Signals and Communication Technology

More information about this series at http://www.springer.com/series/4748

Mathias Wien

High Efficiency Video Coding

Coding Tools and Specification

Springer

Mathias Wien
Institut für Nachrichtentechnik
RWTH Aachen University
Aachen
Germany

ISSN 1860-4862 ISSN 1860-4870 (electronic)
ISBN 978-3-662-44275-3 ISBN 978-3-662-44276-0 (eBook)
DOI 10.1007/978-3-662-44276-0

Library of Congress Control Number: 2014944714

Springer Heidelberg New York Dordrecht London

Printed on acid-free paper

Springer is part of Springer Science+Business Media (www.springer.com)

Preface

Since the first publication of the video coding standard H.264 | AVC in 2003, the video world has changed. Video content has become a major fraction of digital network traffic worldwide and is still growing. The demand for HD and Ultra HD video (with picture resolutions of 4K $\times$ 2K and more) increases, inducing even higher bandwidth needs. The established standard H.264 | AVC is used for HD content while having been developed mainly for resolutions around standard definition television. The standardization effort for "High Efficiency Video Coding" (HEVC) answers the demand for improved compression performance at resolutions of HD and beyond. The design of the included coding tools enables video compression to the desired degree in conjunction with implementation friendliness and options for parallelization at multiple levels. It is expected to be widely adopted for video services at HD and Ultra HD quality, providing Ultra HD video at similar bitrates as used for HD video today. Based on the known concepts of the well-established hybrid coding scheme, new coding structures and better coding tools have been developed and specified for HEVC. The new standard is expected to be taken up easily by established industry as well as new endeavors, answering the needs of today's connected and ever-evolving online world.

This book presents the standard and strives to explain it in a clear and coherent language. It provides a comprehensive and consistently written description of the applied concepts and coding tools. For synchronization of beginners in the field, a chapter on the fundamentals of video coding is included. It provides a general overview on the elements of video coding systems, the representation of color video, and an introduction of the building blocks of the hybrid coding scheme. Another chapter is dedicated to the topic of specification by itself; it deals with requirements on specification text and with imposed fundamental technological guidelines. The understanding of these principles is utile for assessment of the algorithmic design at present as well as for future development of extensions and potential additional coding tools. The following chapters follow the structure of the HEVC specification, giving insight to the design and concepts of the coding tools in the specification.

The book shall help readers to understand the state-of-the-art concepts of video coding with HEVC as their latest instantiation. It shall contribute to the promulgation of the HEVC standard and provide support in adopting it for new and fascinating applications.

Examples in the chapters make use of bitstreams and test sequences according to the common testing conditions of the Joint Collaborative Team on Video Coding (JCT-VC). Bitstreams according to these conditions were frequently made public on the JCT-VC experts mailing list by the reference software coordinators. Different versions of these bitstreams have been used as anchors in numerous tests and extensive experiments throughout the HEVC development. Coding examples in this book are given for the test sequence BasketballDrive, provided courtesy of NTT DOCOMO, Inc., Japan, and ParkScene, provided courtesy of Tokyo Institute of Technology, Nakajima Laboratory, Japan. Examples for the sequences of the JCT-VC common testing conditions further include BlowingBubbles and RaceHorses, provided courtesy of NTT DOCOMO, Inc., Japan; ChinaSpeed, provided courtesy of Shanghai Shulong Computer Technology Co., Ltd., China; FourPeople, provided courtesy of Vidyo Inc., USA; and PeopleOnStreet, provided courtesy of Samsung Electronics Co., Ltd., Korea.

Hundreds of experts in the JCT-VC have worked hard, spending time in never-ending meetings with tremendous numbers of input contributions to achieve the goal of a stable high-quality high-performance specification. Their contribution and commitment are highly appreciated. The standardization work has been carried out under the prudent and stimulating lead of the JCT-VC chairs Gary J. Sullivan and Jens-Rainer Ohm. Their thoughtful and precise guidance is specifically appreciated. It formed and coined this collaborative team.

This book would have not been possible without help and support from several people. I want to thank an uncounted number of JCT-VC experts for numerous conversations and debates on various aspects of the specification. Special thanks go to T. K. Tan and Andrew Segall for manifold and insight-full discussions over the years; and to Benjamin Bross who knew all answers to questions on the specification. Very special thanks go to Rickard Sjöberg for his expertise and helpful comments on picture types and reference picture management; and specifically to Peter Amon for his comprehensive and remarkably careful text review. Peter, I owe you more than a beer. Special thanks for review and support also go to people at Institut für Nachrichtentechnik of RWTH Aachen University: Julian Becker, Max Bläser, Christopher Bulla, Olena Chubach, Christian Feldmann, Iris

Heisterklaus, Cordula Heithausen, Fabian Jäger, Ningqing Qian, Inge Reissel, Christian Rohlfing, and Uday Thakur; very special thanks to Bastian Cellarius who helped generating the example figures.

Most special thanks go to Prof. Jens-Rainer Ohm for providing distinct leadership, kind mentoring, and encouraging guidance. Thank you for providing me room to write this book. It is an outstanding experience and honor to work in your team. The warmest thanks go to my wonderful family: Sabine and our children Frederic, Leonard, and Marlene.

Aachen, May 2014 Mathias Wien

Contents

Symbols and Operators

Mathematical Functions and Operators

$\lvert\cdot\rvert$	Absolute value
$\lfloor \cdot \rfloor$	Round to lower integer value
$\lceil \cdot \rceil$	Round to higher integer value
$\mathrm{Clip}\{a,b,v\}$	$= \begin{cases} a, & \text{if } v < a \\ v, & \text{if } a \le v \le b \\ b, & \text{if } b < v \end{cases}$
mod	Modulo operator
$\mathrm{round}\{v\}$	$= \mathrm{sgn}(v)\left\lfloor \lvert v\rvert + \frac{1}{2} \right\rfloor$
$\mathrm{sgn}(v)$	$= \begin{cases} 1, & \text{if } v > 0 \\ 0, & \text{if } v = 0 \\ -1 & \text{if } v < 0 \end{cases}$

Bit-Wise Operators

&	Bit-wise 'and' operation
|	Bit-wise 'or' operation
$x \gg y$	Bit-shift to the right of binary representation of x by y bits. Integer division by 2^y, $y \geq 0$
$x \ll y$	Bit-shift to the left of binary representation of x by y bits. Integer multiplication with 2^y, $y \geq 0$

Variables

B_{d}	Bit depth
1_{id}	Layer identifier in the NAL unit header

N_L Number of lines in picture
N_P Number of pixels (samples) per line in picture
t_{id} Temporal sub-layer identifier in the NAL unit header

Codes and Binarization

$B_{tp,n}(v)$ Binarization of value v with type 'tp' and parameter n
$C_{tp,n}(v)$ VLC for value v of type 'tp' and parameter n
$C^{i}_{tp,n}(c)$ Return symbol value for code word c of VLC type 'tp' and parameter n

Acronyms

This list collects the acronyms used in this book. It further includes some acronyms which have been established during the work of the JCT-VC on HEVC, and others which are commonly used in the context of video coding. Marked acronyms (*) indicate deprecated naming conventions or refer to tools which have not been adopted into the HEVC specification. If acronyms relate to the H.264 | AVC specification, a corresponding indication is given. For acronyms that are used in this book, a reference to the relevant chapter or section is provided to facilitate access.

3D-HEVC	3D high efficiency video coding (Sect. 12.5)
AAC	Advanced audio coding
AAP	Alternative Approval Process (Sect. 1.3.2)
AhG	Ad hoc group
AI	All intra (JCT-VC CTC, Sect. 3.6.1.1)
AIF*	Adaptive interpolation filter
ALF*	Adaptive loop filter
AMP	Asymmetric motion partitioning (Sect. 7.1.2)
AMVP	Advanced motion vector prediction (Sect. 7.2.3)
ANG*	Angular intra prediction (see Sect. 6.5)
APS	Adaptation parameter set (Sect. 5.3.1)
ASO	Arbitrary slice ordering (H.264 | AVC, Sect. 4.2.7)
AU	Access unit (Sect. 5.2.3)
AUD	Access unit delimiter (Sect. 5.2.2)
AVC	Advanced video coding
BLA	Broken link access [picture] (Sect. 4.1.2.1)
BD	Bjøntegaard delta measurement (Sect. 3.6.1.3)
BL	Base layer (Sect. 12.4)
BO	SAO band offset (Sect. 9.2)
BoG	Break-out group (Sect. 1.4.4)
BP	Bandpass [filter]

BPB	Bitstream partition buffer (Sect. 12.2)
CABAC	Context-based adaptive binary arithmetic coding (Sect. 10.2)
CAVLC	Context adaptive variable length coding (Sect. 2.4.8)
CB	Coding block (Sect. 4.2.4)
CBF*	Coded block flag (see Sect. 8.2.1)
CBP	Coded block pattern (H.264 | AVC, Sect. 4.2.7)
CBR	Constant bitrate (Sect. 5.6.1)
CD	Committee draft (Sect. 1.3.1)
CD	Compact disk
CE	Core experiment (Sect. 1.4.4)
CfE	Call for evidence (Sect. 1.3.1)
CfP	Call for proposals (Sect. 1.3.1)
CIE	Commission internationale de l'éclairage (Sect. 2.3.1)
CIF	Common interchange format 352 × 288 (Sect. 11.2)
CPB	Coded picture buffer (Sect. 5.6.1)
CRA	Clean random access point (Sect. 4.1.2.1)
CRFB*	Compressed reference frame buffer (see Sect. 7.2.1)
CRT	Cathode ray tube
CS	Constraint set (in CfP)
CSS	Coded slice segment (Sect. 5.2.3)
CT	Collaborative team (Sect. 1.4.1)
CTB	Coding tree block (Sect. 4.2.4)
CTC	Common testing conditions (Sect. 3.6.1)
CTU	Coding tree unit (Sect. 4.2.4)
CTX	CABAC context (Sect. 10.2)
CU	Coding unit (Sect. 4.2.4)
CVS	Coded video sequence (Sect. 4.1)
DAM	Draft amendment (Sect. 1.3.1)
DC	Direct current
DCT	Discrete cosine transform (Sect. 2.4.5.2)
DCTIF	DCT interpolation filter (Sect. 7.3.3)
DIF*	DCT interpolation filter, syn. DCTIF (Sect. 7.3.3)
DIF*	Directional interpolation filter
DIS	Draft international standard (Sect. 1.3.1)
DLP*	Decodable leading picture, syn. RADL (Sect. 4.1.2.2)
DMVD	Decoder side motion vector derivation
DPB	Decoded picture buffer (Sect. 5.6.2)
DPCM	Differential pulse code modulation
DST	Discrete sine transform (Sect. 2.4.5.3)
DU	Decoding unit (Sect. 5.2.4)
DUT*	Directional unified transform (see Sect. 8.1.2)
DVD	Digital versatile disk
EG	Exp-Golomb code (Sect. 2.4.8.4)
EL	Enhancement layer (Sect. 12.4)
EO	SAO edge offset (Sect. 9.2)

EOB	End of bitstream (Sect. 5.2.2)
EOS	End of sequence (Sect. 5.2.2)
EOTF	Electro-optical transfer function (Sect. 2.3.3)
FCD	Final committee draft (Sect. 1.3.1)
FD	Filler data (Sect. 5.2.2)
FDAM	Final draft amendment (Sect. 1.3.1)
FDIS	Final draft international standard (Sect. 1.3.1)
FLC	Fixed length code (Sects. 2.4.8.2 and 10.1)
FMO	Flexible macroblock ordering (H.264 I AVC, Sect. 4.2.7)
fps	Frames per second
FRExt	Fidelity range extensions (of H.264 I AVC, Sect. 12.1)
GRD	Gradual decoder refresh (H.264 I AVC)
GOP	Group of pictures (Sect. 4.1)
HD	High definition (Sect. 11.2)
HDTV	High definition television
HEVC	High efficiency video coding
HM	HEVC test model (Sect. 1.2.2)
HP	Highpass [filter]
HRD	Hypothetical reference decoder (Sect. 5.6)
HSS	Hypothetical stream scheduler (Sect. 5.6)
HTM	3D-HEVC test model (Sect. 12.5)
HVC*	High performance video coding (name of HEVC pre-project in MPEG)
HVS	Human visual system (Sect. 2.1.7)
IBDI*	Internal bit-depth increase (see Sect. 3.3.5)
IDCT	Inverse discrete cosine transform (Sect. 2.4.5.2)
IDR	Instantaneous decoder refresh (Sect. 4.1.2.1)
IEC	International Electrotechnical Commission
IEEE	Institute of electrical and electronics engineers
IRAP	Intra random access point (Sect. 4.1.2.1), see also RAP
IS	International standard (Sect. 1.3.1)
ISDN	Integrated services digital network
ISO	International organization for standardization
ITU	International telecommunication union
JCT	Joint collaborative team (of ISO and ITU)
JCT-VC	Joint collaborative team on video coding
JCT-3V	Joint collaborative team on 3D video coding extension development
JM	Joint model (AVC test model)
JPEG	Joint photographic experts group
JTC	Joint technical committee (Sect. 1.3.1)
JVT	Joint video team (Sect. 1.2.1)
KLT	Karhunen–Loéve transform (Sect. 2.4.5)
KTA	Key technical areas (H.264 based exploration software of VCEG)
LCTB*	Largest coded tree block, syn. CTB
LCTB*	Largest coded tree unit, syn. CTU

LCU*	Larges coding unit, syn. CTU
LD	Low delay (JCT-VC CTC, Sect. 3.6.1.1)
LP	Lowpass [filter]
LPS	Least probably symbol (Sect. 2.4.8.5)
LSB	Least significant bit
MAC	Multiplexed analog components (Sect. 5.7.2)
MANE	Media aware network element
MB	Macroblock (H.264 I AVC, Sect. 4.2.7)
MBAFF	Macroblock adaptive frame/field coding (H.264 I AVC, Sect. 4.2.7)
MC	Motion compensation (Sect. 2.4.3)
MDDT*	Mode dependent directional transform (see Sect. 8.1.2)
ME	Motion estimation (Sect. 2.4.4)
MMCO	Memory management control operation (H.264 I AVC, Sect. 5.8)
MP3	MPEG-2 audio layer III
MPEG	Moving picture experts group (Sect. 1.3.1)
MPM	Most probable mode (Sect. 6.6)
MPS	Most probably symbol (Sect. 2.4.8.5)
MSB	Most significant bit
MSE	Mean squared error
MV	Motion vector (Chap. 7)
MVC	Multiview video coding (H.264 I AVC, Chap. 12)
MVD	Motion vector difference (Sect. 7.2.3)
MV-HEVC	Multiview high efficiency video coding (Sect. 12.3)
NAL	Network abstraction layer (Sect. 5.2)
NALU	NAL unit (Sect. 5.2.1)
NB	National body (in ISO)
NGVC*	Next generation video coding (name of HEVC pre-project in VCEG)
NTSC	National television systems committee (Sect. 5.7.2)
NUH	NAL unit header (Sect. 5.2)
NUT	NAL unit type (Sect. 5.2)
PAL	Phase alternating line (Sect. 5.7.2)
PAFF	Picture adaptive frame/field coding (H.264 I AVC, Sect. 4.2.7)
PB	Prediction block (Sect. 4.2.5)
PCM	Pulse code modulation (Chap. 8)
PDAM	Proposed draft amendment (Sect. 1.3.1)
PPS	Picture parameter set (Sect. 5.3.1)
PSNR	Peak signal to noise ratio (Sect. 2.5.1.4)
POC	Picture order count (Sect. 5.4.1)
PU	Prediction unit (Sect. 4.2.5)
QCIF	Quarter common intermediate format 176 × 144 (Sect. 11.2)
QHD	Quarter high definition 960 × 540 (Sect. 11.2)
QP	Quantization parameter (Sect. 2.4.5.6)
RA	Random access (JCT-VC CTC, Sect. 3.6.1.1)
RADL	Random access decodable leading picture (Sect. 4.1.2.2)

RAP	Random access point (Sect. 4.1.2.1)
RASL	Random access decodable skipped picture (Sect. 4.1.2.2)
RBSP	Raw byte sequence payload (Sects. 5.2.1 and 10.1.1)
RD	Rate-distortion (Sect. 2.5.2)
RDO	Rate-distortion optimization (Sect. 2.5.2)
RDOQ	Rate-distortion optimized quantization (Sect. 8.1.7)
RExt	HEVC Range extensions (Sect. 12.1)
RGB	Red green blue, color format (Sect. 2.3.4)
RPL	Reference picture list (Sect. 4.3.2)
RPS	Reference picture set (Sect. 4.3.1)
RQT	Residual quadtree (Sect. 8.2)
RTP	Real-time transport protocol (Sect. 5.1)
SAD	Sum of absolute differences (Sect. 2.5.1.2)
SAO	Sample adaptive offset (Sect. 9.2)
SAR	Sample aspect ration (Sect. 5.7)
SATD	Sum of absolute transformed differences (Sect. 2.5.1.3)
SC	Sub-committee (Sect. 1.3.1)
SD	Standard definition (TV)
SDH	Sign data hiding (Sect. 8.2.4)
SECAM	Séquentiel couleur á mémoire (Sect. 5.7.2)
SEI	Supplemental enhancement information (Sect. 5.5)
SG	Study group (Sect. 1.3.2)
SHVC	Scalable high efficiency video coding (Sect. 12.4)
SODB	String of data bits (Sect. 5.2.1)
SOP	Structure of pictures (Sect. 5.5), see also GOP
SPS	Sequence parameter set (Sect. 5.3.2)
SSD	Sum of squared differences (Sect. 2.5.1.1)
SSE	Sum of squared error (Sect. 2.5.1.1)
STSA	Stepwise temporal sub-layer access (Sect. 4.1.2.3)
SVC	Scalable video coding (H.264 | AVC, Chap. 12)
TB	Transform block (Sect. 4.2.6)
TE	Tool experiment (Sect. 3.6.2.2)
TFD*	Tagged for discard [picture], syn. RASL (Sect. 4.1.2.2)
TMuC	Test model under consideration (Sect. 3.6.2.2)
TMVP	Temporal motion vector predictor (Sect. 7.2.2.2)
TR	Truncated Rice [binarization] (Sect. 10.2.3)
TSA	Temporal sub-layer access (Sect. 4.1.2.3)
TSB	Transform sub-block (Sect. 8.2.1)
TSB	Telecommunication standardization bureau (Sect. 1.3.2)
TU	Transform unit (Sect. 4.2.6)
TV	Television
UHD	Ultra high definition (Sect. 11.2)
VBR	Variable bitrate (Sect. 5.6.1)
VCEG	Visual coding experts group (Sect. 1.3.2)
VCL	Video coding layer (Chap. 5)

VGA	Video graphics array 640 × 480 (Sect. 11.2)
VLC	Variable length code (Sects. 2.4.8.2 and 10.1)
VPS	Video parameter set (Sect. 5.3.1)
VUI	Video usability information (Sect. 5.7)
WD	Working draft (Sect. 1.3.1)
WG	Working group (Sect. 1.3.1)
WPP	Wavefront parallel processing (Sect. 4.2.2.4)
XGA	Extended Graphics Array 1024 × 768 (Sect. 11.2)
XYZ	XYZ color space (Sect. 2.3.1)
YCbCr	Color format with luma and two chroma components (Sect. 2.3)
YUV	Color format with luma and two chroma components (Sect. 2.3)

Chapter 1
Introduction

In this chapter, the general prerequisites for the development of video coding specifications are described, including an overview of previous video coding standards, the formal procedures which are required for standards approval, and the Joint Collaborative Team on Video Coding, which was established for the development of HEVC. In order to immerge in the subject without an overload of definitions, some of the terminology used in this chapter will be taken as is for the moment and will be defined and detailed in the forthcoming sections and chapters.

1.1 How to Read This Book

Before starting with the content of this chapter a brief survey on the structure of the book and how it should be read is provided.

The book is designed to address readers at multiple levels of education in the area of video coding. For an tutorial introduction into the topic of video coding, the reader is invited to study Chap. 2, which includes an overview of the relevant elements of the signal processing chain, the representation of video signals, and an introduction to the elements of the hybrid video coding scheme, which is the basis for all relevant video coding specifications as of today. Chapter 3 details the scope, constraints, and requirements on the specification text as well as fundamental design principles, which have been established in video coding standardization.

The subsequent chapters elaborate on the technical detail of the HEVC specification. Relevant tools and principles, which exist in HEVC as well as in the preceding H.264 | AVC specification, are briefly compared in their previous and new realization, highlighting the main differences between the two schemes. In Chap. 4, the organization of an HEVC coded video sequence in terms of picture types, prediction structures, and the applicable spatial partitioning into the relevant processing units is presented. This terminology introduced in this chapter is used in the following chapters. It is placed before Chap. 5 on high-level syntax to ease the understanding of the high-level structure and the chosen bitstream representation in HEVC. While syntax

M. Wien, *High Efficiency Video Coding*, Signals and Communication Technology,
DOI 10.1007/978-3-662-44276-0_1

and semantics constitute a significant part of the specification, the focus of this book is on the concepts behind the selected design and only some elements of the concrete syntax structure design are discussed here. Since the HEVC specification is publicly available and can be downloaded from the ITU-T website for study, a comprehensive reproduction of the syntax structures has not been considered instrumental. Chapters 6–11 follow the general organization of the clauses of the specification itself. The functionality of the respective building blocks is detailed. Wherever indicated, interconnections between different parts of the design, e.g. prediction tools and entropy coding, are highlighted. The standardization activities towards extension of the HEVC specification, which are ongoing by the time of writing of this book are summarized in Chap. 12.

Like in many other areas in the field of engineering, the video coding community has developed an 'expert language' with lots of acronyms for building blocks, concepts, and schemes. The Appendix provides a comprehensive collection of these acronyms with decryption and additional pointers for reference. Whenever a term or acronym appears unknown while reading, these resources may be utile for a short-hand update.

1.2 A Brief Walk Through the History of Video Coding Standards

Since the early 1990s, the development of video coding standards has been driven by two parallel application spaces: real-time video communication and distribution or broadcast of video content. The corresponding specifications have been published by two main standardization bodies, the International Telecommunications Union (ITU) and the International Standardization Organization/International Electrotechnical Commission (ISO/IEC). The specific organizational structures and processes required for approval of the respective specifications in the two organizations are further discussed in Sect. 1.3. Here, a brief overview of the evolution of video coding standards is provided with a focus on the main corresponding application scenarios and the corresponding main technical achievements. For the sake of simplicity both, ITU recommendations and ISO/IEC standards are referred to as standards in this section. The differences between the two are detailed below.

An overview of the time line of the major standards in the two standardization bodies is shown in Fig. 1.1. For all standards listed in this time line, several corrigenda and extensions have been published over the time. Here, the publication dates of the key versions of the standards have been included. It can be seen that while having started on separate tracks, the two standardization organizations have engaged in increasingly close collaboration, specifically for achieving the latest milestones AVC and HEVC.

In 1984, the first digital video coding standard was published under the name ITU-T H.120 "Codecs for videoconferencing using primary digital group transmission" [1]. It was designed for video conferencing applications at standard television (SD)

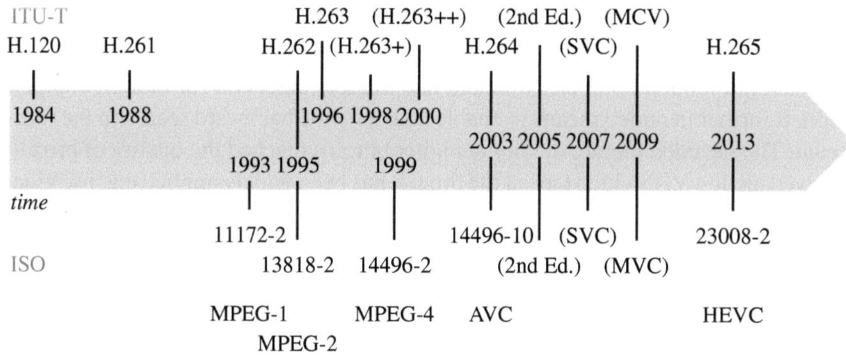

Fig. 1.1 Time line of video coding specifications published by ITU-T and ISO/IEC

resolution with a bitrate of about 1.5–2 Mbit/s but did not reach broad application [2]. The basic concept consists of a DPCM coding structure[1] and 'conditional replenishment' where content from a previous picture is copied to the current picture unless the content differs too much. Variable length code (VLC) tables were used for encoding of the syntax elements. While not being used much, H.120 is still in force today.

The successor ITU-T H.261 "Video codec for audiovisual services at $p \times 64$ kbit/s" was published in its first version in 1988 [3]. The scheme was designed to operate over ISDN [4] in the range of 40 kbit/s–2 Mbit/s. Compared to H.120, it provided improved compression efficiency and a widely extended range of supported bitrates. Targeting at international video conferencing applications, the scheme operated on the video in the Common Intermediate Format (CIF), which provides a reduced picture size relative to SD to enable compatibility with different international standard television resolutions. H.261 was the first video coding standard which is based on the hybrid video coding scheme. It already contained the basic building blocks of all succeeding video coding standards which are detailed in Chap. 2. The most significant compression improvement compared to H.120 was obtained by the introduction of motion compensated prediction which was carried out at sample precision. H.261 has been widely taken up by industry at the time and found long-lasting application for backward-compatibility purposes. Due to its age and fading intellectual property rights protection, the standard recently gained increased attention in the discussion on the establishment of royalty-free video codecs for video communication in the public Internet, see e.g. [5].

In parallel to the deployment of ITU-T H.261, the Moving Picture Experts Group (MPEG) started the development of a video coding standard for video distribution and broadcast on the ISO/IEC side. In 1993, MPEG-1 "Information technology—Coding of moving pictures and associated audio for digital storage media at up to about 1.5 Mbit/s—Part 2: Video" (ISO/IEC 11172-2) was released [6]. While the technical core is based on the hybrid coding scheme and is similar to H.261,

[1] DPCM = Differential Pulse Code Modulation.

MPEG-1 was designed to address the needs of video distribution instead of video communication. On top of the tools specified in ITU-T H.261, this specification added bi-directional motion compensation and half-sample precision for motion compensation. It further included means to enable fast-forward/backward search in the video stream. The decoded picture quality at higher bitrates reached the quality of broadly deployed analog VHS video tape at the time. It has been widely applied e.g. for Video CDs [2]. Besides the video coding specification, the MPEG-1 standard notably comprises a systems specification for transport of audio-visual data and also compression for audio signals, of which the most advanced version (layer 3) gained worldwide attention as the MP3 audio format. MPEG-1 was further developed into MPEG-2 "Information technology—Generic coding of moving pictures and associated audio information —Part 2: Video", which was published in 1995 as ISO/IEC 13818-2 [7]. In terms of video coding tools, the major extension of MPEG-2 to support coding of interlaced video material. Thereby, digital coding of standard television at full picture resolution was enabled. Further, MPEG-2 was the first video coding standard to include tools for spatial and reconstruction fidelity scalability. It was further extended to support coding of multiview (stereo) video. MPEG-2 also included the concept of profiles and levels which can be used to indicate capability requirements for decoders in the encoded stream. Besides the publication by ISO/IEC, MPEG-2 was also adopted by the ITU as H.262 [8] and can therefore be considered as the first joint video coding standard of the two standardization bodies. MPEG-2 found widespread application with DVDs, digital (satellite) television transmission, and also HDTV systems. Due to the immense spread of the mentioned distribution forms, MPEG-2 has been the most-used video coding standard for a long time. By the end of 2012, still about 73 % of satellite TV broadcast in standard resolution were using H.262 | MPEG-2 [9].

In 1996, the ITU-T published H.263 "Video coding for low bit rate communication" as a successor of H.261 for video communication applications [10]. Introducing improved coding tools and VLC design, H.263 was designed for video conferencing applications at very low bit rates. Consequently, it outperformed all previous video coding standards especially at very low bitrates. Over the following years, H.263 was extended to 'H.263+' and 'H.263++' introducing a very large set of negotiable coding tools for improved compression performance or improved functionality (e.g. error resilience features, like forward error correction or data partitioning, and scalability extensions). Furthermore, H.263 included the concept of supplemental enhancement information to be sent in the video bitstream. In order to organize the large amount of options, the concept of profiles and levels was introduced to H.263 in Annex X in the year 2000, which organizes the most commonly used tools into 8 dedicated profiles [10].

In parallel to the further development of H.263 in the ITU-T, MPEG developed MPEG-4 "Information technology—Coding of audio-visual objects—Part 2: Visual" (ISO/IEC 14496-2) [11]. This standard aimed at the specification of a generalized integrated multimedia standard and was published in its first version in 1999. For representation of visual information this standard includes coding of conventional rectangular video and still images. As a new concept it further supported coding of arbitrarily shaped objects and synthetic content. The standard further comprises a scene composition specification for joint presentation of the different types of content. An overview of MPEG-4 is provided e.g. in [12]. For the core video coding part, the basic tools of H.263 were taken over. Later extensions added global motion compensation and quarter-sample motion accuracy with an improved interpolation filter. Despite the immense capabilities for multimedia data representation, the most successful application of MPEG-4 turned out to be the so-called "Advanced Simple Profile", which essentially consisted of the H.263 tools plus global motion compensation and quarter-sample motion accuracy.

1.2.1 Advanced Video Coding

With the increased distribution of video over the Internet, an increased need for improved compression performance was observed. While the target applications for the standards of ITU-T and ISO/IEC were still divergent with real-time video communication on the one side and multimedia delivery and broadcast applications on the other, the largely identical technical building blocks of the specifications for both application spaces indicated a joint specification to be reasonable. This observation may have led to the installation of the Joint Video Team (JVT) of both organization, which developed the "Advanced Video Coding" (AVC) standard. AVC was approved as ITU-T H.264 and MPEG-4 part 10 (ISO/IEC 14496-10) in 2003 [13, 14]. The standardization activity was initiated by the long-term development project 'H.26L' in the ITU-T, which had it's first test model in 1999. Under the impression of the observed improvements in compression efficiency which were achieved by this project, the joint team of both standardization organizations was formed by the end of 2001 [15].

The key requirements for the design of this specification were a simplified and clean design with significantly improved coding performance, bit-exact specification, and network friendliness. The specification allows for a decoder implementation using only 16-bit integer arithmetic. A clear distinction between the video coding layer and the so-called network abstraction layer was introduced to enable integration with different types of transport which the video should be carried in. Over the time, H.264 | AVC was extended and improved by a series of amendments, including "Scalable Video Coding" (SVC) in 2007 and "Multiview Video Coding" (MVC, supporting stereo or more views) in 2009. In 2010, reportedly 66 % of the video viewed on the web were coded with H.264 | AVC [16].

With the main work on H.264 | AVC finished, the Joint Video Team was closed after its last meeting in November 2009.

1.2.2 High Efficiency Video Coding

The wide availability of high definition ('full HD') video content and displays, ongoing developments towards broad application of UHD video services, as well as the increasing deployment of high-quality video services to mobile devices led to the request for a new video coding standard in 2009 [17]. Two key issues were in the focus of the development: increased video resolution and increased use of parallel processing architectures [17]. At the same time, the new standard should provide substantially better compression efficiency than H.264 | AVC at a large range of picture formats. Based on the experience with the installation of the JVT in the previous standardization effort, the two standardization bodies ITU-T and ISO/IEC agreed to form a new collaborative team for the development of the new standard called the Joint Collaborative Team on Video Coding (JCT-VC) [18]. The requirements for the new standard [19] were agreed by the beginning of 2010 when the corresponding joint call for proposals was published [20].

At the first meeting of the JCT-VC in Dresden in April 2010, the responses to the call for proposals were evaluated and the project name "High Efficiency Video Coding" was established. A total of 27 complete proposals was submitted. The compression performance of the proposals was evaluated in a formal subjective viewing assessment where the reconstruction quality of the proposals was compared to the quality of a set of H.264 | AVC anchors [21]. Based on the results of the visual assessment, the software implementation of the proposals and a thorough evaluation of the included coding tools, the first working draft WD1 [22] and test model HM1 were released in October 2010 [23]. The draft was further developed and refined with each meeting of the JCT-VC approximately every 3 month, until with WD10 and HM10, the final draft of HEVC version 1 was approved in January 2013 [24, 25]. HEVC is published by the ITU-T as recommendation H.265, and by ISO/IEC as MPEG-H part 2 [26, 27]. The HEVC verification test report by Tan et al. reveals that HEVC provides bitrate savings of approximately 59 % for the Main profile compared to the H.264 | AVC High profile for the same observed subjective quality [28]. A comparison of the coding efficiency of HEVC and the previous standards H.262 | MPEG-2, H.263, MPEG-4 Visual, and H.264 | AVC is provided in [29].

In this book, the technical details of HEVC as it has been developed in the JCT-VC are presented in Chaps. 4–11. The ongoing standardization work for extensions for professional applications, scalability features, and the support of multiview and 3D video coding which are developed in a second joint collaborative team (JCT-3V) are summarized in Chap. 12.

Before getting into the technical details, the development and standardization procedures as well as the organisational context and background are outlined in the following.

1.3 Evolution of a Specification

The driving force behind the development of a specification is the need for interoperability of products built by different manufacturers and potentially customized by different vendors. For video delivery this applies e.g. for computers, mobile devices, TV sets, or broadcasting institutions. The basic process cycle for the development of a video coding specification is illustrated in Fig. 1.2. Based on envisaged applications, interested parties bring input to the definition of requirements of a new specification to be developed. The defined requirements serve as a guide throughout the development process in the new standardization project. After the definition of an initial draft to start from, the development evolves over several draft stages, including continuous performance evaluation to affirm improvement over successive draft versions. In this phase, the performance evaluation for video coding schemes usually comprises rate-distortion assessment, i.e. the achievable video reconstruction quality subject to the invested bitrate, and complexity measures for monitoring of the hardware and software implementation cost of the draft design. With the final stabilization of the draft design, the final specification can be approved and published for deployment in applications. The deployment of products using the specification may induce the request for additional features or reveal shortcomings of the design which have not been envisaged in the development process. Such indications may lead to new or revised requirements, which are then used for appropriate extension or correction of the existing specification.

The basic process cycle shown in Fig. 1.2 is formalized in various forms in the standardization organizations. In the following, the formal procedures of the ISO/IEC and the ITU, being the relevant standardization bodies in the area of video coding, are described.

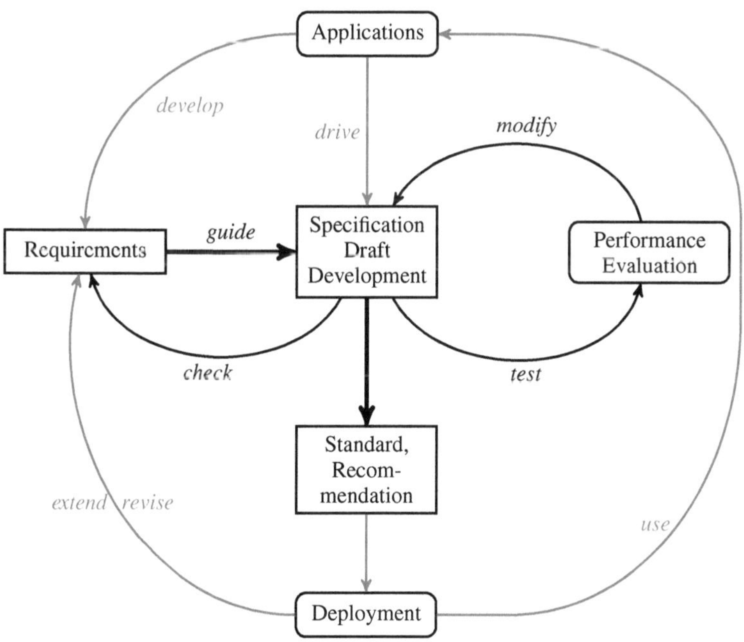

Fig. 1.2 Development course of a specification

1.3.1 Formal Procedure for a Standard in ISO/IEC

While being published by the ISO, the standardization work in this area is tightly connected with the International Electrotechnical Commission (IEC) by the instrument of a Joint Technical Committee (JTC). Standardization work in the area of media coding and transport is carried out in Working Group 11 (WG 11) on coding of moving pictures and audio of the Joint Technical Committee 1 (JCT 1) on information technology. The working group is assigned to Sub-Committee 29 of JCT 1, which comprises the working groups in the area of coding of audio, picture, multimedia and hypermedia information. The described organizational structure results in the official denotation as ISO/IEC JTC 1/SC 29/WG 11. Working Group 11 is also called the Moving Pictures Experts Group (MPEG) and has reached significant visibility by its standards, like MPEG-1, MPEG-2, or MPEG-4, including wide-spread formats like H.264 | AVC video or the MP3/AAC audio as mentioned above. The working group itself is divided into subgroups which are responsible for different parts of the standards.

In ISO/IEC, a working group consists of delegates of national body (NB) institutions (e.g. national standardization committees). Important decisions of the working group are drawn by balloting among the national bodies accredited for the working group. The rules of the ISO/IEC standardization processes are laid down in the ISO/IEC directives and the JTC1 supplement [30, 31]. In the technical work, strong efforts are made to achieve consensus on the specified design. The intention of this consensus building effort is to induce supportive ballots throughout the decision taking process. Unanimity is not required in such ballots. It should be noted that unanimity is also not required in the determination of consent, which can be regularly determined by the absence of strong opposition to a decision. If necessary, pertaining isolated opposition may be overruled in order to proceed [30].

The life-cycle of a work item which is qualified for the standardization process consists of a set of stages which are consecutively passed. The ISO/IEC rules allow for fast track modes, where several stages are skipped. For details the reader is referred to the directives mentioned above. The stages are organized according to increasing stability and maturity of the specification design. The document status changes while passing through the stages with denotation, which indicates increasing stability. Throughout the standardization process, the participating national bodies express their acceptance of the quality of the work by repeated ballots on the draft. This construction of the development process is meant to assert most achievable acceptance of resulting international standard. In the following the procedure and the stages are briefly described. The stages, document states, and the recommended time lines are summarized in Fig. 1.3.

At the *preliminary stage*, a potential new work item is evaluated by the working group. This stage can be started by a call for evidence (CfE), which brings developments outside of the working group to the attention of the committee. If the preliminary evaluation reveals sufficient substance, a new work item can be proposed. The national bodies in the working group decide by simple majority on the acceptance

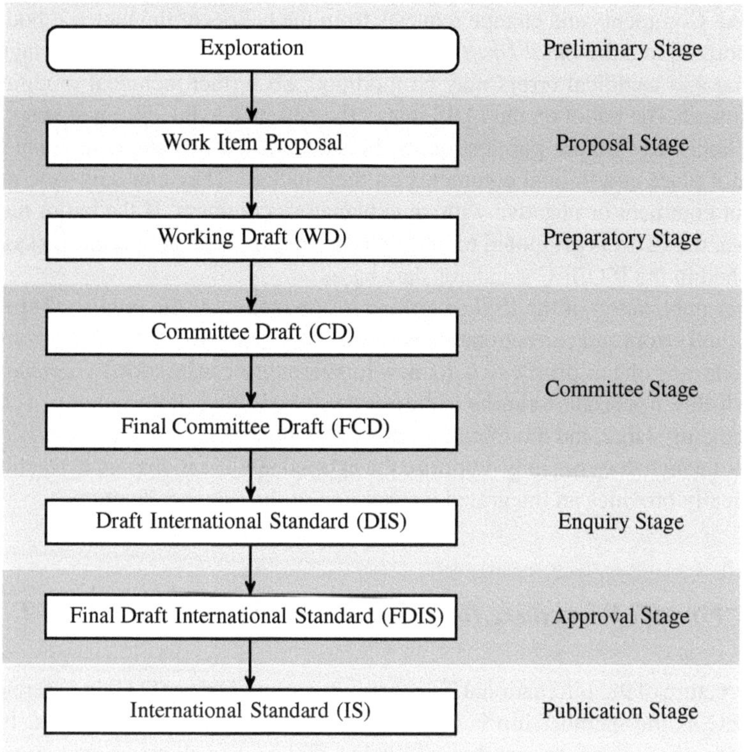

Fig. 1.3 Illustration of the process stages and document states of the ISO/IEC standardization process

of the new work item. The new work item is registered as a project of the technical committee and a time line for the standardization process is defined. At this point, the project enters the *proposal stage*. In MPEG, this step is usually accompanied by a *Call for Proposals* (CfP), where the starting point of the standardization work is selected from the competing responses. As an accepted project and with the result of the CfP, the work item then enters the *preparatory stage* where a *Working Draft* (WD) is generated. The initial working draft is refined and improved over multiple meeting cycles until sufficient maturity is determined. At this stage, the project is promoted to *committee stage* and the working draft is turned into the *Committee Draft* (CD), which is balloted by the national bodies. Comments and requests for changes which are raised in the ballot have to be resolved until consensus on the specification design is reached. The comments to the balloting vote of national bodies may contain conditions under which an originally negative vote could be turned into a positive vote. The convergence of the CD is then expressed by promotion to the *Final Committee Draft* status which is the final status to be reached at the committee stage. When the FCD status is reached, the working group decides on entering the *enquiry stage* where the promotion of the FCD to a *Draft International Standard* (DIS) is balloted. Comments and change requests from the ballots of the national bodies are incorporated into the *Final Draft International Standard* (FDIS). At this stage, only editorial and technical errors may be modified, no further technical modifications are allowed. The ballot on the FDIS marks the entrance to the *approval stage* which is the final stage before publication of the standard. At this stage, national bodies must not place conditional comments on their ballots. They can only vote positive without comment or negative with an explanatory comment. If the ballot turns out positive, the FDIS is promoted to the status of an *International Standard* (IS) and is published in the ISO/IEC standards data base.

After publication of the IS the working group maintains the published specification. Small errors and corrections are issued as *corrigenda* to the IS. Extensions, such as the addition of new profiles e.g. for new functionality, can be added as *amendments*, which follow a procedure similar to the one described above. If the amount of changes is sufficiently large, the modifications can be incorporated into a new *edition* of the standard which does not only comprise the applied modifications and corrections but specifically provides an integrated text version of the full specification.

1.3.2 Formal Procedure for a Recommendation in the ITU-T

The structure of the International Telecommunications Union (ITU) is different from ISO/IEC as the membership is not strictly organized in national bodies. Instead, countries can register as members (Member States) as well as organizations (Sector Members). Both types of membership allow for full participation in decision processes of the ITU.

In the ITU, normative specifications are called recommendations instead of standards. A recommendation is considered to be 'a set of guidelines which should be followed'. As such, the claim of a recommendation can be seen to be lower than the claim of a standard. In practice the recommendations in the area of video coding are followed equivalently to ISO/IEC standards. In the ITU, the standardization work is organized in study groups. The work inside a study group is organized in questions which are redefined every 4 years. Depending on the size of the study group, further partitioning into working parties is arranged as needful. The work of a question is lead by a rapporteur who is responsible for the organization of the question work and for the output of recommendations of the respective question. By the time of writing of this book, the development of recommendations in the area of video coding is performed in Question 6 on visual coding which is part of Working Party 3 on media coding and signal processing of Study Group 16 on Multimedia of the standardization sector (ITU-T). The group is called the Visual Coding Experts Group (VCEG) with the official denotation ITU-T Q6/16, indicating the question and the study group.[2] It is responsible e.g. for H.261, H.263, and H.264 | AVC.

[2] Initially, the acronym stood for Video Coding Experts Group. This was modified when VCEG was appointed responsibility for still image coding standards as well [32].

The approval process applicable to video coding recommendations is specified in ITU-T A.8 [33]. This approval process is called the *alternative approval process* (AAP), which since 2001 is applied to all recommendations that have no policy or regulatory implications. Such recommendations follow the traditional approval process specified in ITU-T Resolution 1 [34]. A summary of the AAP is shown in Fig. 1.4. As a concept of the organization, ITU works on a consensus basis. This means that all decisions are taken without substantial opposition. Balloting is not applied. Similar to ISO/IEC, consensus does not necessarily imply unanimity in all cases but does require that verbalized opposition is sufficiently resolved such that it is not further brought up in the debate.

If the work on the specification in the question has reached a sufficient level of maturity, the respective recommendation text is presented for review at a study group or working party meeting. At this meeting *consent* is sought on the draft recommendation. If no substantive opposition is indicated the draft recommendation is made ITU-T public for a *Last Call* period of four weeks where Member States or Sector Members may submit comments on the draft. Comments that are received have

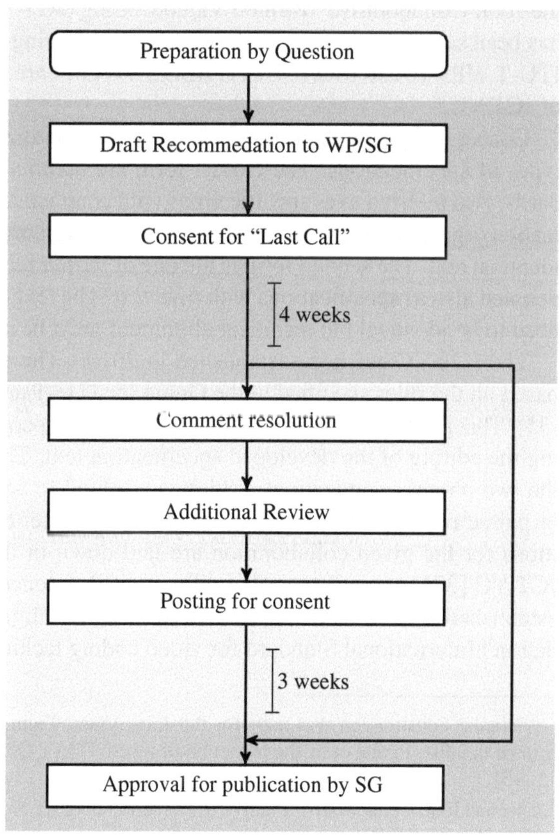

Fig. 1.4 Illustration of the alternative approval process (AAP) in the ITU-T

to be resolved by the rapporteur. If the comments only address editorial issues, those are addressed and the recommendation is considered to be approved. In turn it gets published by the ITU-T Telecommunication Standardization Bureau (TSB). Thereby, a recommendation can be set into force in about five weeks after the corresponding study group or working party meeting in the best case. If substantive comments are received the draft is sent back to the rapporteur for resolution in an *Additional Review* period. For extensions and amendments of the recommendation, the same formal process as for new recommendations is followed.

1.4 The Joint Collaborative Team on Video Coding

Since both, ISO/IEC and ITU-T are defining standards and recommendations in a very similar application space, collaboration of these organizations can provide significant synergy for resulting specifications. As mentioned before, cooperation or collaboration has been sought on multiple projects in the past. Among the available modes of cooperation, setting up a so-called collaborative team is the closest form. In the Joint Collaborative Team on Video Coding (JCT-VC), this type of collaboration has been selected for the project of jointly specifying HEVC [18]. For the JCT-VC, ITU-T WP3/16 and ISO/IEC JTC 1/SC 29/WG 11 are referred to as the parent bodies of JCT-VC.[3]

Conceptually, a collaborative effort on a standardization project can result in two types of specifications: The closest form are *identical* recommendations and standards, also referred to as specifications with *common text*. In this case, both standardization organizations work to approve all editions, corrigenda, and amendments with identical text. The second form is the one of *paired* recommendations and standards, denoted also as specifications with *twin text*. The text of such specifications does not need to be identical but technical alignment must be achieved [35].

The JCT-VC has been established in 2010.[4] The working rules for JCT-VC are based on the rules specified in the Guide for ITU-T and ISO/IEC JTC 1 cooperation [35]. This guide specifies the available modes of operation and requirements regarding the editing of the developed specification text. The aligned approval process in the two parent organizations, which is required to achieve publication of identical or paired recommendations and standard, is further detailed below. Specific definitions for the given collaboration are laid down in the Terms of Reference of the JCT-VC [37]. According to these Terms of Reference "the JCT on video coding is established to collaboratively develop technically aligned twin text for a Recommendation | International Standard for video coding technology more advanced than the

[3] A similar construction was used for the Joint Video Team (JVT) on H.264 | AVC which was formed in 2001. In that case, the parent bodies were ITU-T Q6/16 and ISO/IEC JTC 1/SC 29/WG 11 [15].

[4] A second Joint Collaborative Team on 3D Video Coding Extension Development has been founded in 2012 for the development of advanced 3D video coding capabilities [36]. See Sect. 12.3 for a brief overview of the planned multiview and 3D extensions of HEVC.

current AVC standard (ITU-T Recommendation H.264 | ISO/IEC 14496-10)" [37]. Hence, for HEVC as has been done for H.264 | AVC, identical text is not mandated for JCT-VC though divergence of the approved text in the parent bodies is expected to be avoided by strongest efforts.

1.4.1 Approval Process

The synchronized approval process of both parent organizations of a collaborative team is visualized in Fig. 1.5. The process for approval of a new specification is shown. For amendments, the same process is applied accordingly. In the figure, the

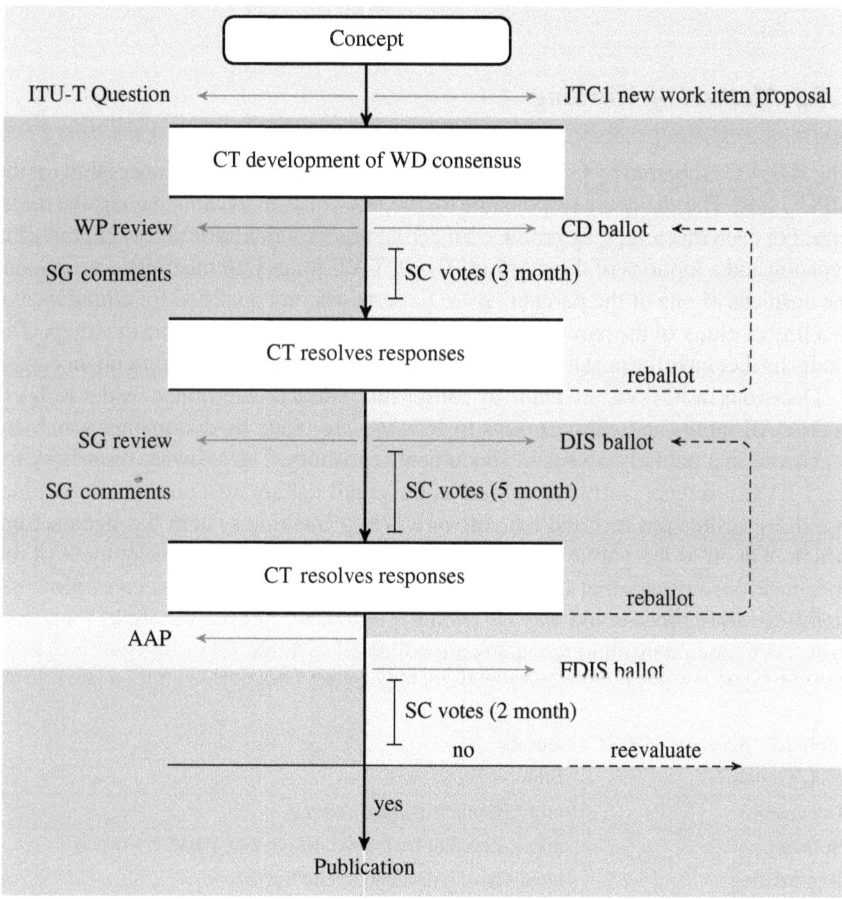

Fig. 1.5 Illustration of the approval process for a collaborative team (CT) of ISO/IEC JTC 1 and ITU-T [35]

collaborative team is denoted as CT according to [35]. This approval process is originally targeting the collaboration towards common text. However, it is closely followed for twin text specifications as it secures technical alignment of the approved text in both organizations. At the stage where the CT has reached consensus on sufficient maturity of the draft specification to start the approval process, a time line for the approval process is established and the ISO/IEC JTC 1 balloting for the CD status is conducted. In parallel, the ITU-T WP3/16 is provided the draft text and comments are formally requested. Based on the received comments, the draft text is revised and, if substantial modifications were necessary, the balloting on the CD is repeated. If the committee draft stage is successfully passed (FCD), the text is prepared to be balloted DIS in ISO/IEC JTC 1. The resulting text is first submitted to ITU-T SG 16 for initiation of the AAP. The output of the AAP including comments received from the Study Group are passed to ISO/IEC JTC 1 and the FDIS ballot is performed. Upon a positive result, both organization proceed to publication of the new specification.

1.4.2 Method of Working

The JCT-VC is chaired by Gary Sullivan for the ITU side and Jens-Rainer Ohm for the MPEG side. The chairs are responsible for the work of JCT-VC and manage the meetings. For each meeting, they produce a meeting report which particularly captures the decision and adoptions of the meeting. The JCT-VC holds four meetings a year under the auspices of one of the parent bodies. Experts who are qualified for attendance of meeting of either of the parent bodies are qualified to participate in the meetings. The chairs further have the possibility to personally invite other experts to attend meetings.

Decisions in JCT-VC are made by consensus, which is determined by the JCT-VC chairs. All inputs and contributions to JCT-VC are made by documents which are registered in a publicly accessible document repository. The software repository for the HEVC reference software as well as the email list are also publicly accessible. For the specification text and the software, a bug tracking system has been set up, which is open to the public as well. This open approach in the development of the specifications ensures that any interested organizations and experts can follow the standardization process and may contribute if indicated. The relevant links for access to JCT-VC documents and resources are collected in Table 1.1.

Table 1.1 Access to JCT-VC resources

JCT-VC item	Link
Documents	http://phenix.it-sudparis.eu/jct/
Software	https://hevc.hhi.fraunhofer.de/svn/svn_HEVCSoftware/
Bug tracking	https://hevc.hhi.fraunhofer.de/trac/hevc
Email list	https://mailman.rwth-aachen.de/mailman/listinfo/jct-vc

1.4.3 Deliverables

The development work for the new international standard and recommendation is reflected in a set of deliverables, which turn to become normative or remain to be supplemental in their final form. These comprise the specification text itself, the reference software, a conformance specification, and the test model. Furthermore, a verification report is produced which documents and demonstrates the achieved performance increase compared to the previous standard H.264 | AVC.

1.4.3.1 Specification

The draft specification is developed as a working draft document or draft amendment, depending on the state of the working phase. Since this document represents the current state of the main deliverable of the group, it has highest priority and receives most attention of the involved experts. While the specification of the first version of HEVC has been finalized, ongoing JCT-VC work on maintenance and extensions is reflected in corresponding specification drafts [25]. Depending on the scale of the introduced changes, they may be published as an amendment or as a new edition of the standard. While amendments only include the applicable changes and extensions of the specification, a new edition would imply the publication of a complete integrated version of the specification text.

A new version of the draft text is released after every meeting, integrating the adoptions of the meeting. Before a target release date which is established from meeting to meeting, the editors usually update the pre-versions of the document in the document repository such that the JCT-VC experts can study the state of the specification in a close manner.

1.4.3.2 Test Model

Besides the development of the specification text, a test model document is maintained which describes the encoder control and algorithms implemented in the reference software. The test model text complements the reference software which implements the reference decoder and a rate-distortion optimized encoder. It is referred to as the "High Efficiency Video Coding Test Model" (HM) for HEVC [24]. The document aids experts in the work with and the analysis of the reference software, including the integrated normative tools. By describing the encoder decisions for application of the specified tools, the test model text serves as a tutorial example on how to implement an encoder control for the tool set in the specification.

1.4.3.3 Reference Software

The reference software implements the decoding process as specified in the (draft) specification and an example encoder which generates bitstreams complying to the specification. It is referred to as the HM software [38].

A new version of the reference software is released after every meeting, integrating the adoptions of the meeting. The integration work is documented in the version control system of the software repository and can thereby be closely monitored by involved and interested experts. In the development phase, the reference software specifically serves as the platform to test and analyze proposed tool changes. Simulations which are performed using the reference software confirm the expected rate-distortion performance along the development of the specification draft.

The software reference decoder can be used by encoder manufacturers for testing if their encoded bitstreams comply to the specification. Since the reference software does not necessarily include all restrictions specified in the text, successful decoding of a bitstream by this software may give a good indication but not a final proof for compliance. The reference software is maintained and developed to meet the goal of compliance as closely as possible.

The encoder of the reference software provides a rate-distortion optimized implementation, which aids in comparing the performance impact of tools in the context of the reference model. It should be noted that the reference software commonly does not include sophisticated rate control for real encoding tasks nor does it include significant error concealment in the decoder in the case that e.g. corrupted bitstreams are fed to the decoder. Such tools are up to the implementers of encoders and decoders for their respective target applications.

1.4.3.4 Conformance Specification

With the approval of the final version of the specification text, the task to assert conformance to the specification gains high attention. The conformance specification is developed to provide means to manufacturers of encoders and decoders to test their product for compliance to the specification text. An important means for conformance testing of decoders is a suite of conformance bitstreams which are generated by the contributing experts of the JCT-VC. These bitstream are designed to include a complete as possible test set for all tools included in the specification. The design task for the conformance specification is to approach completeness as much as possible.

1.4.4 Structure of the JCT-VC

During the meetings, the review and decisions on input contributions is done either in plenary or in parallel tracks chaired by one of the two JCT-VC chairs. If consensus

Table 1.2 Tasks and appointed JCT-VC experts

Function	Name
Chairs	Gary J. Sullivan
	Jens-Rainer Ohm
Editing team	Benjamin Bross
	Woo-Jin Han
	Jens-Rainer Ohm
	Gary J. Sullivan
	Ye-Kui Wang
	Thomas Wiegand
Software coordinators	Frank Bossen
	Karsten Sühring
	David Flynn
Conformance coordinators	Teruhiko Suzuki
	Wade Wan
Verification coordinators	T.K. Tan
	Vittorio Baroncini
	Marta Mrak
	Marta Karczewicz
	Wade Wan
	Jiangtao Wen

on a decision is determined in one of the parallel tracks, the subject is reviewed in plenary in order to update all participants and search affirmation of the consensus by the whole group.

Some of the duties in JCT-VC require permanent engagement of qualified experts. This specifically concerns the chairs but also the editors, test model software coordinators and the coordinators for the establishment of a conformance test set. A list of the experts appointed for the relevant tasks in JCT-VC is provided in Table 1.2.

1.4.4.1 Break-Out Groups

Depending on the topic, informal break-out groups (BoGs) are frequently installed. The chair of a break-out group is appointed by the JCT-VC chairs with the consensus of the participants. In the BoG, input documents and proposals are reviewed, analyzed, or e.g. merged, and decisions of the JCT-VC are prepared. The BoG chair summarizes the activity in a BoG report which is registered as an input document during the meeting. Decisions on the outcome of a BoG are made by consensus of the JCT-VC.

1.4.4.2 Editing Team

An editing team is appointed which is responsible for drafting the specification text. It receives input from proponents of adopted proposals and integrates these into the specification. When doing so, it assures that conventions that are defined for the specification text are followed and that the tool specification integrates with the previously available text.

1.4.4.3 Software Coordinators

Similar to the editing team, the software coordinators task is to handle the integration of adoptions into the reference software. They manage the software code base and ensure a sufficient quality of the software integration of the tools. By extensive testing according to established JCT-VC common testing conditions, the performance impact of integrated tools is monitored in order to maintain and confirm the performance of the reference software over subsequent development cycles.

1.4.4.4 Ad Hoc Groups

JCT-VC ad hoc groups (AhGs) are instantiated to work on certain topics in between meetings. These AhGs follow the formal definition in the ISO/IEC JTC 1 [31]. The working topic is defined by the AhG mandates, which are set up at a JCT-VC meeting. The appointed AhG chairs coordinate and moderate the activity and provide a written report as input to the next meeting. This report includes a summary of the activity since the last meeting, an overview of related input documents and a set of recommendations to the JCT-VC. While ad hoc groups on certain topics may continuously be initiated from meeting to meeting, the groups formally only exist for one meeting cycle and may be restructured according to the needs of the JCT-VC at each meeting.

1.4.4.5 Core Experiments

The regular process for a tool to be adopted into the draft specification is the definition of a core experiment which verifies the validity of a proposed change. A core experiment can be defined at a meeting on technology that has been proposed by input to the same or previous meetings. Conceptually, the proponent must provide a precise text description of the proposed technology as well as a proposed implementation of the technology into the current reference software. No further development of the technology is allowed in the course of a core experiment. Such development must be reported as separate input and is not jointly treated with the core experiment contributions. Conceptually, a successful core experiment can be considered as the last step before adopting a proposal into the draft specification. The successful conduction of a core experiment does not imply guaranteed adoption, though.

While the proponent reports the performance results of the tool under investigation in this test, the most important task is on the core experiment participants who provide a *cross-check* of the proposed technology. Cross-checkers must be from a different participant company or organization and must be independent from the proponent. In the most thorough form, the cross-check is performed by an independent implementation of the proposal based on the proposed text description and a corresponding independent verification of the simulation results reported by the proponent.[5] In practice, the verification task is often performed by assessing the proponents implementation into the reference software, asserting that no other than the proposed changes have been implemented, and verifying the simulation results.

The described core experiment process is well suited for testing and verification of proposed addition or modification of coding tools. Studies of changes in structures above the coding layer (the high-level syntax) do not easily allow for verification of the benefit of proposed changes. In such cases, assessment by qualified experts is obligatory.[6]

References

1. Codecs for videoconferencing using primary digital group transmission. ITU-T Rec. H.120. http://www.itu.int/rec/TREC-H.120/en (1993). Accessed 14 Apr 2014
2. Sullivan, G.J.: Overview of international video coding standards (preceding H.264/AVC). Pres. ITU-T VICA Workshop, ITU Headquarter, ITU-T, Geneva. http://www.itu.int/ITU-T/worksem/vica/docs/presentations/S0_P2_Sullivan.pdf (2005). Accessed 14 Apr 2014
3. Video codec for audiovisual services at p×64 kbit/s. ITU-T Rec. H.261. http://www.itu.int/rec/T-REC-H.261/en (1993). Accessed 14 Apr 2014
4. Décina, M., Scace, E.L.: CCITT recommendations on the ISDN: A review. IEEE J. Sel. Areas Commun. **4**(3), 320–325 (1986). doi:10.1109/JSAC.1986.1146333
5. Rtcweb Status Pages. IETF. http://tools.ietf.org/wg/rtcweb/charters (2013). Accessed 14 Apr 2014
6. Information technology—Coding of moving pictures and associated audio for digital storage media at up to about 1.5 Mbit/s—Part 2: Video. ISO/IEC 11172-2:1993 (MPEG-1). http://www.iso.org/iso/home/store/catalogue_tc/catalogue_detail.htm?csnumber=22411 (2006). Accessed 14 Apr 2014
7. Information technology—Generic coding of moving pictures and associated audio information—Part 2: Video. ISO/IEC 13818-2:2013 (MPEG-2). http://www.iso.org/iso/home/store/catalogue_tc/catalogue_detail.htm?csnumber=61152 (2013). Accessed 14 Apr 2014
8. Information technology—Generic coding of moving pictures and associated audio information: Video. ITU-T Rec. H.262 (MPEG-2). http://www.itu.int/rec/T-REC-H.262/en (2012). Accessed 14 Apr 2014
9. Del Rosario, J.: Reality check on MPEG-2. Northern Sky Research. http://www.nsr.com/new_sresources/the-bottom-line/reality-check-on-mpeg-2/ (2013). Accessed 14 Apr 2014
10. Video coding for low bit rate communication. ITU-T Rec. H.263. http://www.itu.int/rec/T-REC-H.263/en (2005). Accessed 14 Apr 2014

[5] In the development process of MPEG-4 Visual, even two complete independent reference implementations were maintained by the working group to assert the validity of the reported results.

[6] This assessment sometimes has been referred to as a ‘mental cross-check’.

11. Information technology—Coding of audio-visual objects—Part 2: Visual. ISO/IEC 14496-2:2004 (MPEG-4). http://www.iso.org/iso/home/store/catalogue_tc/catalogue_detail.htm?csnumber=39259 (2009). Accessed 14 Apr 2014
12. Pereira, F., Ebrahimi, T.: The MPEG-4 Book. Prentice Hall, Upper Saddle River (2002)
13. Advanced video coding for generic audiovisual services. ITU-T Rec. H.264 (AVC). http://www.itu.int/rec/T-REC-H.264/en (2014). Accessed 14 Apr 2014
14. Information technology—Coding of audio-visual objects—Part 10: Advanced video coding. ISO/IEC 14496-10:2012 (AVC). http://www.iso.org/iso/home/store/catalogue_tc/catalogue_detail.htm?csnumber=61490 (2012). Accessed 14 Apr 2014
15. MPEG and VCEG. Terms of reference for Joint Video Team (JVT) activities. Doc. ITU-T Q.6/SG16 and ISO/IEC JTC1/SC29/WG11, Pattaya, Thailand. http://www.itu.int/oth/T3401000001/en (2001). Accessed 14 Apr 2014
16. Website on the Joint Video Team. International Communication Union. http://www.itu.int/en/ITU-T/studygroups/com16/video/Pages/jvt.aspx (2010). Accessed 14 Apr 2014
17. Sullivan, G.J., et al.: Overview of the high efficiency video coding (HEVC) standard. IEEE Trans. Circ. Syst. Video Technol. **22**(12), 1649–1668 (2012). doi:10.1109/TCSVT.2012.2221191
18. VCEG and MPEG. Terms of reference of the Joint Collaborative Team on Video Coding Standard Development. Doc. VCEG-AM90. ITU-T SG16/Q6 VCEG, Kyoto, JP (2010)
19. VCEG. Draft requirements for next-generation video coding project. Doc. VCEG-AL96. ITU-T SG16/Q6 VCEG, London, GB (2009)
20. VCEG and MPEG. Joint call for proposals on video compression technology. Doc. VCEG-AM91. ITU-T SG16/Q6 VCEG, Kyoto, JP (2010)
21. JCT-VC. Report of subjective testing of responses to Joint Call for Proposals (CfP) on video coding technology for high effciency video coding (HEVC). Doc. JCTVC-A204. 1st meeting: Joint Collaborative Team on Video Coding (JCT-VC) of ITU-T SG 16 WP 3 and ISO/IEC JTC 1/SC 29/WG 11, Dresden, Germany (2010)
22. Wiegand, T., et al.: WD1: Working draft 1 of high effciency video coding. Doc. JCTVC-C403. 3rd Meeting: Joint Collaborative Team on Video Coding (JCT-VC) of ITU-T SG 16 WP 3 and ISO/IEC JTC 1/SC 29/WG 11, Guangzhou, CN (2010)
23. McCann, K., et al.: High effciency video coding (HEVC) test model 1 (HM 1) encoder description. Doc. JCTVC-C402. 3rd Meeting: Joint Collaborative Team on Video Coding (JCT-VC) of ITU-T SG 16 WP 3 and ISO/IEC JTC 1/SC 29/WG 11, Guangzhou, CN (2010)
24. Kim, I.-K., et al.: High efficiency video coding (HEVC) test model 10 (HM10) encoder description. Doc.JCTVC-L1002. 12th Meeting: Joint Collaborative Team on Video Coding (JCT-VC) of ITU-T SG 16 WP 3 and ISO/IEC JTC 1/SC 29/WG 11, Geneva, CH (2013)
25. Bross, B., et al.: High efficiency video coding (HEVC) text specification draft 10 (for FDIS & Last Call). Doc. JCTVC-L1003. 12th Meeting: Joint Collaborative Team on Video Coding (JCT-VC) of ITU-T SG 16 WP 3 and ISO/IEC JTC 1/SC 29/WG 11, Geneva, CH (2013)
26. High efficiency video coding. ITU-T Rec. H.265 (HEVC). http://www.itu.int/rec/T-REC-H.265/en (2013). Accessed 14 Apr 2014
27. Information technology—High efficiency coding and media delivery in heterogeneous environments—Part 2: High efficiency video coding. ISO/IEC 23008-2:2013 (HEVC). http://www.iso.org/iso/home/store/catalogue_tc/catalogue_detail.htm?csnumber=35424 (2013). Accessed 14 Apr 2014
28. Tan, T.K., et al.: HEVC verification test results. Doc. JCTVC-Q204. 17th Meeting: Joint Collaborative Team on Video Coding (JCT-VC) of ITU-T SG 16 WP 3 and ISO/IEC JTC 1/SC 29/WG 11, Valencia, ES (2014)
29. Ohm, J.-R., et al.: Comparison of the coding efficiency of video coding standards—including high efficiency video coding (HEVC). IEEE Trans. Circ. Syst. Video Technol. **22**(12), 1669–1684 (2012). doi:10.1109/TCSVT.2012.2221192
30. ISO/IEC. ISO/IEC Directives, Part 1: Consolidated ISO supplement—procedures specific to ISO. Technical report, Fourth edition, ISO/IEC. http://www.iso.org/iso/home/standards_development/resources-for-technical-work/iso_iec_directives_and_iso_supplement.htm (2013). Accessed 14 Apr 2014

31. ISO/IEC JTC1. ISO/IEC Directives, Part 1: Consolidated JTC 1 Supplement Procedures for the technical work. Technical report, ISO/IEC. http://isotc.iso.org/livelink/livelink/open/jtc1 (2012). Accessed 14 Apr 2014
32. Question 6/16–Visual coding, ITU-T. http://www.itu.int/ITU-T/2005-2008/com16/sg16q6.html (2008). Accessed 14 Apr 2014
33. Alternative approval process for new and revised ITU-T Recommendations. ITU-T Rec. A.8. http://www.itu.int/rec/T_REC-A.8/en (2008). Accessed 14 Apr 2014
34. ITU-T. Resolution 1—Rules of procedure of the ITU telecommunication standardization sector (ITU-T). Technical report, ITU-T, Dubai. http://www.itu.int/en/ITU-T/wtsa12/Documents/resolutions/Resolution%2001.pdf (2012). Accessed 14 Apr 2014
35. Guide for ITU-T and ISO/IEC JTC 1 cooperation. ITU-T and ISO/IEC JTC 1 Rec. A.23 Annex A. http://www.itu.int/rec/TREC-A.23/en (2010). Accessed 14 Apr 2014
36. Terms of reference of the Joint Collaborative Team on 3D Video Coding Extension Development (JCT-3DVE). ITU-T and ISO/IEC JTC 1, Geneva. http://www.itu.int/en/ITU-T/studygroups/com16/video/Documents/ToR-JCT-3V-TD-PLEN-0532-16.pdf (2012). Accessed 14 Apr 2014
37. Terms of reference of the Joint Collaborative Team on Video Coding Standard Development. ITU-T and ISO/IEC JTC 1, Kyoto. http://www.itu.int/oth/T4601000001/en (2010). Accessed 14 Apr 2014
38. JCT-VC. HEVC Reference Software. https://hevc.hhi.fraunhofer.de/svn/svn_HEVCSoftware/ (2014). Accessed 14 Apr 2014

Chapter 2
Video Coding Fundamentals

This chapter provides an overview of the fundamentals of the video coding, including the tool chain, from acquisition of the video sequence via coding and transmission to display. The premier focus is on the aspects that concern the encoding and decoding process. These include the representation format of video sequences including the representation of color. The fundamental concept and the main building blocks of hybrid video coding are presented. The aim of the presentation is to provide a conceptual overview of the components and how they interact. In the following chapters, the realization of the building blocks of this scheme in HEVC will be presented and analyzed in greater detail.

2.1 Video Coding Systems

The setup of a complete video coding system can be organized according to the block diagram in Fig. 2.1. The building blocks of the video coding system are summarized below:

- Video Acquisition—Source of the video sequence which is output in a digital form. The acquisition process may be temporally and locally decoupled from the following processing steps.
- Pre-Processing—Operations on the raw uncompressed video source material, such as trimming, color format conversion, color correction, or de-noising.
- Encoding—Transformation of the input video sequence into a coded bitstream. The aim of encoding is to generate a compact representation of the input video sequence which is suitable for the transmission method in the given application scenario.
- Transmission—Packaging the bitstream into an appropriate format and transmission over the channel. Transmission includes sending and delivery of the video data to the receiver side, as well as potential methods for loss protection and loss recovery. In some scenarios, like in conversational applications with real-time

M. Wien, *High Efficiency Video Coding*, Signals and Communication Technology,
DOI 10.1007/978-3-662-44276-0_2

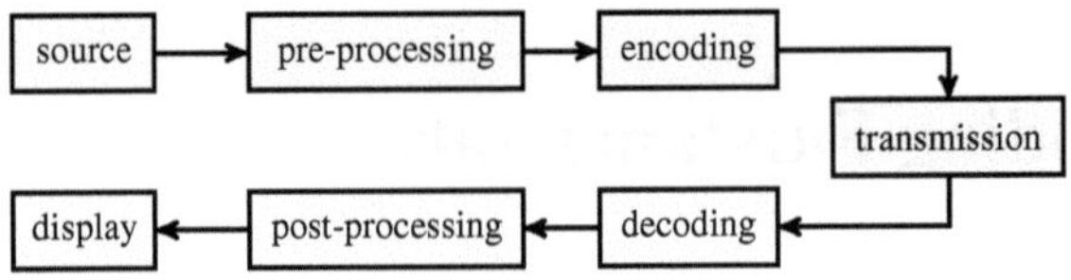

Fig. 2.1 General block diagram of a video coding system

requirements, feedback from the receiver side to the sender side is available and can be used to control the encoder configuration. Such feedback can further be used e.g. to request re-transmission in case of lost information.

- Decoding—Transformation of the received bitstream into a reconstructed video sequence. Since video encoding usually requires lossy compression to achieve the target transmission bitrate constraints, the decoded video constitutes an approximation of the original source video. If unrecoverable transmission losses have occurred, the decoder applies concealment strategies to recover the corrupted video sequence as much as possible.
- Post-Processing—Operations on the reconstructed video sequence for enhancement or for adaptation of the sequence for display. These operations can e.g. include color correction, trimming, or re-sampling. Also special effects may be applied as determined by the application.
- Display—Presentation of the video sequence for viewing. The video sequence needs to be transferred into the appropriate color format for display. The output timing of the pictures of the video sequence is an important aspect to achieve the intended visual impression.

The processing steps described above represent a simplified view on the tool chain that is used. In real applications, e.g. multiple pre-/post-processing steps in conjunctions with iterative de- and re-encoding and transmission may be used. In many applications, transcoding techniques are applied, where the incoming bitstream is transformed into a bitstream with different properties (e.g. lower bitrate), or into a bitstream following a different video coding specification.

2.1.1 Video Acquisition

Video sequence can originate from a variety of sources. These include e.g. capture of natural scenes, scanning of photographed film material, analogue recording of video, or synthetically generated video material like computer generated animation. Furthermore, mixtures and combinations of the above mentioned sources may be possible. Depending on the origin of the video material, the signal reveals different characteristic properties concerning the represented scene content as well as the presence of other signal components which interfere with the content, such as camera noise, film grain, motion blur, or artifacts from digitized analogue or physical source material. In a simple signal model, an acquired video sequence can be described as an undistorted video signal which is additively distorted by a noise component. Further, non-additive artifacts in the video sequence may for example relate to geometrical distortions or color alignment issue due to camera lense properties. When natural scenes are captured, camera properties like the construction and the resolution of the sensor chip or the shutter speed play an important role for the resulting video sequence properties and the picture quality.

2.1.2 Pre-processing

As indicated in the previous section, the incoming video source material may have various degradations and show undesirable artifacts, which should be removed. Further pre-processing operations include trimming, geometrical format conversion, and color format conversion. Furthermore, the picture rate at which the video sequence is captured might be subject to modifications. Since different distribution methods operate with different picture rates, picture rate conversion is an important task.

In the pre-processing stage, the incoming raw video material may therefore undergo multiple signal processing and signal enhancement operations in preparation for the next steps in the video tool chain. For video coding, specific pre-processing may be applied in order to adapt the signal to the requirements of the target application. This includes e.g. the picture rate, the picture resolution, and the color format. Further dedicated pre-processing may be applied to the source material to support the encoder for an improved compression efficiency or adjusted visual impression. Applicable operations include e.g. local lowpass filtering or denoising filtering.

2.1.3 Encoding

The encoder stage transforms the input video sequence into the coded representation of a bitstream. The bitstream in contemporary video coding standards has a packet oriented structure where the encoded information is hierarchically organized according to the employed specification.

The encoder must generate a bitstream to the application requirements. These include e.g. the selection of tools applied for encoding (which correspond to a certain profile in a standard video coding scheme) or constraints on decoder-side buffer requirements and the resulting bitrate for transmission of the coded video.

The optimization of the encoder control decisions is a crucial aspect of encoder design. On the picture level, the selection of coding tools as well as the bit allocation over the regions of the picture has to be considered. Two independent encoder implementations which are conforming to the same video coding specification may produce very different reconstructed video quality.

Depending on the time that is available for encoding the video, the amount and effort spent for optimization varies. In real-time application, the encoder can only use past information of the video it is encoding to adapt to the content. Furthermore, transmission requirements and other unpredictable events influence the encoding process.

In production systems that prepare the video for storage and media distribution, the video is processed for large scale distribution to a potentially very high number of customers. Here, off-line processing of the video can be applied and serious effort for encoder optimization and parameter tuning may be invested.

In preparation for transmission to the decoder side, the encoded bits are encapsulated into a transport format according to the application requirements.

2.1.4 Transmission

The transmission building block in the diagram of Fig. 2.1 represents the delivery of the encoded video sequence to the receiving side. This transmission may be in real-time and even bi-directional, e.g. in conversational applications. Here, the end-to-end delay between the sender and the receiver side is crucial. In live broadcast applications (like sports events) a certain delay may be tolerable as usually no direct feedback between receiver and sender is required. In storage and video distribution applications, the transmission may be temporally completely decoupled from the production process. Such applications include video-on-demand systems or storage on optical discs like DVD or Blu-ray.

Design aspects of the transmission system include the accessibility of the desired signal. This is e.g. important in multi-channel applications like television where the ability of entering to or switching between different channels is an important feature. In conversational applications where more than two participants are connected in a joint video call, the organization and redirection of the video signals to the correct receiver sides is required. In streaming and storage applications, features like fast-forward or backward search operations are desirable.

A further important aspect is the robustness of the transmission signal against disruptions, distortions, and losses. In packet-oriented transmission scenarios, an error-free transmission of the packets is often required. Distortions in the transmission signal may induce the loss of the affected packets in such systems. Therefore, mechanisms for protection against losses, and potentially the recovery of lost packets are desirable.

2.1.5 Decoding

The decoder stores the received bitstream in a buffer and reconstructs the encoded data into a video sequence in the format indicated by the encoder. In a standard video coding system, the decoding process follows the normative procedures as defined in the underlying video coding specification. If no losses occur, the reconstructed video at the output of the decoder exactly matches the video as held available for reference at the encoder side. The bitstream may include information that only parts of the decoded video are displayed. It may even be indicated that complete decoded pictures are suspended from display (while potentially being used for prediction).

An important aspect of decoder control is the buffer management for the decoded pictures as well as for the incoming bitstream. Decoded pictures in the decoded picture buffer may be further used for reference in the decoding process. Further, in systems that operate at a defined picture rate for presentation, it must be asserted that the decoding process makes the decoded pictures available for display according to the timing constraints imposed by the employed format.

In case of transmission losses, the decoder must be able to act on missing data and potentially unforeseen conditions. In such cases, error resilience is required and the decoder must be able to apply error concealment strategies.

2.1.6 Post-processing

The pictures that are output by the decoder may be fed into a post-processing stage where operations for image enhancement and for display adaptation may be performed. Also post-decoder error concealment methods may be applied.

If the format of the decoded video material does not correspond to the format required for display, application of conversion operations similar to those listed for pre-processing is indicated. These include e.g. the color format, the picture resolution or the picture rate.

2.1.7 Display

The display device is the interface to the human visual system (HVS) as the final sink for the video signal.

As indicated before, not all samples of the coded video sequence are necessarily used or even available for display. For example, parts of the left or right boundary of the picture (or both) may be excluded from the area for display. If the remaining part of the picture is to be displayed on the full screen, this mode of operation is called *overscan*. If number of coded samples is smaller than the number of samples available on the display, the video may either be 'stretched' to match the display resolution, or the video is displayed without modification and the display area is not completely covered by the video. This mode of operation is called *underscan*.

The video format, the color space of the source material, and other helpful information that may be utile for adequate preparation of the video for display can be signaled with the video data. Such information is called *Video Usability Information*and is further described in Sect. 5.7. While it does not necessarily have an impact on the normative decoding process, keeping the information associated with the video itself ensures that the intended parameters for display are conveyed to the decoding side. In some application scenarios, the video usability information may be fixed and therefore be hard-coded for the system.

2.2 Structure of a Video Sequence

A *video sequence* consists of a series of pictures, which are presented with constant or variable time intervals between successive pictures. In order to achieve the impression of motion, a picture rate of at least about 24 pictures per second is required. The minimum picture rate depends on the lighting conditions and the content to be displayed [1]. The unit of the picture rate is Hertz (Hz), with $1\,\text{Hz} = 1\,\text{s}^{-1}$. The picture rate is also referred to as the frame rate using the acronym 'fps' (frames per second). High definition (HD) video today applies picture rates of 50–60 Hz. For Ultra HD formats, picture rates of up to 120 Hz are specified [2]. Even higher picture rates are discussed, see e.g. [3].

In the context of video coding, an input video sequence is encoded into a bitstream which contains basically all information and data that is necessary to decode the bitstream and reconstruct the output video sequence for display at the receiving side. In the vast majority of video coding applications, encoding the video sequence into a bitstream corresponds to a lossy compression, i.e. the reconstructed video sequence at the decoder only approximates the original input video sequences. The transmission between the sender, where the original sequence is encoded, and the receiver, which includes the decoder, may be instantaneous or close to real-time, e.g. in video communication or live broadcast scenarios. It may also be completely decoupled, e.g. if the data is stored on a Blu-ray disc, DVD, or in streaming applications.

2.2.1 *Pictures, Frames, and Fields*

A *picture* is an array or a set of arrays of *samples* with intensity values. A sample in the array is also denoted as *pixel* (derived from *pic*ture *el*ement, or a *pel*). If the picture is monochrome, i.e. it has one single color component, the picture consists of a single sample array. For representation of color in video or pictures, usually three color components are employed as detailed below. Correspondingly, a color picture consists of three intensity arrays, with one array for each component. In some applications a fourth array is included which represents the 'opaqueness' of the pixels. This array is often called the α component. Such information can be used when mixing multiple videos into one scene, e.g. to add a foreground speaker to a background video.

Historically, the video format for capture, distribution, and display of television programs was interlaced. In interlaced video, the even and odd sample lines are collected in the *top* and *bottom field* pictures, respectively. The two fields are alternately

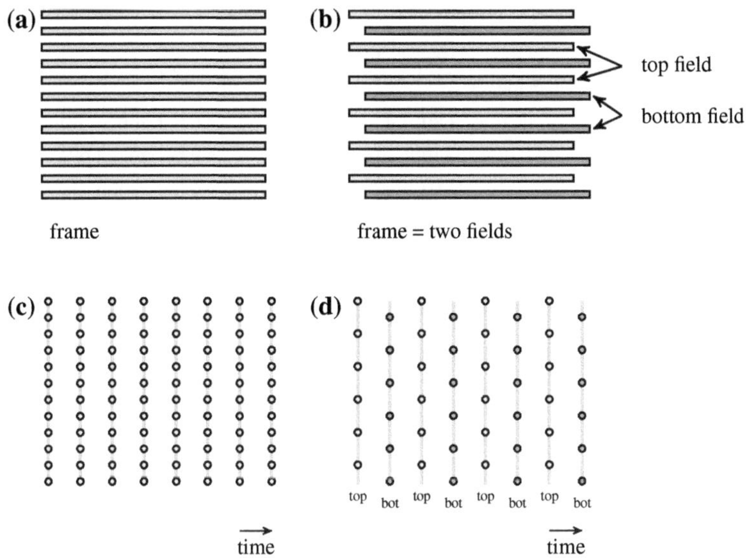

Fig. 2.2 Frames and fields in progressive and interlaced video. **a** Lines of a progressive frame. **b** *Lines* of the *top* and *bottom* fields of an interlaced frame. **c** One-dimensional representation of the scanning of progressive frames over time. **d** One-dimensional representation of the scanning of interlaced frames over time

displayed. The displays were built with CRT screens[1] with relatively slow fading characteristics. The alternating display of the even and odd lines on these screens was the state of the art in display technology until the end of the 20th century.

In interlaced video, a pair of a top and a bottom field constitutes a *frame*. For non-interlaced material, a frame holds a single picture which has been acquired in *progressive scan*. The progressive and interlaced scans are visualized in Fig. 2.2, including an illustration of the relation of the fields over time. For details on the interlaced and progressive video formats, the reader is referred to [1].

2.2.2 Sample Shape

When capturing video, each picture is represented by a set of lines and a set of samples (i.e. samples) per line. The shape of the samples in a picture is not necessarily square. There exists a variety of other aspect ratios of width and height of the samples that are commonly used in video coding systems. Different sample aspect ratios can be used to adapt the display size of the video to the aspect ratio of the screen in dependency of the number of lines and the number of samples per line in a picture.

[1] CRT = Cathode Ray Tube

Table 2.1 Common sample aspect ratios

Index[a]	Sample aspect ratio	Pixel shape	Usage examples	
1	1:1	□	7680 × 4320 frame	No horizontal overscan
			3840 × 2160 frame	No horizontal overscan
			1280 × 720 frame	No horizontal overscan
			1920 × 1080 frame	No horizontal overscan
			640 × 480 frame	No horizontal overscan
2	12:11	□	720 × 576 4:3 frame	With horizontal overscan
			352 × 288 4:3 frame	No horizontal overscan
3	10:11	□	720 × 480 4:3 frame	With horizontal overscan
			352 × 240 4:3 frame	No horizontal overscan
4	16:11	□	720 × 576 16:9 frame	With horizontal overscan
			528 × 576 4:3 frame	No horizontal overscan
5	40:33	□	720 × 480 16:9 frame	With horizontal overscan
			528 × 480 4:3 frame	No horizontal overscan
6	24:11	□	352 × 576 4:3 frame	No horizontal overscan
			480 × 576 16:9 frame	With horizontal overscan
7	20:11	□	352 × 480 4:3 frame	No horizontal overscan
			480 × 480 16:9 frame	With horizontal overscan
8	32:11	□	352 × 576 16:9 frame	No horizontal overscan
9	80:33	□	352 × 480 16:9 frame	No horizontal overscan
10	18:11	□	480 × 576 4:3 frame	With horizontal overscan
11	15:11	□	480 × 480 4:3 frame	With horizontal overscan
12	64:33	□	528 × 576 16:9 frame	No horizontal overscan
13	160:99	□	528 × 480 16:9 frame	No horizontal overscan
14	4:3	□	1440 × 1080 16:9 frame	No horizontal overscan
15	3:2	□	1280 × 1080 16:9 frame	No horizontal overscan
16	2:1	□	960 × 1080 16:9 frame	No horizontal overscan

[a] Index indication according to the HEVC Video Usability Information

A set of commonly used sample aspect ratios is presented in Table 2.1. The table provides the aspect ratio and visualizes the shape of the samples under the assumption of identical height. For each aspect ratio, typical application formats and the corresponding application of horizontal overscan are indicated [4].

2.3 Representation of Color

The perception of color in the eye is stimulated by electromagnetic waves. The visible range of the electromagnetic waves includes the approximate range from 380 to 780 nm. The impression of color corresponds to the perception of an intensity density distribution over the visible spectral range. Colors corresponding to a single wavelength are called spectral colors or *primary colors*. These cover a range from violet over blue, green, yellow, to red. A large set of observed color impressions are created by the combination of multiple spectral colors (e.g. magenta or brown).

The human visual system has three color receptors (cone cells) with maximum sensitivity in the wavelength areas of red, green, and blue. It further comprises 'gray-scale' receptors (rod cells) which are especially responsive in low lighting conditions, e.g. at night. The impression of different colors is generated by the combined stimulus of the three color receptors. The human visual system has the feature that the same color impression can be achieved by multiple different combinations of spectral colors. Stimuli leading to the same color impressions are characterized to be metameric. This metamerism is utilized for presentation of color on displays, where usually a mixture of three light sources (red, green, and blue) is used to generate the impression of color.

2.3.1 The CIE Standard Observer

Visual perception can be split into perception of brightness (light and dark) and chromaticity (the color impression). While the brightness is driven by the summarized intensity of the observed spectrum, the color impression is driven by the shape of the intensity distribution. A functional expression to represent perceived color by a mathematical description was first established and standardized in the CIE 1931 Standard Observer.[2] The model assigns each color a point in a three-dimensional XYZ space. The X, Y, Z values are derived from the observed spectrum using three color matching functions $\bar{x}(\lambda)$, $\bar{y}(\lambda)$, and $\bar{z}(\lambda)$ which simulate the sensitivity of the human color receptors, as shown in Fig. 2.3. As can be seen from the figure, $\bar{x}(\lambda)$ has a large peak in the range of the color red and a small peak in the range of blue, $\bar{y}(\lambda)$ has a peak in the range of green and $\bar{z}(\lambda)$ has its peak in the range of blue.

The color impression translates into the contribution of each color matching function. Let $I(\lambda)$ be the spectral intensity at wavelength λ. The X, Y, Z values are derived as

$$X = \int_{380\,\text{nm}}^{780\,\text{nm}} I(\lambda)\bar{x}(\lambda)\,\mathrm{d}\lambda, \quad Y = \int_{380\,\text{nm}}^{780\,\text{nm}} I(\lambda)\bar{y}(\lambda)\,\mathrm{d}\lambda, \quad Z = \int_{380\,\text{nm}}^{780\,\text{nm}} I(\lambda)\bar{z}(\lambda)\,\mathrm{d}\lambda. \tag{2.1}$$

[2] Defined in 1931 by the Commission internationale de l'éclairage (CIE), specified in ISO 11664-1 [5].

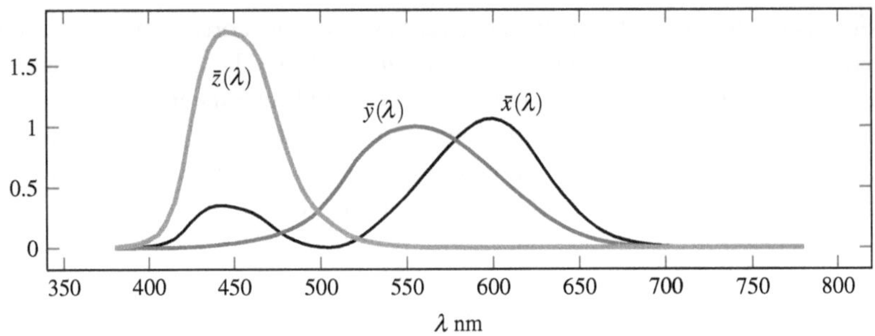

Fig. 2.3 CIE1931 standard observer color matching functions

In order to get an expression of the chromaticity, which is independent of the observed brightness, the X, Y, Z values are normalized as

$$x = \frac{X}{X+Y+Z}, \quad y = \frac{Y}{X+Y+Z}, \quad z = \frac{Z}{X+Y+Z}. \tag{2.2}$$

Since $x + y + z = 1$, the chromaticity can be specified by the (x, y)-pair. These (x, y) values are used to specify reference colors in various standards which deal with coding and display of images and video, see e.g. Table E-3 in HEVC Annex E.

This color representation further allows for the definition of a 'white point', i.e. the combination of x, y, z values which corresponds to a defined color impression of white. Several definitions of white points exist, including 'white C' specified by the CIE with the standard observer in 1931 and now deprecated, and 'white D65' which was specified by the CIE later on. See the references for further reading [1, 6].

2.3.2 Color Primaries

As stated above, display systems usually employ a mixture of red, green, and blue light sources to generate the impression of color on the screen. The mixture of red, green, and blue spans the RGB color space. As these colors are spectral or primary colors, they are called the *color primaries*.

In order to specify the perceived color impression generated from an input signal, color primaries are specified in all relevant video transmission and video display specifications. A set of commonly used color primaries is provided in Table 2.2.

2.3.3 Display Transfer Characteristics

The relation between the optical intensity and the corresponding signal representation is not necessarily linear. In order to deal with this effect, opto-electronic transfer

Table 2.2 Example color primaries used in ITU-R recommendations

Primary	Red		Green		Blue	
	x	y	x	y	x	y
ITU-T BT.601 625	0.640	0.330	0.290	0.600	0.150	0.060
ITU-T BT.601 525	0.630	0.340	0.310	0.595	0.155	0.070
ITU-T BT.709	0.640	0.330	0.300	0.600	0.150	0.060
ITU-T BT.2020	0.708	0.292	0.170	0.797	0.131	0.046

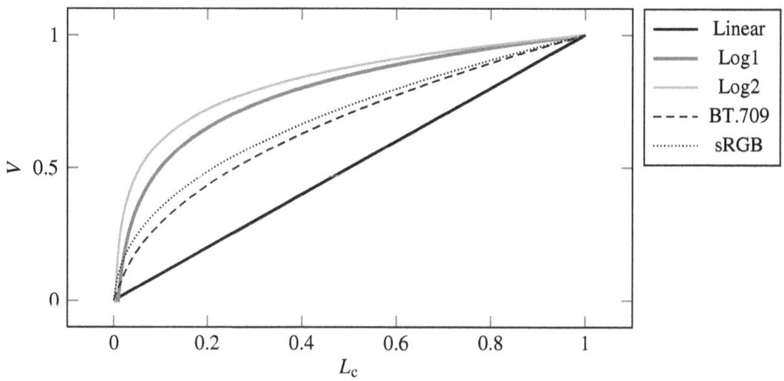

Fig. 2.4 Visualization of transfer characteristics between linear intensity input L_c and the value V of the signal component

characteristics are used, which describe the relation between a linear optical intensity input L_c and the nominal value V.[3]

Figure 2.4 shows some representative transfer characteristics according to Table E-4 in HEVC Annex E. A linear transfer characteristic is compared to logarithmic and other characteristics. Other characteristics very similar to the ITU-T BT.709 curve are specified in SMPTE-240M or ITU-T BT.2020. The curves for IEC 61966-2-4 and the extended gamut system in ITU-T BT.1361 are not shown. These specifications represent a much wider value range and include negative values for L_c.

2.3.4 Color Conversion

While capture and display usually are operating in the RGB color space, the usage of a *luminance* and two *chrominance* signals is established for coding and

[3] The reference electro-optical transfer function (EOTF) for flat panel displays used in HDTV production is specified in ITU-T BT.1886 [9].

transmission.[4] The luminance component represents the gray level intensity while the two chrominance components are required to specify the chromaticity as described in Sect. 2.3.1. Together with the color primaries and the transfer characteristics, the conversion parameters between the two color spaces define the complete settings of the color processing chain. If information on the parameters used in the generation process of the coded video sequence is available, the receiver is enabled to reproduce (or at least approximate) the intended color impression for display.

Let the linear intensity of the RGB components based on the color primaries as e.g. listed in Table 2.2 be denoted by E_R, E_G, and E_B. The direct conversion of these values would turn these intensities into a luminance and two chrominance signals. Since the system transfer characteristics for the color impression are not linear in general, these have to be taken into account by applying a transfer characteristics function as shown in Fig. 2.3. The resulting modified values are denoted by E'_R, E'_G, and E'_B. The conversion result from these modified intensity values is denoted by *luma* and *chroma* to indicate the different conversion source. The luma component is indicated by Y or L while the two chroma components are indicated by Cb and Cr. The derivation of the luma and chroma signals from the modified intensities is performed as follows.

Let the components E'_R, E'_G, E'_B be real valued and have a value range of [0, 1] each, where $E'_R = E'_G = E'_B = 0$ corresponds to nominal black and $E'_R = E'_G = E'_B = 1$ corresponds to nominal white. Let k_R, $k_G = 1 - k_R - k_B$, and k_B be weighting factors for the three components E'_R, E'_G, and E'_B, respectively. The real valued luma and chroma signals are derived as

$$E'_Y = k_R \cdot E'_R + (1 - k_R - k_B) \cdot E'_G + k_B \cdot E'_B, \tag{2.3}$$

$$E'_{PB} = \frac{1}{2} \cdot \frac{1}{1 - k_B} \cdot (E'_B - E'_Y), \tag{2.4}$$

$$E'_{PR} = \frac{1}{2} \cdot \frac{1}{1 - k_R} \cdot (E'_R - E'_Y). \tag{2.5}$$

The luma signal E'_Y has a range from nominal black ($E'_Y = 0$) to nominal white ($E'_Y = 1$). The two chrominance components have nominal range from −0.5 to 0.5. Example values for k_R and k_B are listed in Table 2.3 [2, 7, 8].

Table 2.3 Example color conversion coefficients used in ITU-R recommendations

	k_R	k_B
ITU-T BT.601	0.299	0.114
ITU-T BT.709	0.2126	0.0722
ITU-T BT.2020	0.2627	0.0593

4 Analog television started off with presentation of luminance only (black and white). By additional transmission of two chrominance signals, a backward compatible transmission of color and monochrome television signals was enabled [1].

For digital representation, the signals need to be quantized to a predefined precision which corresponds to a specified bit depth. In many applications, the bit depth of the component signals is set to 8 bit. In HD applications, 10 bit or even higher bit depths like 12 bit, 14 bit, or even 16 bit may be employed.

Since signal processing like filtering and prediction may induce some signal overshoot or undershoot, the quantization of the real-valued E'_{Y}, E'_{PB}, E'_{PR} signals to integer values of the specified bit depth B_{d} is performed using a slightly reduced value range, see HEVC, Annex E [4][5]:

$$Y = \mathrm{round}\left\{219 \cdot E'_{\mathrm{Y}} + 16\right\} \tag{2.6}$$

$$C_{\mathrm{b}} = \mathrm{round}\left\{224 \cdot E'_{\mathrm{PB}} + 128\right\} \tag{2.7}$$

$$C_{\mathrm{r}} = \mathrm{round}\left\{224 \cdot E'_{\mathrm{PR}} + 128\right\} \tag{2.8}$$

The quantization process is denoted for 8 bit signals here. For higher bit depth B_{d}, the (2.6)–(2.8) are scaled by $2^{B_{\mathrm{d}}-8}$. The resulting integer Y, C_{b}, and C_{r} components are denoted as the luma component and the two chroma components, respectively. The color space defined by these components is denoted as the YCbCr color space.

2.3.5 Chroma Sub-sampling

The human visual system is less sensitive to color than it is to structure and texture information. Therefore, in many application scenarios, it is more important to provide a high resolution luma component than to provide such detail for the chroma components. In high-definition high-quality applications like production, full resolution components at a high bit depth may be advisable. In consumer applications, sub-sampling of the chroma components is commonly applied.

Various types of chroma sub-sampling exist. Here, two chroma sub-sampling formats explicitly considered in the HEVC specification are briefly summarized. A comprehensive treatment on the subject is provided e.g. in [1]. The chosen color sub-sampling is commonly expressed in the relation between the number of luma samples compared to the number of chroma samples. The common notation is

$$\mathrm{YCbCr}\ Y{:}X_1{:}X_2.$$

The first value Y denotes the number of luma samples, the value X_1 and X_2 describe the sub-sampling format of the chroma components relative to the luma value. In the most common formats, $Y = 4$ is used. The X_1 value specifies the horizontal sub-sampling. The value $X_2 = 0$ indicates that the same X_1 sub-sampling factor is applied for the vertical direction. $X_2 = X_1$ indicates that no vertical sub-sampling

[5] Using $\mathrm{round}\{v\} = \mathrm{sgn}(v)\left\lfloor |v| + \frac{1}{2}\right\rfloor$.

Fig. 2.5 Chroma locations in typical video formats as used in the specification of the HEVC decoding process. **a** YCbCr 4:2:0 video. **b** YCbCr 4:2:2 video

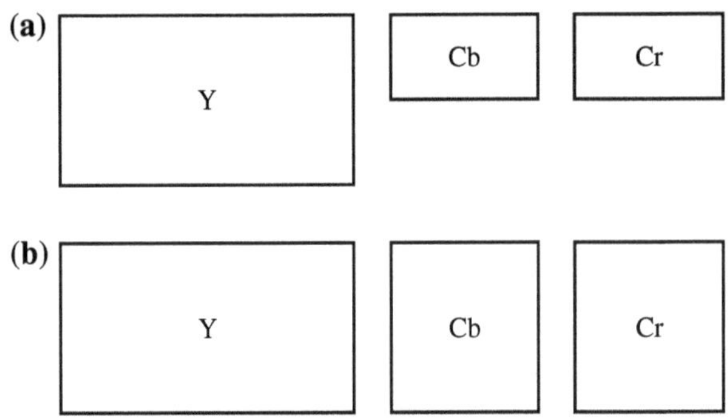

Fig. 2.6 Illustration of the YCbCr sample arrays for pictures using the chroma sub-sampling shown in Fig. 2.5 for video sequences with 16:9 picture size. **a** YCbCr 4:2:0. **b** YCbCr 4:2:2

is performed and both chroma components apply the same horizontal sub-sampling factor. A visualization for the two most common formats YCbCr 4:2:0 and YCbCr 4:2:2 is shown in Fig. 2.5. The corresponding sample arrays as used in the decoding process are shown in Fig. 2.6.

The locations of the chroma samples shown in Fig. 2.5 corresponds to the locations that are used in the specification of the HEVC decoding process (as it was done for H.264 | AVC). While these chroma locations are used for coding, the consideration of the correct chroma locations as present in the original source material is required for display. Otherwise, display of the video may include an observable shift between the luma component and the chroma component.

A variety of chroma locations may be considered for this purpose. For a coded video sequence, the applicable locations can be indicated in the Video Usability Information, see Sect. 5.7. The possible locations are summarized in Fig. 2.7. The numbering used in the Figure corresponds to the numbering used for signaling of the chroma locations in HEVC Annex E. In the specification of the decoding process for YCbCr 4:2:0 video, chroma location "0" is employed. The presentation includes the

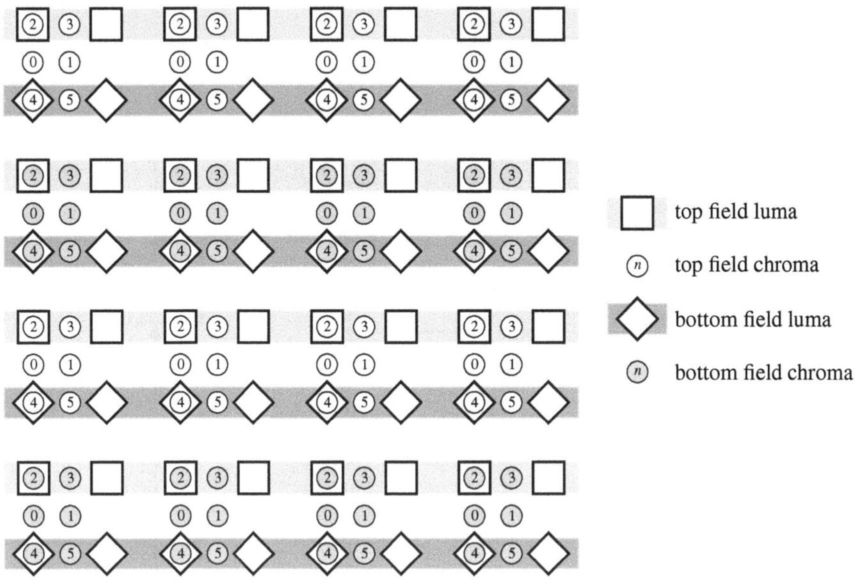

Fig. 2.7 Possible locations for chroma samples relative to the luma samples in video source material

distinction between top field and bottom field locations for both luma and chroma. In case of interlaced video material, the top field chroma samples are associated with the top field luma samples in a top field picture, while the bottom field chroma samples are associated with the bottom field luma samples in a bottom field picture. The chroma locations shown in Fig. 2.5 correspond to chroma location "0" for YCbCr 4:2:0 video and "2" and "4" for YCbCr 4:2:2 video. For each of these locations, a half way shift of the chroma location between two neighboring luma locations may have been used (e.g. in H.261, or JPEG). Such shifted chroma locations can be indicated by a "+1" operation on the indices mentioned previously.

2.4 The Hybrid Video Coding Scheme

Since H.261, the hybrid video coding scheme has been the basic structure for all video coding standards and recommendations of ITU-T and ISO/IEC MPEG. While the structure has not been changed, the algorithms represented by the building blocks have been refined and the applicable configuration for the algorithms has become more and more flexible over the last 25 years. The coding scheme is called *hybrid* as it combines temporal prediction between pictures of the video sequence with transform coding techniques for the prediction error [10].

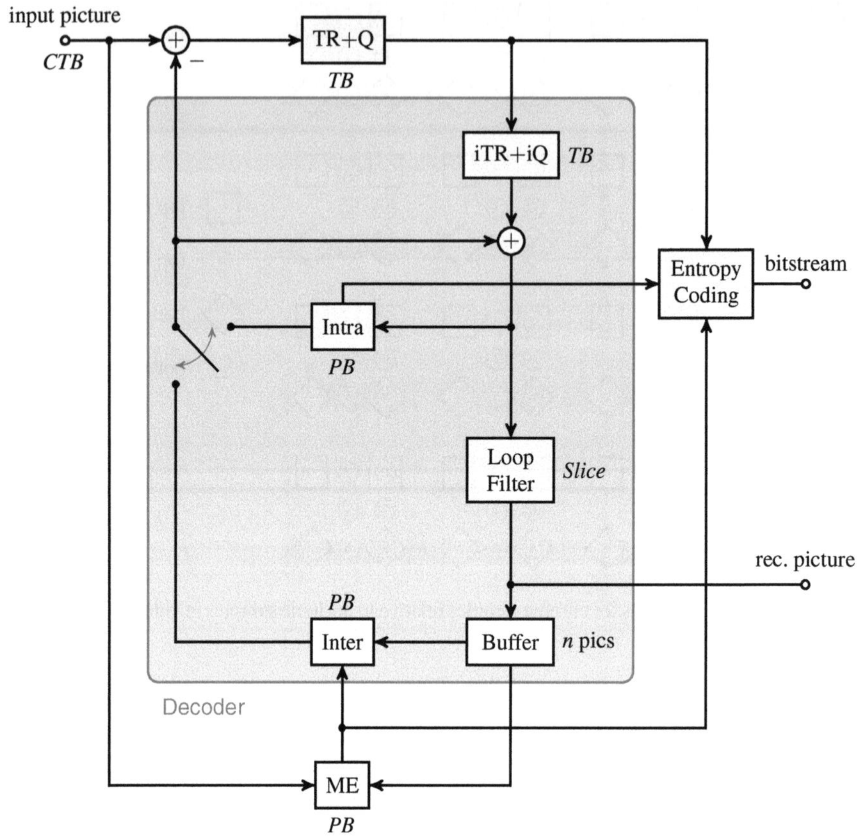

Fig. 2.8 Encoder block diagram for the hybrid video coding scheme (*CTB* coding tree block; *ME* motion estimation; *PB* prediction block; *Q* quantization; *TB* transform block; *TR* transform)

The hybrid video coding scheme has proved to be a well-suited tool to efficiently compress a video signal into a bitstream of smallest possible size. By prediction and transformation of the prediction error signal, it eliminates *redundant* information from the data. With the application of quantization after the transform step, *irrelevant* parts of the information are removed.[6]

The basic structure of the hybrid video coding scheme is shown in Fig. 2.8. The figure is simplified to a great extent to make the structure and the interdependencies visible. It corresponds to a classical DPCM loop[7]: The pictures of the input video sequence are fed to the encoder. A prediction signal generated from information available both, encoder and decoder is subtracted from the input signal. The residual, rep-

[6] If too strong quantization is applied, also relevant parts of the video signal content may be affected.

[7] DPCM: Differential Pulse Code Modulation.

resenting the resulting prediction error, is transformed, quantized, and encoded into the bitstream. The prediction parameters needed to reproduce the prediction signal at the decoder side are encoded into the bitstream as well. Under the assumption of error-free transmission, the encoder and decoder sides are synchronized since the encoder includes the complete prediction structure of the decoder. In Fig. 2.8, the building blocks which are included in both encoder and decoder are marked by a gray box.

Previously decoded pictures can be used for *inter* prediction in the current picture, which is to be encoded. The encoder can also choose to employ already coded neighboring samples within the current picture for *intra* prediction. The first coded picture of a video sequence can only apply intra prediction as no previous pictures are available. For the following pictures, the encoder decides between the two prediction options based on a decision criterion (e.g. rate-distortion optimization).

The process operates in a block-based fashion where the complete input picture is segmented into non-overlapping blocks, which are processed one after the other. For construction of the prediction signal, the transformed and quantized prediction error is first reconstructed and added with the available prediction. The signal is then processed by the loop filter (e.g. a deblocking filter). If the full picture has been processed, the reconstructed picture is available, as it also is at the decoder side. The reconstructed picture is stored in the decoded picture buffer to make it available for prediction.

For inter prediction, the motion estimation stage (ME) searches for the best prediction available for the current picture block in the decoded picture buffer. For intra prediction, sample values from already reconstructed neighboring blocks of the current picture can be used for prediction. Depending on the encoder decision which prediction mode has been selected, either the intra or the inter prediction signal is used for prediction of the current block.

The block diagram shown in Fig. 2.8 does not include the building blocks required for encoder control. An encoder implementation requires a control engine which decides on the applicable prediction modes, prediction and filtering parameters, as well as the applicable quantization parameters. Control information to inform the decoder on the selected prediction tools and configurations is included in the bitstream as well.

In the following, the building blocks of the hybrid video coding scheme are briefly described. The acronyms used in Fig. 2.8 are explained. For simplification purpose, the description focuses on the luma component. Prediction and transform coding are performed similarly for the chroma components.

2.4.1 Picture Partitioning

Each picture is partitioned into a complete set of non-overlapping blocks. An illustration of the basic partitioning of a picture is shown in Fig. 2.9. In previous video coding specifications, the basic block structure was the so-called *macroblock* of 16×16 luma samples and the corresponding chroma samples. As a generalization

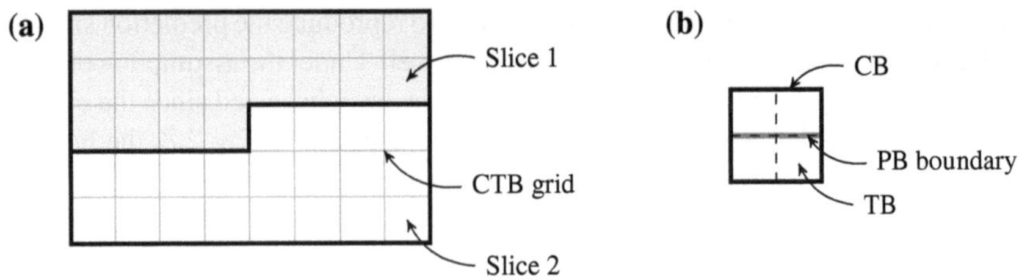

Fig. 2.9 Partitioning. **a** Picture into slices and coding tree blocks (*CTBs*). **b** Partitioning of a coding block into prediction blocks (*PBs*) and transform blocks (*TBs*)

of the macroblock concept, the coding tree block (CTB) is introduced in HEVC. It can be split into multiple coding blocks (CBs). For prediction, each coding block is partitioned into a set of one or more prediction blocks (PB). In parallel, the coding block can be partitioned into transform blocks (TB).

The picture can further be partitioned into one or more slices, which consist of an integer number of CTBs and are independently decodable.

2.4.2 Intra Prediction

Intra prediction is used to remove correlation within local regions of a picture. The basic assumption for intra prediction is that texture of a picture region is similar to the texture in the local neighborhood and can thus be predicted from there.

Intra prediction is performed on a prediction block basis. In the block diagram of Fig. 2.8, it is denoted by the building block "intra" For intra prediction, samples from the reconstructed neighborhood of the prediction block under consideration are employed to form the prediction signal. Intra prediction is applied when no pictures for inter prediction are available, or if inter prediction would be less efficient or more expensive (in the sense of an applicable cost criterion) than an applicable inter prediction.

The direct neighbor samples are commonly employed for prediction, i.e. samples from the sample line above the current block and samples from the last column of the reconstructed blocks to the left of the current block. The values of the available neighboring samples are combined to form a e.g. directional or planar prediction signal. An illustration of directional intra prediction is shown in Fig. 2.10.

The quality and the frequency of usage of intra prediction depends on various aspects. Criteria are the available intra prediction directions, the method of potential pre-filtering of the prediction samples before application, but also the availability of other prediction methods that are not directional (like DC or planar prediction). As for inter prediction, an efficient signaling of the applicable intra prediction mode is essential. The more variants are available for use, the more efficient signaling is needed in order not to trade the prediction benefit for the signaling cost.

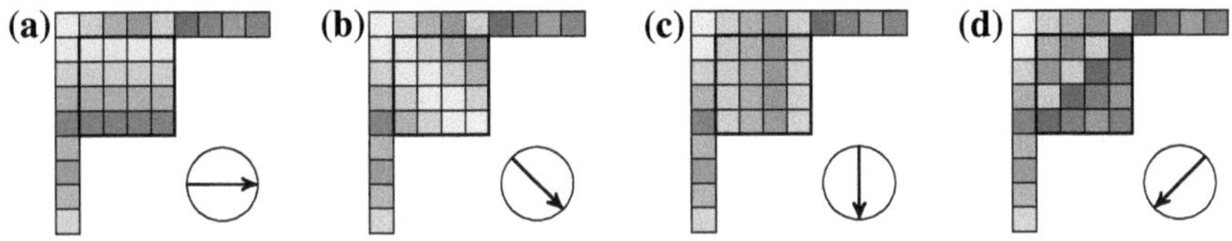

Fig. 2.10 Examples for directional intra prediction of the current block from neighboring samples. **a** Horizontal. **b** Diag. down right. **c** Vertical. **d** Diag. down left

2.4.3 Inter Prediction

A basic assumption in the context of inter-picture prediction is that a significant part of the content of the pictures in a video sequence consists of objects which move in the scene. From picture to picture, only small differences between the scene content are observed and these differences are mostly due to motion. If this motion is used for prediction, the scene can be efficiently represented by motion vectors and a prediction error signal. The motion vectors indicate how picture regions should be moved from the reference picture to the current picture to form the prediction. The prediction error signal contains the part of the scene content that could not be described by the applied motion model.

Inter prediction is performed on a prediction block basis. In the block diagram of Fig. 2.8, it is denoted by the building block "inter". Usually, inter prediction is performed by selecting an available reference picture from the decoded picture buffer and indicate a displacement relative to the location of the current prediction block in the selected picture.

This displacement is motivated by the assumption of planar motion of rigid objects in the observed scene, see Fig. 2.11a. More complex motion, like movement in direction of the camera or rotation, or motion of non-rigid objects, cannot be represented by this very simple model. Examples for pure rotational motion of the rigid object

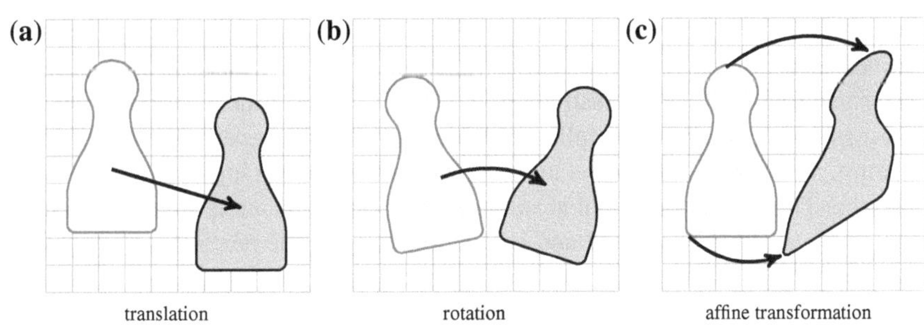

Fig. 2.11 Illustration of motion models for objects between successive pictures. **a**, **b** rigid transformation. **c** non-rigid transformation

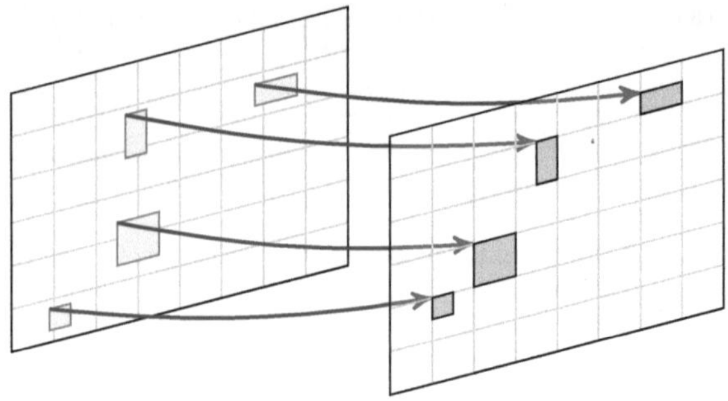

Fig. 2.12 Block displacement using a fixed grid for inter prediction from the reference picture

and affine transformation where the shape of the object is affected are shown in Fig. 2.11b, c. More complex objects, e.g. like waves or smoke, but also effects like shadow, reflections, or transparency reduce the possibility to predict from picture to picture by simple displacement even more. However, by adapting the applicable prediction block size, the prediction quality using these displacements has been considered to be sufficient and suitable for video coding applications.

In terms of prediction precision, motion parameters which e.g. describe the rotational or affine motion might be the an appropriate solution for some of the example scenarios. However, the signaling cost for the required parameters, the estimation effort needed to find the right model parameters, and not least the complexity impact when implementing higher order motion models in the decoder have so far prevented the specification of higher order motion models in video coding specifications. Since only the translational motion model, using displacement vectors, is available, more complex motion has to be (somewhat) approximated by adapting the applicable size and shape of the prediction blocks.

For inter prediction, the current picture is partitioned into non-overlapping rectangular blocks. For each of the blocks, a displacement vector is applied to get the prediction for the block from the displaced area of the reference picture, see the illustration in Fig. 2.12. For each prediction block in a fixed block raster in the current picture, a related area in the reference picture is determined. While the blocks in the current picture follow the given block grid, the reference blocks in the reference picture are located arbitrarily subject to the optimization criterion for motion estimation.

In terms of motion vector precision, the most simple realization would allow for integer displacements on the sample grid. Since motion in the captured scene will not necessarily correspond to full sample deltas between successive pictures in the sequence, sub-sample motion vector precision is required for high compression performance. The prediction signal at sub-sample locations is interpolated using interpolation filters. In H.264 | AVC and HEVC, quarter-sample precision is applied for motion vectors in the luma component. The usage of higher precision luma motion vectors, as well as the usage of adaptive interpolation filters, has been investigated during the development of HEVC but has not been adopted into a video coding specification due to the observed trade-off between the compression gain improvement and the implementation complexity cost.

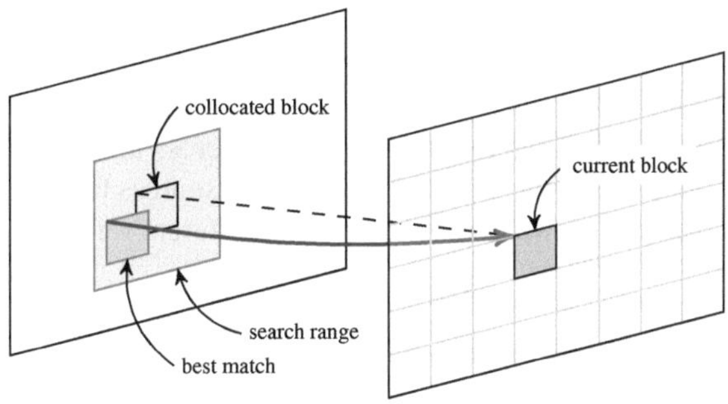

Fig. 2.13 Motion estimation in a search range around the collocated block in the reference picture

2.4.4 *Motion Estimation*

The motion estimation stage operates on a prediction block level and is only part of the encoder.[8] The estimator takes the current prediction block to be used and tries to find the best matching area in an available reference picture. The determination of what the best match would be is subject to the employed cost criterion.

A traditional search method is to shift the current prediction block over a search area around the collocated block position in the reference picture, and to determine the cost criterion for each position, see the illustration in Fig. 2.13. If all available positions are tested, the method is called a full search. Fast search methods may only test a subset of the available positions, or e.g. use criteria from the neighborhood to generate a candidate set of motion vectors to be considered during the search operation. Since motion estimation is one of the computationally most demanding tasks in the encoder, the implementation of an efficient search algorithm is crucial for a successful and fast encoder.

2.4.5 *Residual Coding*

Intra or inter prediction remove correlation within pictures and between pictures of a video sequence. The subtraction of the prediction signal from the current block generates the residual signal, containing the part of the original signal which could not be predicted by the selected prediction method. While prediction already reduces the correlation of the residual signal, it still contains information which can be further compressed. This decorrelation is performed by the application of a transformation, which is applied to represent the correlated parts of the residual signal in the residual block by a potentially small number of transform coefficients. These coefficients are then quantized and coded into the bitstream.

[8] Motion estimation at the decoder side to circumvent the transmission of motion vector information has been evaluated [11], but so far has not become part of a video coding specification.

The transform is applied on a transform block basis. In the block diagram of Fig. 2.8, forward transform and quantization are denoted by the building block "TR+Q". The inverse quantization and inverse transform block is denoted by "iTR+iQ". Usually the transform block size is smaller or equal to the prediction block size. Thereby, transformation across artificial boundaries is prevented. Such boundaries may occur if the prediction sources for two neighboring prediction blocks diverge (e.g. when predicting from different reference pictures).

2.4.5.1 Transform Type

By concept, the Karhunen-Loàve Transform (KLT) provides optimum decorrelation and energy compaction of the residual signal [12]. As it would not be practical to compute and signal the applicable KLT for a given residual characteristics, the application of transforms approximating the features of the KLT is advisable. For this purpose, parametric models of the prediction error signal have been used. Jain [13] has shown that for an important parametric model which includes the first order Markov process, a family of sinusoidal unitary transforms constitutes a complete orthonormal set of bases (eigenvectors). This family comprises the type I–VIII Discrete Cosine Transform (DCT) and the type I–VIII Discrete Sine Transform (DST). The different types of the DCT and DST vary in the way the transform matrices are constructed from cosine or sine functions. Details can be found e.g. in [14]. Here, only the transforms which are used in the context of video compression are detailed, see below. For prediction error signals whose characteristics can be approximated by a process following this parametric model, a corresponding sinusoidal transform can hence be considered as a suitable choice for decorrelation.

In all standardized video coding specifications which are based on the hybrid video coding scheme, the DCT-II/DCT-III transform pair is used for transformation of the prediction error, where the DCT-II is the forward and the DCT-III the corresponding inverse transform. Since this is the most commonly used DCT pair, it is often referred to as *the* DCT. In the context of HEVC, also the DST-VI / DST-VII pair is applied as further detailed in Chap. 8. The suitability of this transform pair for application with intra prediction error signals has been presented in [15]. The DST-VI will be referred to as *the* DST in this book.

2.4.5.2 The DCT Matrix

The DCT is a block transform and can be represented in matrix notation. A DCT matrix of size $N \times N$ is denoted by $\mathbf{T}_{\mathrm{DCT},N}$ with base vectors $\mathbf{t}_n$ of length N,

$$\mathbf{T}_{\mathrm{DCT},N} = \begin{bmatrix} \mathbf{t}_0 \\ \mathbf{t}_1 \\ \vdots \\ \mathbf{t}_{N-1} \end{bmatrix}. \tag{2.9}$$

The row vector $\mathbf{t}_n$ contains the n-th base function or base vector of the transform. The elements $t_n(m)$ of this vector constitute the coefficients of the $N \times N$ DCT matrix $\mathbf{T}_{\mathrm{DCT},N}$. For the DCT-II, these are defined by

$$t_n^{\mathrm{II}}(m) = \sqrt{\frac{2}{N}} \cdot a(n) \cdot \cos\left(\frac{\pi(2m+1)n}{2N}\right), \quad m, n = 0, ..., N-1, \tag{2.10}$$

with

$$a(n) = \begin{cases} 1/\sqrt{2} & : \quad n = 0, \\ 1 & : \quad n = 1, ..., N-1. \end{cases} \tag{2.11}$$

This is the orthonormal definition of the DCT-II, i.e. $\mathbf{T}_{\mathrm{DCT},N} \cdot \mathbf{T}_{\mathrm{DCT},N}^{-1} = \mathbf{I}_N$, where $\mathbf{I}_N$ is the $N \times N$ identity matrix. Other not normalized definitions of the DCT-II matrix are sometimes used as well [16]. As can be seen from (2.10), the base vectors are sorted with respect to increasing frequency of the cos() function in $\mathbf{T}_{\mathrm{DCT},N}$.

The DCT-II has symmetric base functions on the even positions and anti-symmetric base functions on the odd positions, i.e.

$$t_n^{\mathrm{II}}(N-m-1) = \begin{cases} t_n^{\mathrm{II}}(m) & : \quad m = 0, \ldots, N/2-1, \ n \text{ even}, \\ \\ -t_n^{\mathrm{II}}(m) & : \quad m = 0, \ldots, N/2-1, \ n \text{ odd}. \end{cases} \tag{2.12}$$

Hence, each DCT base function can only have $N/2$ different coefficient values at maximum.

Furthermore, the even base functions of a $2N \times 2N$ DCT can be generated from the base functions of the $N \times N$ DCT by scaling and mirroring,

$$\mathbf{t}_{2n,2N}^{\mathrm{II}} = \frac{1}{\sqrt{2}} \cdot \left[\mathbf{t}_{n,N}, \quad \mathbf{J}_N \cdot \mathbf{t}_{n,N}^{\mathrm{II}}\right], \tag{2.13}$$

where $\mathbf{J}_N$ is the opposite identity or reversal matrix of size $N \times N$. From these symmetry properties it can be seen that overall, a $N \times N$ DCT-II matrix contains $N-1$ different coefficient values.

The elements of the DCT-III transform matrix are defined as

$$t_n^{\mathrm{III}}(m) = \sqrt{\frac{2}{N}} \cdot a(m) \cdot \cos\left(\frac{\pi m(2n+1)}{2N}\right), \quad m, n = 0, ..., N-1, \tag{2.14}$$

with $a(m)$ given in (2.11). As stated above, the DCT-III is the inverse of the DCT-II. Furthermore, the matrices are unitary, i.e.

$$\mathbf{T}_{\mathrm{DCT-II},N}^{-1} = \mathbf{T}_{\mathrm{DCT-II},N}^{\mathrm{T}} \left(= \mathbf{T}_{\mathrm{DCT-III},N}\right). \tag{2.15}$$

The contribution of each transform coefficient to the reconstruction of the corresponding signal block can be visualized by plotting the set of DCT base pictures

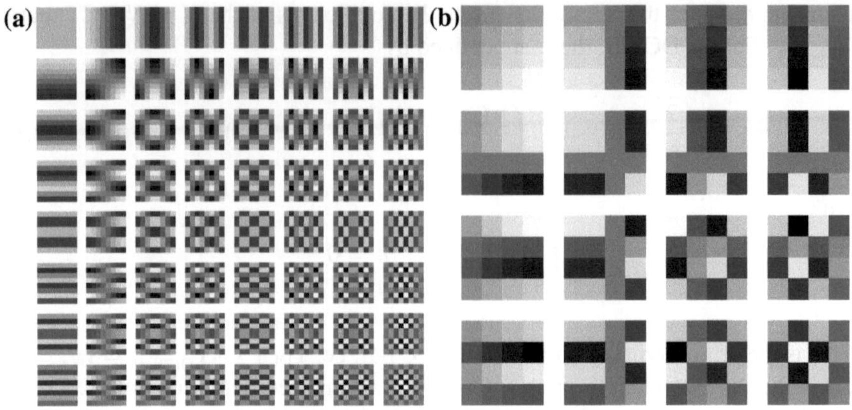

Fig. 2.14 Transform base pictures. *Black* indicates large negative values, *white* indicates large postive values. **a** 8 × 8 DCT. **b** 4 × 4 DST

for each transform coefficient. In Fig. 2.14a, the set of DCT base pictures is shown for the DCT of block size $N = 8$. These are generated by reconstructing a $N \times N$ block of transform coefficients with a single non-zero coefficient each. Thereby, the contribution of each coefficient in the block can be visualized. In the figure, the DC base image is located in the top-left corner. The horizontal and vertical frequencies increase to the right and to the bottom, respectively.

2.4.5.3 The DST Matrix

Similar to the DCT-II/DCT-III transform matrices, the matrices for the DST-VI/DST-VII pair can be defined. The DST matrix is denoted as

$$\mathbf{T}_{\mathrm{DST},N} = \begin{bmatrix} \mathbf{t}_0 \\ \mathbf{t}_1 \\ \vdots \\ \mathbf{t}_{N-1} \end{bmatrix}. \tag{2.16}$$

The elements of the $N \times N$ DST-VI matrix are defined as

$$t_{n-1}^{\mathrm{VI}}(m-1) = \frac{2}{\sqrt{2N+1}} \cdot \sin\left(\frac{\pi(2m-1)n}{2N+1}\right), \quad m, n = 1, \ldots, N. \tag{2.17}$$

The elements of the $M \times M$ DST-VII matrix are defined as

$$t_{n-1}^{\mathrm{VII}}(m-1) = \frac{2}{\sqrt{2N+1}} \cdot \sin\left(\frac{\pi m(2n-1)}{2N+1}\right), \quad m,n = 1,\ldots,N. \tag{2.18}$$

The DST matrices do not show the same symmetry properties as the DCT. However, some factorization is possible. Approaches for factorization and fast computation have been proposed [17].

Similar to the DCT matrices, the DST-VII is the inverse of the DCT-VI. The matrices are unitary as well, i.e.

$$\mathbf{T}_{\mathrm{DST-VI},N}^{-1} = \mathbf{T}_{\mathrm{DST-VI},N}^{\mathrm{T}} \left(= \mathbf{T}_{\mathrm{DST-VII},N}\right). \tag{2.19}$$

The contribution of the transform coefficients to the reconstructed block can be visualized by base pictures. In Fig. 2.14b, the set of DST base pictures is shown for the DST of block size $N = 4$. The lowest frequency base image is located in the top-left corner. The horizontal and vertical frequencies increase to the right and to the bottom, respectively.

2.4.5.4 The Hadamard Transform Matrix

Due to its low complex implementation (basically only additions) the Hadamard transform is often applied in video encoders for calculation of the sum of absolute transformed difference (SATD), see Sect. 2.5.1.3.

The Hadamard transform matrix is orthogonal with identical elements that only differ in their sign. Omitting normalization, the $N \times N$ matrix with $N = 2^n$ can be recursively derived as[9]

$$\mathbf{T}_{\mathrm{H}1} = \begin{bmatrix} 1 & 1 \\ 1 & -1 \end{bmatrix}, \quad \mathbf{T}_{\mathrm{H}n} = \begin{bmatrix} \mathbf{T}_{\mathrm{H}n-1} & \mathbf{T}_{\mathrm{H}n-1} \\ \mathbf{T}_{\mathrm{H}n-1} & -\mathbf{T}_{\mathrm{H}n-1} \end{bmatrix}. \tag{2.20}$$

The resulting matrix has only entries with value ± 1. For normalization, the $\mathbf{T}_{\mathrm{H}n}$ must be multiplied with $\sqrt{2}^{-n}$.

2.4.5.5 Transform and Quantization Scheme

Let $\mathbf{T}$ and $\mathbf{B}$ denote the transform matrix and a residual block, both of size $N \times N$. The transform is unitary, i.e.

$$\mathbf{T}^{\mathrm{T}} \cdot \mathbf{T} = \mathbf{I}, \tag{2.21}$$

[9] The Hadamard transform shares the base vectors with the Walsh transform. In the Walsh matrix, the base vectors are sorted according to increasing 'frequency', i.e. increasing number of sign changes within one base vector, which is comparable to the organization of the DCT base vectors. Omitting normalization, the Walsh transform matrix could also be derived as $\mathbf{T}_{\mathrm{W}} = \mathrm{sgn}\{\mathbf{T}_{\mathrm{DCT}}\}$.

with $\mathbf{I}$ denoting the identity matrix. The block $\mathbf{B}$ is transformed into the transform coefficients $\mathbf{C}$ by horizontal and vertical application of the transform,

$$\mathbf{C} = \mathbf{T} \cdot \mathbf{B} \cdot \mathbf{T}^{\mathsf{T}}. \tag{2.22}$$

With (2.21), the picture block can be perfectly reconstructed applying the inverse transformation procedure,

$$\begin{aligned} \mathbf{B} &= \mathbf{T}^{-1} \cdot \mathbf{C} \cdot \left(\mathbf{T}^{-1}\right)^{\mathsf{T}} \\ &= \mathbf{T}^{\mathsf{T}} \cdot \left(\mathbf{T} \cdot \mathbf{B} \cdot \mathbf{T}^{\mathsf{T}}\right) \cdot \mathbf{T} = \mathbf{I} \cdot \mathbf{B} \cdot \mathbf{I}. \end{aligned} \tag{2.23}$$

In the compression application, the transform coefficients are quantized and encoded for transmission in the bitstream. Since the transform is orthonormal, the Parsival theorem can be applied [10]. Thereby, the quantization error energy introduced by quantization of the transform coefficients is identical to the reconstruction error energy after inverse transform. The quantization aims at the removal of the most irrelevant information for a given bitrate budget. At the decoder side, the quantized transform coefficients $\mathbf{C}_{\mathrm{q}}$ are transformed back to form the reconstructed residual signal $\tilde{\mathbf{B}}$,

$$\tilde{\mathbf{B}} = \mathbf{T}^{\mathsf{T}} \cdot \mathbf{C}_{\mathrm{q}} \cdot \mathbf{T}.$$

The inverse transformation distributes the quantization error induced by a single transform coefficient over the whole reconstructed block. This effect helps to conceal the introduced quantizer distortion for the observer.

2.4.5.6 Quantization

The quantization process maps signal amplitudes to a predefined set of representative values. If individual values are quantized, the process is called scalar quantization. For multidimensional values vector quantization can be applied [10]. In HEVC, as in previous video coding specifications, scalar quantization is used.

Quantization is an inherently non-linear lossy operation, which cannot be inverted. The quantization process is the part of the hybrid video coding scheme which inserts signal degradation and distortion by removing signal information from the coded representation. By careful control of the quantizer settings within the coded picture and over the coded video sequence, the amount of removed irrelevant information (i.e. information which can be removed without noticeable degradation) compared to the amount of removed relevant information, which leads to visible distortions, can be optimized.

The design of the quantizer is driven by the probability distribution of the observed signal amplitudes, balancing the rate needed to encode the quantized values and the distortion introduced by mapping amplitude intervals to a defined reconstruction

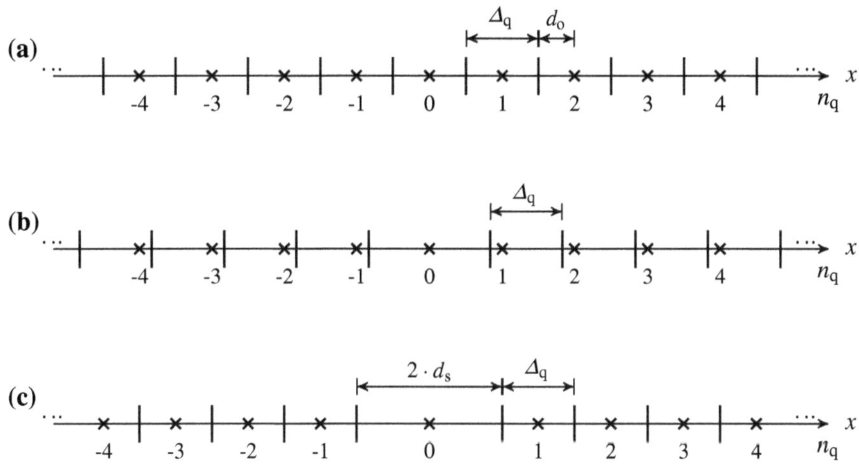

Fig. 2.15 Example scalar quantizer types. An input value x is mapped to the reconstruction value of the n_qth interval. **a** Uniform quantizer (uniform distribution). **b** Uniform quantizer with shifted interval boundaries (Laplacian distribution). **c** Dead-zone quantizer with extended interval around 0

value. Solutions for this optimization problem are described in literature [18]. For a set of known amplitude probability distributions like the uniform or Laplacian distributions, optimum quantizers can be derived and described in a closed form [19]. DCT transform coefficients can be modelled to follow a Laplacian distribution, which lead to the present quantizer design used in DCT based hybrid video coding schemes [20, 21].

Three classical scalar quantizer schemes are shown in Fig. 2.15. For quantization, an input value x is mapped to the n_qth reconstruction value representing the interval which the input value x falls into. n_q is called the quantizer *level*. The intervals are scaled to achieve finer or coarser quantization according to the application needs. The quantizer scaling is expressed by the quantizer step size Δ_q.

In a uniform quantizer, the level n_q for a quantized value x is calculated by

$$n_q = \operatorname{sgn}(x) \cdot \left\lfloor \frac{|x|}{\Delta_q} + d_o \right\rfloor, \tag{2.24}$$

where d_o is an offset which determines the position of the interval boundaries. For example in Fig. 2.15a, b, $d_o = \frac{1}{2}$ and $d_o = \frac{1}{6}$, respectively. These correspond to optimized quantizers for the uniform distribution and the Laplacian distribution, respectively. Reconstruction of the quantized value is trivially achieved by scaling with the quantizer step size,

$$x_q = \Delta_q \cdot n_q. \tag{2.25}$$

The corresponding equations for a dead-zone quantizer as shown in Fig. 2.15c can be given as follows

$$n_\mathrm{q} = \begin{cases} 0, & |x| < d_\mathrm{s} \\ \mathrm{sgn}(x) \cdot \left\lfloor \frac{|x| - d_\mathrm{s}}{\Delta_\mathrm{q}} + 1 \right\rfloor, & \text{otherwise} \end{cases} \tag{2.26}$$

and

$$x_\mathrm{q} = \begin{cases} 0, & n_\mathrm{q} = 0 \\ \mathrm{sgn}(n_\mathrm{q})\left[d_\mathrm{s} + \Delta_\mathrm{q} \cdot \left(\left|n_\mathrm{q}\right| - 1\right) + d_\mathrm{o}\right], & \text{otherwise.} \end{cases} \tag{2.27}$$

In the example of Fig. 2.15c, the parameters were set to $d_\mathrm{s} = \Delta_\mathrm{q}$ and $d_\mathrm{o} = \frac{\Delta_\mathrm{q}}{2}$.

The number and values of the applicable quantizer step sizes in a video coding system is limited. The applicable quantizer step size is usually indicated by the *quantization parameter* (QP), which serves as an index to a predefined set of applicable quantization step sizes. Commonly, low quantization parameters correspond to fine quantization and high quantization parameters correspond to coarse quantizer step sizes. A high granularity of the quantizer step sizes is beneficial to allow for precise rate control in the encoded bitstream. On the other hand, the signaling for a high number of available quantizer step sizes induces additional coding cost, which needs to be considered. The specification has to balance granularity and coding cost.

For the orthonormal transform, application of the same quantizer step size to the transform coefficients regardless of their frequency index induces the minimum mean squared error for the reconstructed block [12]. However, this approach does not consider the subjective impact of quantization on the human visual system if applied to coefficients which correspond to different spectral frequencies in the block. In order to enable adaptation of the quantization scheme in this regard, a quantization weighting matrix $\mathbf{W}_\mathrm{q}$ of transform block size $N \times N$ can be applied. This matrix defines an additional scaling of the quantizer step size in dependence of the transform coefficient which is quantized,

$$\mathbf{W}_\mathrm{q} = \begin{bmatrix} w_{\mathrm{q}0,0} & w_{\mathrm{q}0,1} & \cdots & w_{\mathrm{q}0,N-1} \\ w_{\mathrm{q}1,0} & w_{\mathrm{q}1,1} & \cdots & w_{\mathrm{q}1,N-1} \\ \vdots & & \ddots & \vdots \\ w_{\mathrm{q}N-1,0} & w_{\mathrm{q}N-1,0} & \cdots & w_{\mathrm{q}N-1,N-1} \end{bmatrix}. \tag{2.28}$$

For the transform coefficient $c_{i,j}$, the applicable quantizer step size is scaled to

$$\Delta'_\mathrm{q}(QP, i, j) = w_\mathrm{q}(i, j) \cdot \Delta_\mathrm{q}(QP) \tag{2.29}$$

for both, quantization and inverse quantization. Quantization weighting matrices can e.g. be designed to provide the finest quantization for the low frequencies while increasing the quantizer step size with increasing frequency index.

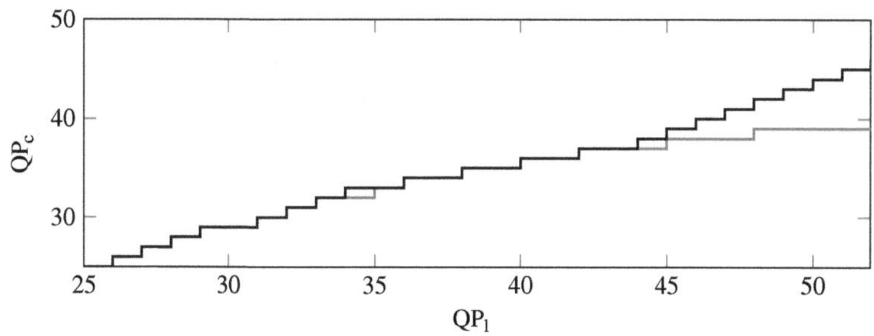

Fig. 2.16 Schematic presentation of the mapping of the luma quantization parameter QP_l to the chroma quantization parameter QP_c for HEVC (*black*) and H.264 | AVC (*gray*)

For YCbCr 4:2:0 video, the signal characteristics of luma and the two chroma components are quite different. Specifically, chroma often exhibits a strong lowpass character. If strong quantization is applied, the chroma information may be completely quantized to zero, which would lead to the complete loss of color. Accordingly, in order to reduce this the quantizer step size for chroma is adapted by reducing the chroma quantizer step size for high QP values.

As an example, the correspondence between the luma quantization parameter QP_l and the chroma quantization parameter QP_c is shown for the standards H.264 | AVC and HEVC in Fig. 2.16. For QP values below $QP_l = 30$ in the QP range $0 \leq QP_l \leq 51$, the chroma quantization parameter is aligned to luma. At high QP values, the chroma is kept at a finer quantization than the luma component.

2.4.5.7 Integer Transforms

Before H.264 | AVC, the transformation in video coding standards was specified in floating point operations. Since deviations in different floating point implementations of encoders and decoders may easily lead to differences in the inverse transform result, a significant effort was spent to define conformance conditions such that it could be ensured that two independent DCT implementations would induce the same (or almost the same) reconstructed block. The normative requirements for this task were specified by the IEEE in [22]. After the withdrawal of this standard in 2003, MPEG approved the replacement specification ISO/IEC 23002-1 in 2006 [23].

In order to dispose this transform conformance issue, the specification of integer transforms has been introduced. In H.264 | AVC and HEVC, the inverse transforms are specified in integer arithmetic. In fact, all mathematical operations specified in the decoding process are in integer arithmetic in these standards. As a recommended solution for the previously existing standards, a fixed-point 8×8 IDCT and DCT have been specified in ISO/ICE 23002-2 [24]. This transform pair fulfills the requirements of ISO/IEC 23002-1 and is recommended for implementation as it enables bit-exact reconstruction if used for both, encoder and decoder implementations.

Integer DCT Matrices

The integer approximations of the DCT are called integer DCTs in the following. Several design aspects rule into the development of these transforms. These include the required dynamic range in the transformation process, the approximation precision, and the implementation friendliness in terms of required multiplications, additions, and bit-shifts.

Three example 8×8 integer DCT matrices which follow different design criteria are given below:

$$\mathbf{T}_{8,a} = \begin{bmatrix} 17 & 17 & 17 & 17 & 17 & 17 & 17 & 17 \\ 24 & 20 & 12 & 6 & -6 & -12 & -20 & -24 \\ 23 & 7 & -7 & -23 & -23 & -7 & 7 & 23 \\ 20 & -6 & -24 & -12 & 12 & 24 & 6 & -20 \\ 17 & -17 & -17 & 17 & 17 & -17 & -17 & 17 \\ 12 & -24 & 6 & 20 & -20 & -6 & 24 & -12 \\ 7 & -23 & 23 & -7 & -7 & 23 & -23 & 7 \\ 6 & -12 & 20 & -24 & 24 & -20 & 12 & -6 \end{bmatrix} \quad (2.30)$$

$$\mathbf{T}_{8,b} = \begin{bmatrix} 8 & 8 & 8 & 8 & 8 & 8 & 8 & 8 \\ 12 & 10 & 6 & 3 & -3 & -6 & -10 & -12 \\ 8 & 4 & -4 & -8 & -8 & -4 & 4 & 8 \\ 10 & -3 & -12 & -6 & 6 & 12 & 3 & -10 \\ 8 & -8 & -8 & 8 & 8 & -8 & -8 & 8 \\ 6 & -12 & 3 & 10 & -10 & -3 & 12 & -6 \\ 4 & -8 & 8 & -4 & -4 & 8 & -8 & 4 \\ 3 & -6 & 10 & -12 & 12 & -10 & 6 & -3 \end{bmatrix} \quad (2.31)$$

$$\mathbf{T}_{8,c} = \begin{bmatrix} 64 & 64 & 64 & 64 & 64 & 64 & 64 & 64 \\ 89 & 75 & 50 & 18 & -18 & -50 & -75 & -89 \\ 83 & 36 & -36 & -83 & -83 & -36 & 36 & 83 \\ 75 & -18 & -89 & -50 & 50 & 89 & 18 & -75 \\ 64 & -64 & -64 & 64 & 64 & -64 & -64 & 64 \\ 50 & -89 & 18 & 75 & -75 & -18 & 89 & -50 \\ 36 & -83 & 83 & -36 & -36 & 83 & -83 & 36 \\ 18 & -50 & 75 & -89 & 89 & -75 & 50 & -18 \end{bmatrix} \quad (2.32)$$

Matrix $\mathbf{T}_{8,a}$ in (2.30) is an orthogonal, single-norm integer approximation of the 8×8 DCT matrix [25]. The matrix $\mathbf{T}_{8,b}$ in (2.31) corresponds to the inverse 8×8 transform in the High profiles of H.264 | AVC [26, 27]. The coefficient values of this matrix allow for an implementation of the inverse transform process with 16-bit dynamic range. The base vectors of this matrix are orthogonal but have diverging norms. These are dealt with in the design of the corresponding integer quantization process. The last matrix $\mathbf{T}_{8,c}$ in (2.32) corresponds to the inverse 8×8 transform in

HEVC [4]. The base vectors of this transform have approximately the same norm but are not exactly orthogonal, i.e. the product $\mathbf{T}_{8,c} \cdot \mathbf{T}_{8,c}^{\mathsf{T}}$ does not result in a diagonal matrix. The deviation from orthogonality however is so small that it is leveled out by normalization during the quantization and reconstruction processes. The HEVC transform and quantization design is further detailed in Chap. 8.

Normalization and Quantization

By definition, integer DCT matrices can not be orthonormal. Therefore, normalization must be implemented into the coding scheme in this case. For the discussion, a uniform quantizer and a $N \times N$ orthogonal integer DCT matrix are assumed, i.e.

$$\mathbf{T} \cdot \mathbf{T}^{\mathsf{T}} = \|\mathbf{T}\|^2 \cdot \mathbf{I}. \tag{2.33}$$

The transform and quantization operation to generate the quantizer levels $\mathbf{n}_q$ for a residual block $\mathbf{X}$ at the encoder side can be written as

$$\mathbf{n}_q = \text{round}\left\{ \frac{1}{\|\mathbf{T}\|^2 \cdot \Delta_q} \cdot \mathbf{T} \cdot \mathbf{X} \cdot \mathbf{T}^{\mathsf{T}} \right\}, \tag{2.34}$$

where Δ_q is the quantizer step size. Accordingly, the normalization at the decoder side has to be included in the reconstruction $\mathbf{X}_r$ of the residual block:

$$\mathbf{X}_r = \text{round}\left\{ \frac{\Delta_q}{\|\mathbf{T}\|^2} \cdot \mathbf{T}^{\mathsf{T}} \cdot \mathbf{n}_q \cdot \mathbf{T} \right\}. \tag{2.35}$$

In order to enable the implementation of the divisions in (2.33) and (2.35) as bit-shift operations, the quotient factors can be designed as

$$\frac{f_q}{2^{N_e}} \approx \frac{1}{\|\mathbf{T}^2\| \cdot \Delta_q} \quad \text{and} \quad \frac{g_q}{2^{N_d}} \approx \frac{\Delta_q}{\|\mathbf{T}\|^2}, \tag{2.36}$$

with appropriate bit-shift values N_e and N_d at the encoder and decoder side, respectively. For integer DCTs with base vectors of different norm (e.g. in H.264 | AVC), separate f_q and g_q values have to be specified according to the applicable normalization of the respective transform coefficients.

2.4.6 In-loop Filtering

Loop filtering is a means to improve the reconstruction quality of the picture for display. Since the filter is located within the loop, the enhancement not only affects the quality of the output pictures but also the reference pictures, which are available for prediction when coding succeeding pictures. Thereby, loop filters have a strong impact on the overall performance of the video coding scheme. Optimization criteria can be in terms of visual quality but also in terms of objective fidelity measures such as the minimization of the mean squared error between the original picture and the reconstructed picture.

Conceptually, two classes of loop filter operations can be distinguished: The first class comprises linear filters which are applied to an area of a picture or even a complete picture. For the selected area, the filtering operation can be performed as a convolution with the filter impulse response in the spatial domain or as a multiplication with the filter transfer function in the frequency domain. The filter configuration can be flexible or static over the video sequence and the applicable filter parameters may be determined at the encoder side according to a selected optimization criterion, e.g. using a Wiener filter approach [10, 12].[10]

The filters of the second class of loop filters operate on a local spatial domain of the picture. Based on local features, the applicable filtering operation for a small set of samples is determined. The H.264 | AVC and HEVC deblocking filters are representatives of this class. A deblocking filter deals with block boundary structures which are induced by block-wise motion compensated prediction and block-wise processing of the residual signal. The edges of these blocks become visible in the reconstructed pictures due to quantization of the residual signal such that perfect reconstruction of the original picture is not achieved. In order to mitigate this effect, the deblocking filter operates in a local area across prediction and transform block boundaries. The strength of the deblocking filter is locally controlled by the detected amount of 'blockiness' at the processed boundary. The realization of the HEVC deblocking filter and the corresponding deblocking filter control are discussed in Sect. 9.1.

Another representative of local loop filters is sample-adaptive offset (SAO) filtering, which is newly specified in HEVC and has not been included in a ITU-T or ISO/IEC video coding standard before. SAO operates on a sample basis, independent from the block structures of the reconstructed picture. Two types of SAO filtering are defined in HEVC: Derivation of a sample level correction based on the local neighborhood, or derivation of a sample level correction based on the intensity level of the sample itself. Conceptually, sample-adaptive offset filtering can be considered as a picture restoration operation. While such filters may commonly be applied to the reconstructed video in a post-processing stage, it is integrated into the coding loop in this case. SAO can be used e.g. to effectively reduce ringing artifacts or to correct level values which have been shifted due to quantization. The operation of the HEVC sample-adaptive offset filter is further detailed in Sect. 9.2.

The applicable loop filters are jointly denoted by "Loop Filter" in the block diagram of Fig. 2.8. The filters operate on a slice basis, as the filtering is applied across

[10] During the development of HEVC, the specification of an adaptive loop filter was evaluated. This filter partitioned the picture into filtering blocks on a quadtree basis and applied adaptive filters to the partitions. In the final design, the overall trade-off between compression improvement and implementation cost was considered a too high burden and this loop filter type was not included in the HEVC specification [28].

block boundaries within slices. In H.264 | AVC and HEVC, the deblocking filter can also be configured to be applicable across slice boundaries.

A filter process within the coding loop has already been present in the H.261 specification [29]. There, a separable three-tap averaging filter can be applied on an 8×8 block basis. The filtering is activated on a macroblock basis. In contrast to H.264 | AVC and HEVC, which apply the loop filter after reconstruction as shown in Fig. 2.8, the H.261 loop filter is located after the motion compensation stage. Thereby, the inter prediction signal before addition of the reconstructed residual is filtered instead of the reconstructed picture. Since H.261 only specifies full-sample motion compensated prediction, this application of the loop filter can be interpreted as a replacement for missing sub-sample motion vector precision. If a block does not fit the full-sample motion compensation, the prediction is improved by lowpass filtering the motion compensated block.

2.4.7 The Decoded Picture Buffer

The decoded picture buffer holds the decoded reconstructed pictures until they are scheduled for display. This is specifically necessary if the decoding order of the pictures diverges from the output order of the pictures. In addition to the storage for display, pictures can be stored for use as reference pictures in inter prediction. In the block diagram of Fig. 2.8, the decoded picture buffer is denoted by "Buffer".

The size of the buffer defines limits for the applicable temporal coding structures applied to a video sequence. Conceptually, two types of reference pictures are differentiated in the buffer: short-term and long-term reference pictures. Short-term reference pictures denote reference pictures from the temporal proximity of the current picture to be predicted. Long-term reference pictures are explicitly marked as such. These pictures can be used to store a picture with certain scene content for reference usage on a larger temporal scale, e.g. if some scene content appears repeatedly after interruptions by other content. The handling of reference pictures and the decoded picture buffer in HEVC are detailed in Sects. 4.3 and 5.6.2.

2.4.8 Entropy Coding

The syntax elements which represent the quantized transform coefficients, as well as the applicable prediction modes, motion vectors, intra prediction directions, etc. need to be coded into the bitstream. The values of the decoded syntax elements are used in the decoding process to derive the intended reconstructed video sequence. The operation of mapping the syntax elements into the bitstream is done by the building block "Entropy Coding" in the block diagram of Fig. 2.8. The entropy coding stage maps the incoming syntax elements, such as flags, indices, vectors, modes, or coefficients, to binary code words and bit-string representations. By the design and layout of the assigned coding methods, the entropy coding stage has significant impact on the overall compression efficiency of a video coding scheme.

Depending on the type of coded information, the design criteria for the entropy coding stage differ. Syntax elements indicating high-level properties of the bitstream are usually coded using code word representations which are easy to access not only by the decoder but also by other applications which may want to acquire information regarding general features of the bitstream. Here, *fixed length codes* are often employed. Such information is often placed at prominent positions in the bitstream for easy access. Byte aligned positions relative to defined starting points alleviate access to relevant syntax elements in the bitstream. Since such information usually comprises only a very small portion of the bitstream, fixed-length code words can be used without much penalty in the coding efficiency. Syntax elements which convey information at the picture or slice level are coded using *variable length codes* (VLCs) as appropriate. For these code words, a trade-off regarding decoding complexity and compactness of the representation has to be found during the design of the coding scheme. The highest effort in terms of an efficient entropy coding representation is spent on the slice and block level. This part of the coded information typically covers the vast majority of the bitstream. State-of-the-art video coding schemes apply context-dependent adaptive coding at this level.[11]

In H.264 | AVC, a context adaptive VLC based scheme (CAVLC) and a context adaptive binary arithmetic coding scheme (CABAC) were specified and employed on the block level, depending on the selected standard profile. In HEVC, only context-based adaptive binary arithmetic coding has been selected after thorough investigation on the performance, computational complexity, and description complexity impact [31].

In the following, the basic principles of coding with variable length codes and arithmetic coding are summarized. For further study, the reader is referred to literature, e.g. starting from [10, 32, 33]. VLCs and CABAC as used in HEVC are detailed in Chap. 10.

2.4.8.1 Entropy and Information Content

Let $S = \{s_0, s_1, \ldots, s_{N-1}\}$ be an alphabet of N symbols s_n. Let the symbols be sent by a source, let the symbols be independent and identically distributed, and let each symbol s_n occur with a probability $p_S(n)$. When a source sends symbols of this alphabet, the information content of each symbol can be computed as

$$I(s_n) = \log_2\left(\frac{1}{p_S(n)}\right) = -\log_2\left(p_S(n)\right). \tag{2.37}$$

[11] For the HEVC Random Access configuration according to the JCT-VC common testing conditions [30], the portion of the bitstream which is not encoded with CABAC is in the range of 0.1–1.0 %.

The logarithm to the base 2 applies for a coded representation with binary values. For such a binary code, the information content can be interpreted as the minimum required code word length to represent the given symbol s_n. The entropy of the source is defined as the average information content of the source,

$$H = \sum_{n=0}^{N-1} p_S(n) \cdot I(s_n) = - \sum_{n=0}^{N-1} p_S(n) \cdot \log_2 \left(p_S(n)\right). \tag{2.38}$$

According to the foundations of information theory, the entropy corresponds to the least number of bits per symbol that need to be spent for encoding, if the symbols shall be uniquely decodable.

When designing a coding scheme for a given source, the entropy of the source thereby provides the lower bound on the achievable compression. The assumption of independent or uncorrelated symbols, which was used in the definition above, does not generally hold. Instead, determining and utilizing the correlation between successive symbols as well as the impact on the coding context on the symbol probability distribution is crucial for the derivation of a (close-to) optimum coding scheme. However, the achievable gain in terms of representation compactness always has to be balanced with the accompanying algorithmic effort that has to be spent.

2.4.8.2 Fixed and Variable Length Codes

As stated above, fixed length codes are used for coding high-level syntax in a video coding scheme. Trivially, a flag which has two states can be coded with 1 bit. For syntax elements with more values, the applicable number of bits for the fixed-length code must be specified. Typically, the value of the syntax element is directly coded as its binary representation. Depending on the probability of the available syntax element values and observing (2.38), this representation can be more or less efficient.

When designing VLCs, various strategies can be applied. Under the assumption that the probability distribution of the symbols in the alphabet is known (i.e. here the probability for the values of the syntax element under consideration), the VLC design method proposed by Huffman can be applied [34]. The method guarantees a prefix-free set of resulting code words, i.e. a bitstream consisting of these code words can be unambiguously decoded. The length of the resulting code words increases with decreasing probability of the corresponding symbol. It can be shown that by design, the average code length of a Huffman code is bound by $H + 1$ [33].

While Huffman codes provide an efficient coded representation for a given source alphabet, the resulting code words are not systematic. Therefore, code word tables have to be stored at both encoder and decoder side in the coding application. An alternative method is the usage of *systematic codes*, which do not require the storage of code word tables at the decoder side but are rather generated on the fly by a specified construction rule, see e.g. [35].

The construction of the applicable VLCs can further be made adaptive to better fit variable source statistics. Strategies for the design of context adaptive VLCs include

Table 2.4 Unary code

v	$C_{\mathrm{u}}(v)$
0	1
1	0 1
2	0 0 1
3	0 0 0 1
$\vdots$	$\vdots$

context-based selection of VLC tables or adaptive assignment of code words to syntax element values, see e.g. [36].

In the following, some systematic VLCs are presented which play a role in the context of HEVC. The presentation focuses on unsigned integer values. A codeword that is assigned to a value $v \geq 0$ is denoted by $c = C(v)$. The value is decoded from the codeword by $v = C^{\mathrm{i}}(c)$.

2.4.8.3 Unary Code

A *unary* code $C_{\mathrm{u}}(v)$ is a systematic code that represents an non-negative integer value v by v bits of one parity and a stop-bit of the other parity. A schematic presentation of the code is given in Table 2.4. Given a value v, the code word consists of v 0-bits and a terminating 1-bit. The length of the code word is

$$n_{\mathrm{c}} = v + 1. \tag{2.39}$$

Accordingly, the value is decoded from the code word by

$$\begin{aligned} v &= C_{\mathrm{u}}^{\mathrm{i}}(c) \\ &= n_{\mathrm{c}} - 1. \end{aligned} \tag{2.40}$$

2.4.8.4 Golomb Codes

Golomb codes are a family of systematic codes that can be adapted to the source statistics and are thereby well suited for coding applications [35, 37]. Golomb codes are generally constructed by a prefix and a suffix part.

Golomb-Rice Codes

A Golomb-Rice code $C_{\mathrm{gr}\,k}(v)$ of grade k is constructed by a unary coded prefix and k suffix bits [38]. An example is given in Table 2.5 for $k = 4$. In the table and in the following $x_0, x_1, \ldots, x_n$ denote bits of the code word with $x_i \in \{0, 1\}$.

Table 2.5 Golomb-Rice code of order $k = 4$

v	$C_{\mathrm{gr}4}(v)$
$0, \ldots, 15$	$1\ x_3\ x_2\ x_1\ x_0$
$16, \ldots, 31$	$0\ 1\ x_3\ x_2\ x_1\ x_0$
$32, \ldots, 47$	$0\ 0\ 1\ x_3\ x_2\ x_1\ x_0$
$\vdots$	$\vdots$

Let the code be used for unsigned integer values and the suffix be the k-bit binary representation of an integer $0 \leq i < 2^k$. The number of prefix bits is denoted by n_{p}, the number of suffix bits is denoted by n_{s}. For the Golomb-Rice code, the number of suffix bits is $n_{\mathrm{s}} = k$. When encoding a value v, the number of prefix bits is determined by

$$n_{\mathrm{p}} = 1 + \left\lfloor \frac{v}{2^k} \right\rfloor . \tag{2.41}$$

The suffix then is the n_{s}-bit binary representation of

$$v_{\mathrm{s}} = v - 2^k (n_{\mathrm{p}} - 1). \tag{2.42}$$

Accordingly, the value v can be reconstructed from the code word c by

$$\begin{aligned} v &= C^{\mathrm{i}}_{\mathrm{gr}\,k}(c) \\ &= 2^k (n_{\mathrm{p}} - 1) + \sum_{i=0}^{k-1} x_i \cdot 2^i . \end{aligned} \tag{2.43}$$

Exp-Golomb Codes

While the Golomb-Rice codes use a suffix of fixed length, it is also possible to determine the length of the suffix by the length of the prefix. Exponential Golomb codes (Exp-Golomb) follow this approach [39].

A kth-order Exp Golomb code $C_{\mathrm{eg}\,k}(v)$ is constructed by a unary prefix code and a suffix of configurable length. The number of bits in the suffix n_{s} is determined by the value n_{p} of the prefix code, where

$$n_{\mathrm{s}} = k + n_{\mathrm{p}} - 1. \tag{2.44}$$

The number of prefix bits n_{p} of $C_{\mathrm{eg}\,k}(v)$ is determined from the value v by

$$2^k \left(2^{n_{\mathrm{p}}-1} - 1\right) \leq v < 2^k \left(2^{n_{\mathrm{p}}} - 1\right) . \tag{2.45}$$

Table 2.6 Exp-Golomb codes of order $k = 0$ and $k = 1$

v	$C_{\text{eg}\,0}(v)$	v	$C_{\text{eg}\,1}(v)$
0	1	0,1	$1\,x_0$
1,2	$0\,1\,x_0$	$2,\ldots,5$	$0\,1\,x_1\,x_0$
$3,\ldots,6$	$0\,0\,1\,x_1\,x_0$	$6,\ldots,13$	$0\,0\,1\,x_2\,x_1\,x_0$
$7,\ldots,14$	$0\,0\,0\,1\,x_2\,x_1\,x_0$	$14,\ldots,29$	$0\,0\,0\,1\,x_3\,x_2\,x_1\,x_0$
$\vdots$	$\vdots$	$\vdots$	$\vdots$

The suffix is then the n_s-bit binary representation of $v_\text{s} = v - 2^k\left(2^{n_\text{p}-1} - 1\right)$.

A value v can be reconstructed from the code word $c = C_{\text{eg}\,k}(v)$ code as

$$v = C^{\text{i}}_{\text{eg}\,k}(c) = 2^k\left(2^{n_\text{p}-1} - 1\right) + \sum_{i=0}^{n_\text{s}-1} x_i \cdot 2^i. \tag{2.46}$$

Examples for Exp-Golomb codes with $k = 0$ and $k = 1$ are given in Table 2.6. The 0th-order Exp-Golomb code $C_{\text{eg}\,0}(v)$ is used for coding of signed and unsigned values in HEVC, see Sect. 10.1.

2.4.8.5 Arithmetic Coding

With *arithmetic coding*, the direct connection between the symbol and the bitstream representation is abolished. In contrast to variable length codes, symbols are represented by coded number intervals instead of explicit code words. A sequences of symbols can be represented by a single value from a defined interval in a given number range. Thereby, arithmetic coding conceptionally allows for encoding symbols at fractional numbers of bits. This is particularly useful for symbols which occur with a very high probability. According to (2.37), such symbols have an information content of less than 1 bit, which would lead to an inefficient representation by a VLC code word (which by nature must have a length of at least 1 bit). Arithmetic coding is used in H.264 | AVC and HEVC at the slice level for highest compression efficiency. It has been specified and used in image and video compression standards such as H.263 Annex E or JPEG2000 before [35, 40]. The basic concept of arithmetic coding is illustrated here. The arithmetic coding scheme using integer arithmetic as specified in HEVC is detailed in Chap. 10.

Let the available number range be [0, 1], and let the symbols of an alphabet {A, B} occur with probability $p_\text{A} = 0.7$ and $p_\text{B} = 1 - p_\text{A} = 0.3$, respectively. Since the probability for symbol A is higher than for symbol B, symbol A is called the most probable symbol (MPS). Symbol B is denoted the least probable symbol (LPS). For encoding a sequence of these symbols, a representative interval is derived and the

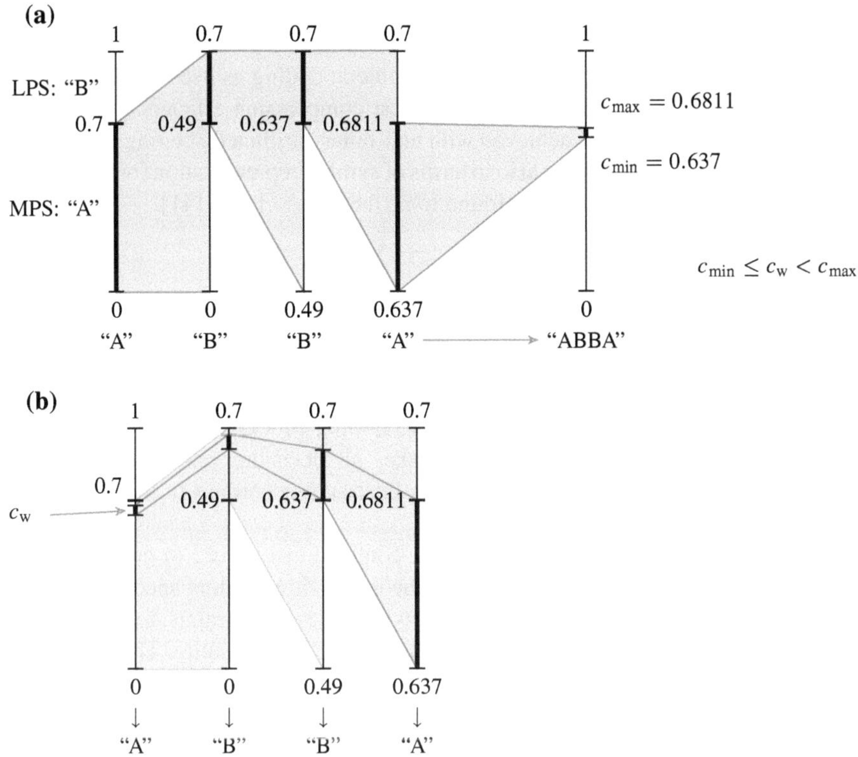

Fig. 2.17 Example for arithmetic coding of a sequence of binary symbols "A, B, B, A" with probabilities $p_A = 0.7$ and $p_B = 0.3$. **a** Arithmetic encoding corresponds to an interval subdivision. The sequence can be represented by a code word c_w from the last interval. **b** Arithmetic decoding compares the code word to the interval boundaries to successively reconstruct the sequence

sequence can be identified by any number from that interval. The process of interval determination is illustrated in Fig. 2.17a. The symbols are assigned to intervals of the available range according to their probability. In the coding step for each symbol, the corresponding interval is selected and taken as the available range for the next symbol. With each step, the corresponding interval is refined. If a pre-defined minimum interval size is reached, a rescaling operation is applied and the encoding process is continued. This is also the point in time where the bits for the previous interval indicators are written into the bitstream. For decoding, a corresponding set of steps has to be performed as illustrated in Fig. 2.17b. The decoded interval indicator is successively compared to the interval boundaries and the symbols corresponding to the intervals are thereby successively decoded. If the successively reduced interval size indicates the need for rescaling, further bits are read from the bitstream and the decoding process continues. The binary scheme as shown here can be similarly extended to alphabets with more than two symbols, e.g. as used in H.263.

A substantial advantage of arithmetic coding is the possibility to update the relation of the interval boundaries for each symbol after each coding step. Thereby, the size of the interval can be adapted to the estimated statistics of the encoded symbols. Context-based adaptive binary arithmetic coding as used in H.264 | AVC and HEVC utilizes this possibility for highest compression efficiency. Conceptually, similar results could be achieved with non-binary arithmetic coding. In terms of implementation complexity but also in terms of symbol representation (binarization), binary arithmetic coding has been found to be the better choice [41].

2.5 Encoder Control

The block diagram in Fig. 2.8 does not include the notion of encoder control to simplify the diagram. In fact, virtually all building blocks in the block diagram are affected by it. The encoder control takes all decisions related to coding of the pictures of the video sequence into the bitstream according to the application requirements.

As a fundamental property, the encoder control must ensure to generate a bitstream conforming to the requirements of the given video coding specification.The most important task in most applications is, within these constraints, to optimize the bitrate (or the bitstream size) and the reconstructed video quality. This could e.g. mean to minimize the bitrate given that the reconstruction quality does not fall below a predefined level, or to maximize the reconstructed video quality subject to given bitrate constraints. Another very important aspect of encoder control is to ensure that the amount of bits sent to the decoder side is regulated such that the decoder buffer for the incoming bitstream does neither overfill nor run empty. This task is e.g. achieved by application of a hypothetical reference decoder model as further discussed in Sect. 5.6. In low-delay and live applications, a fast encoder operation must be achieved which satisfies real-time constraints. In such application scenarios, the video is usually directly processed upon availability to the encoder. If the video material is processed offline for later usage, the encoding speed is of less importance. In such application scenarios, elaborate encoder optimization strategies can be applied and multi-pass encoding may be used. Multi-pass encoding allows the encoder control to make use of 'look-ahead' knowledge when encoding the picture of the video sequence. Thereby, bitrate can be efficiently invested for better prediction in preparation of forthcoming content.

The decisions the encoder control needs to take include the applicable partitioning of the picture as well as the partitioning of the coding blocks. Furthermore, the applicable inter and intra prediction modes, motion vectors, loop filtering modes, the applicable transforms, and the reference picture management are concerned. The selection of the applicable motion vectors and the corresponding reference pictures are a major task. The motion vector determination by the motion estimation stage usually covers a significant portion of the encoder processing time. A variety of fast search strategies have been developed to reduce the computational complexity burden of this task while keeping the prediction quality as high as possible.

The fundamental decision criterion at the encoder side is to balance the bitrate to be spent against the corresponding reconstruction quality that is achieved. The criterion is used on the block level for decision on prediction modes as well as on a larger level, e.g. when deciding on the applicable picture structure. The bitrate cost can be directly measured by counting the bits that would be spent to encode a block in a given mode. The more complete this analysis is, the more precise and reliable the result will be (e.g. including the encoding cost of the quantized transform coefficients). In a VLC-based solution this can be achieved by corresponding table look-ups. With arithmetic coding, determining the precise number of bits to be spent requires more effort as trial encoding and appropriate coder state management needs to be implemented for a precise measurement.

For determination of the reconstruction quality, the objective distortion of the reconstructed picture compared to the original picture is often measured. This measurement is commonly taken on a sample by sample basis. The corresponding distortion measures as used in the HEVC reference software are summarized in the following subsection. Conceptually, also more elaborate measures that e.g. take into account the modelled subjective impression by an observer can be applied. Such measures might be able to better describe the visual impact of a encoder decision. For effective application in an encoder implementation such measures must be operable on a block basis as well as a picture basis. Due to their effectiveness, the simple difference measures as presented below are widely used.

2.5.1 Distortion Measures

In the following, the most commonly used sample-based distortion measures are summarized. Let the current block be denoted by $\mathbf{B}_c$ with $N \times M$ samples $b(n,m)$, and let the prediction block be $\mathbf{P}$ with $N \times M$ prediction samples $p(n,m)$.

2.5.1.1 SSD—Sum of Squared Difference

The sum of squared differences (SSD) or sum of squared error (SSE) is an evaluation measure commonly used in signal processing. The sum of the squared differences between $\mathbf{B}_c$ and $\mathbf{P}$ corresponds to the squared Euclidian norm or the squared L_2-norm in a vector space. It is defined as

$$D_{\mathrm{SSE}} = \sum_{n=0}^{N-1} \sum_{m=0}^{M-1} \left| b(n,m) - p(n,m) \right|^2. \tag{2.47}$$

The measure corresponds to the energy of the error between **P** and $\mathbf{B}_c$. Considering **P** as an approximation of $\mathbf{B}_c$, the difference signal can be interpreted as noise that has been added to the original signal block.

2.5.1.2 SAD—Sum of Absolute Differences

The sum of absolute differences is defined similarly to the SSE but omits squaring the summation elements. The SAD corresponds to the L_1 norm in a vector space and is defined as

$$D_{\text{SAD}} = \sum_{n=0}^{N-1} \sum_{m=0}^{M-1} \left| b(n,m) - p(n,m) \right|. \tag{2.48}$$

The SAD can be seen as the simplest similarity metric. It is used for motion estimation and also in image analysis applications.

2.5.1.3 SATD—Sum of Absolute Transformed Differences

The sum of absolute transformed differences applies a Hadamard transform before calculating the sum of absolute differences. While the SAD itself serves as a measure for the difference between the samples in the blocks, the preceding transform in the SATD provides an additional estimate for the effort needed for transform coding the difference signal, the residual. Thereby, the SATD serves as a combined metric, for the block difference as well as for the coding cost of the residual.

For calculation of the SATD, the prediction residual is transformed by the Hadamard transform as

$$\mathbf{R}_{\text{H}} = \mathbf{T}_{\text{H}} \cdot (\mathbf{B} - \mathbf{P}) \cdot \mathbf{T}_{\text{H}}^{\text{T}}. \tag{2.49}$$

The SATD is then calculated as

$$D_{\text{SATD}} = \sum_{n=0}^{N-1} \sum_{m=0}^{M-1} \left| r_{\text{H}}(n,m) \right|. \tag{2.50}$$

Though having an increased complexity compared to the simple SAD, the features described above make the SATD a very suitable metric for distortion measurement in the context of motion estimation.

2.5.1.4 PSNR—Peak Signal-to-Noise Ratio

Since the conduction of formal subjective tests is an expensive and tedious process, *objective* testing methods are employed, which are considered to sufficiently approximate results of corresponding subjective testing campaigns. Objective testing methods have the advantage of being exactly reproducible by independent evaluators for a given set of test material.

The most commonly used objective quality measure in video coding is the peak signal-to-noise ratio, PSNR. The PSNR measured the relation of the mean squared error between the original signal (i.e. picture) p_{orig} and the reconstructed signal p_{rec} to the available maximum amplitude of the original signal,

$$\begin{aligned}\text{PSNR} &= 10 \cdot \log_{10}\left(\frac{A_{\max}^2}{\frac{1}{N_{\text{P}}N_{\text{L}}}\sum_{n=0}^{N_{\text{P}}-1}\sum_{m=0}^{N_{\text{L}}-1}\left|p_{\text{orig}}(n,m) - p_{\text{rec}}(n,m)\right|^2}\right)\\ &= 10 \cdot \log_{10}\left(N_{\text{P}}N_{\text{L}} \cdot \frac{A_{\max}^2}{D_{\text{SSE}}}\right)\ \text{dB},\end{aligned} \tag{2.51}$$

with $A_{\max} = 255 = 2^8 - 1$ for 8-bit video.

Commonly, the PSNR is measured on a picture basis and the average is taken as the PSNR value for a full sequence. This measurement method is valid if quality fluctuations between successive pictures of the decoded sequence do not specifically need to be taken into account. Otherwise, the D_{SSE} for the full sequence and the corresponding PSNR can be evaluated.

For evaluation, rate-distortion (RD) plots are used, showing the sequence PSNR over the bitrate of the corresponding bitstream. The bitrate is usually calculated from the file size of the coded bitstream divided by the duration of the decoded sequence in seconds and is measured in kbit/s. Depending on the measurement task, the bitrate can also be calculated only from parts of the bitstream, e.g. if specific side information like supplemental enhancement information shall not be taken into account.

2.5.2 Rate-Distortion Optimization

Let each coding tree block (or each macroblock) be coded using a specific coding mode. The coding mode comprises the applicable prediction mode, the corresponding prediction parameters, and the applicable block partitioning for transform coding. With all available coding modes at hand, the encoder has to decide which coding mode with what prediction parameters to apply for each block (e.g. each coding tree block or each macroblock) in each picture of the video sequence, in order to optimize the output in terms of the used bitrate R and observed distortion D induced by the selected coding modes. This problem can be solved by using classical rate-distortion optimization, applied on a picture basis [42].

Let the complete picture be partitioned into a set of N coding tree blocks, $P = \{\text{CTB}_n\}$, $n = 0, \ldots, N-1$, and let the coding mode for coding tree block CTB_n be denoted by a coding mode index M_n. The set of applied coding modes for the picture is denoted by $M = \{M_n\}, n = 0, \ldots, N - 1$. The optimization task for the encoder is to find the best coding mode set M_{opt} for the current picture subject to a given rate constraint R_c, such that the distortion in the picture is minimized,

$$\min_M \big(D(P, M)\big), \text{ with } R(P, M) < R_c. \tag{2.52}$$

Using the Lagrangian weighting factor λ, this constrained problem is transformed into an unconstrained problem,

$$M_{\text{opt}} = \arg\min_M \big(J(P, M|\lambda)\big), \tag{2.53}$$

with the joint cost criterion

$$J(P, M|\lambda) = D(P, M) + \lambda \cdot R(P, M). \tag{2.54}$$

When encoding a picture, it is assumed that the cost for the coding tree blocks is additive. Under this assumption, the total minimum joint cost for coding the picture can be determined by selecting the minimum joint cost for each coding tree block,

$$\min_M \sum_{n=0}^{N-1} J(\text{CTB}_n, M|\lambda) = \sum_{n=0}^{N-1} \min_{M_n} \big(J(\text{CTB}_n, M_n|\lambda)\big). \tag{2.55}$$

This assumption does not take into account potential dependencies of the coding decision for coding tree blocks on future blocks. Also, potentially existing inter-dependencies between pictures are neglected. At an increased computational effort, such information could be included in the coding decision e.g. if look-ahead coding strategies are applied.

The applicable weighting factor λ depends on the employed distortion measure and also on the coding configuration for the given video sequence. It has to be determined by the application.

2.6 Compression Artifacts

Prediction, decorrelation, and entropy coding of a signal can be used to generate a compact coded representation which allows for perfect reconstruction of the signal. The quantization operation removes information from the signal which cannot be recovered. This results in typical compression artifacts which are briefly reviewed in this section.

In terms of artifacts potentially occurring within a single picture, three types are present which occur in the context of predictive coding. These are *ringing*, *blurring*, and *blocking*. These artifacts can be considered to be 'local' to a picture as they are

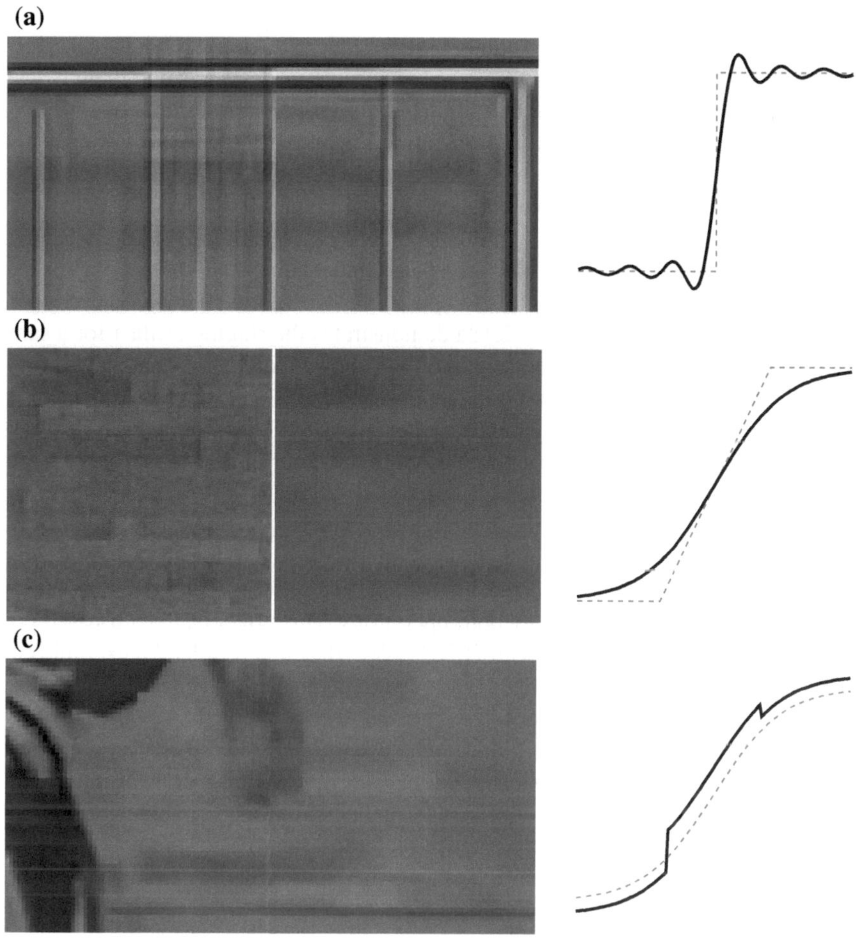

Fig. 2.18 Examples and illustrations for compression artifacts in video coding. **a** Ringing. **b** Blurring. **c** Blocking

induced by prediction and reconstruction of the quantized prediction error of the picture under consideration and do appear in still-image coding schemes as well as in video coding schemes. For video sequences, additional artifacts may be introduced which can severely harm the visual impression of the reconstructed video. *Motion jerkiness* can be introduced when the coded motion vectors do not match the motion observed in the scene.[12] *Pumping* of the reconstruction quality over time

[12] This effect may e.g. be observed with rate-distortion optimized H.264 | AVC encoding. Here, the rate-distortion optimization favours the skip mode, which is very cheap in terms of coding cost while it omits an update of the motion information according to the scene motion. Thereby, skip-coded

may occur, e.g. if the quantizer control periodically switches between fine and coarse quantization.

Cause and effect for each of ringing, blurring, and blocking are briefly described here. An example with a snapshot from a reconstructed video and a one-dimensional illustration for each of the artifacts is shown in Fig. 2.18.

Ringing occurs in the context of transformation and quantization. If a steep edge in the signal is transformed and the coefficients are quantized, some of the transform coefficients—specifically at high frequencies—are quantized to zero. As a consequence, the missing frequency components induce artifacts in the reconstructed signal, which are characterized as overshoots and undershoots. This effect occurs also when filtering a step function with a lowpass filter where the effect is referred to as the Gibbs phenomenon [43]. Fig. 2.18a demonstrates the ringing artifact for a sample area of a reconstructed video sequence and shows the Gibbs phenomenon for a lowpass filtered one-dimensional step response.

Blurring is observed if strong lowpass filtering is applied to a signal. The lowpass filtering removes the detail and smoothes out the contour of the signal. This artifact is induced by strong quantization of the signal. It can be even enhanced by the operation of the deblocking filter, which adds a smoothing operation at otherwise visible block boundaries. A combination of small transform block sizes and strong deblocking filtering abets a blurred impression of the reconstructed video as few detail information is preserved by both operations. This effect can be observed e.g. for video coded at low bitrates with H.264 | AVC Baseline profile. An example can be found in Fig. 2.18b, providing a blurred sample area of a reconstructed picture and an illustration for a one-dimensional signal.

Blocking is an artifact which is induced by two different sources. These are block-wise prediction and block transform coding. The blocks of a picture from a video sequence are either intra predicted from a spatial neighborhood or are inter predicted using motion compensation of a reference picture. Both types of prediction operate on a block basis where the sample values at the boundaries of the neighboring blocks do not necessarily relate to each other. As a consequence, the block structure of the prediction operation becomes visible. The second source of blocking artifacts is the block transform nature of the transform coding stage in the video coding scheme. With an increasing amount of quantization, the block boundaries between neighboring transform blocks become more and more visible as the reconstructions of the neighboring blocks are independent. If the residual signal is even quantized to zero, the blocking artifact of the prediction operation comes to full effect. These blocking effects can be mitigated by deblocking filtering. Other approaches such as overlapped block motion compensation [44] and overlapped transforms [45] alleviate the effect as well but induce an increased computational complexity and integration issue in the overall design. Fig. 2.18(c) shows an example for blocking artifacts in a reconstructed picture. The one-dimensional illustration exemplifies the transition discontinuities if shifted parts of the signal are attached for approximation.

---

regions in a scene may appear to 'jump' back and forth in successive pictures, depending on how coarse the motion approximation by the skip modes has been.

References

1. Poynton, C.: Digital Video and HD: Algorithms and Interfaces. Morgan Kaufman Publishers, Waltham (2012)
2. Parameter values for ultra-high definition television systems for production and international programme exchange. ITU-R Rec, BT.2020-0. http://www.itu.int/rec/R-REC-BT.2020/en (2012). Accessed 14 Apr 2014
3. Salmon, R., et al.: Higher Frame Rates for more Immersive Video and Television. British Broadcasting Corporation. http://www.bbc.co.uk/rd/publications/whitepaper209 (2011). Accessed 14 Apr 2014
4. High efficiency video coding. ITU-T Rec. H.265 (HEVC). http://www.itu.int/rec/T-REC-H.265/en (2013). Accessed 14 Apr 2014
5. Colorimetry—Part 1: CIE standard colorimetric observers. ISO 116641:2007. http://www.iso.org/iso/home/store/catalogue_tc/catalogue_detail.htm?csnumber=52495 (2007). Accessed 14 Apr 2014
6. Hunt, R.G.: The Reproduction of Colour, 6th edn. Whiley-VCH, Chichester (2004)
7. Studio encoding parameters of digital television for standard 4:3 and wide screen 16:9 aspect ratios. ITU-R Rec, BT.601-7. http://www.itu.int/rec/R-REC-BT.601/en (2011). Accessed 14 Apr 2014
8. Parameter values for the HDTV standards for production and international programme exchange. ITU-R Rec, BT.709-5. http://www.itu.int/rec/R-REC-BT.709/en (2002). Accessed 14 Apr 2014
9. Reference electro-optical transfer function for flat panel displays used in HDTV studio production. ITU-R Rec, BT.1886-0. http://www.itu.int/rec/R-REC-BT.1886/en (2011). Accessed 14 Apr 2014
10. Ohm, J.-R.: Multimedia Communication Technology. Springer, Berlin, Heidelberg (2004)
11. Kamp, S., Wien, M.: Decoder-side motion vector derivation for block-based video coding. IEEE Trans. Circ. Syst. Video Technol. **22**(12), 1732–1745 (2012). doi:10.1109/TCSVT.2012.2221528
12. Jain, A.K.: Fundamentals of Digital Image Processing. Prentice-Hall, Englewood Cliffs (1989)
13. Jain, A.K.: A sinusoidal family of unitary transforms. IEEE Trans. Pattern Anal. Mach. Intell. **1**(4), 356–365 (1979). doi:10.1109/TPAMI.1979.4766944
14. Britanak, V., et al.: Discrete Cosine and Sine Transforms: General Properties, Fast Algorithms and Integer Approximations. Academic Press, New York (2006)
15. Han, J., et al.: Towards jointly optimal spatial prediction and adaptive transform in video/image coding. In: Proceedings of IEEE International Conference on Acoustics, Speech and Signal Processing (ICASSP '10), pp. 726–729 (2010). doi:10.1109/ICASSP.2010.5495043
16. Rao, K.R., Yip, P.: Discrete Cosine Transform. Academic Press, San Diego, CA (1990)
17. Chivukula, R.K., Reznik, Y.A.: Fast computing of discrete cosine and sine transforms of types VI and VII. In: Tescher, A.G. (ed) Applications of Digital Image Processing XXXIV, vol. 8135 (2011). SPIE, San Diego, CA. doi:10.1117/12.903685
18. Gray, R.M., Neuhoff, D.L.: Quantization. IEEE Trans. Inf. Theory **44**(6), 2325–2383 (1998). doi:10.1109/18.720541
19. Sullivan, G.J.: Efficient scalar quantization of exponential and Laplacian random variables. IEEE Trans. Inf. Theory **42**(5), 1365–1374 (1996). doi:10.1109/18.532878
20. Reininger, R.C., Gibson, J.D.: Distributions of the two-dimensional DCT coefficients for images. IEEE Trans. Commun. **31**(6), 835–839 (1983). doi:10.1109/TCOM.1983.1095893
21. Lam, E.Y., Goodman, J.W.: A mathematical analysis of the DCT coefficient distributions for images. IEEE Trans. Image Process. **9**(10), 1661–1666 (2000). doi:10.1109/83.869177
22. IEEE Standard Specifications for the Implementations of 8×8 Inverse Discrete Cosine Transform. IEEE 1180 (1991). doi:10.1109/IEEESTD.1991.101047
23. Information technology—MPEG video technologies—Part 1: Accuracy requirements for implementation of integer-output 8×8 inverse discrete cosine transform. ISO/IEC

23002-1:2006 (MPEG-C). http://www.iso.org/iso/home/store/catalogue_tc/catalogue_detail.htm?csnumber=42030 (2006). Accessed 14 Apr 2014
24. Information technology—MPEG video technologies—Part 2: Fixed-point 8 × 8 inverse discrete cosine transform and discrete cosine transform. ISO/IEC 23002-2:2008 (MPEG-C). http://www.iso.org/iso/home/store/catalogue_tc/catalogue_detail.htm?csnumber=45433 (2008). Accessed 14 Apr 2014
25. Wien, M., et al.: Integer transforms for H.26L using adaptive block transforms. Doc. 11th meeting: ITU-T SG16/Q15 VCEG, Q15-K-24, Portland (2000)
26. Advanced video coding for generic audiovisual services. ITU-T Rec. H.264 (AVC). http://www.itu.int/rec/T-REC-H.264/en (2014). Accessed 14 Apr 2014
27. Information technology—Coding of audio-visual objects—Part 10: Advanced video coding. ISO/IEC 14496–10:2012 (AVC). http://www.iso.org/iso/home/store/catalogue_tc/catalogue_detail.htm?csnumber=61490 (2012). Accessed 14 Apr 2014
28. Sullivan, G.J, Ohm, J.-R.: Meeting report of the tenth meeting of the Joint Collaborative Team on Video Coding (JCT-VC), Stockholm, SE. Doc. JCTVC-J1000. 10th Meeting: Joint Collaborative Team on Video Coding (JCT-VC) of ITU-T SG 16 WP 3 and ISO/IEC JTC 1/SC 29/WG 11, Stockholm, SE (2012)
29. Video codec for audiovisual services at p × 64 kbit/s. ITU-T Rec. H.261. http://www.itu.int/rec/T-REC-H.261/en (1993). Accessed 14 Apr 2014
30. Bossen, F.: Common test conditions and software reference configurations. Doc. JCTVC-K1100. 11th Meeting: Joint Collaborative Team on Video Coding (JCT-VC) of ITU-T SG 16 WP 3 and ISO/IEC JTC 1/SC 29/WG 11, Shanghai, CN (2012)
31. Sullivan, G.J., Ohm, J.-R.: Meeting report of the seventh meeting of the Joint Collaborative Team on Video Coding (JCT-VC), Geneva, CH. Doc. JCTVC-G1100. 7th Meeting: Joint Collaborative Team on Video Coding (JCT-VC) of ITU-T SG 16 WP 3 and ISO/IEC JTC 1/SC 29/WG 11, Geneva, CH (2011)
32. Cover, T.M., Thomas, J.A.: Elements of Information Theory. Wiley, New York (1991)
33. Sayood, K.: Introduction to Data Compression, 3rd edn. Morgan Kaufmann Series in Multimedia Information and Systems. Morgan Kaufmann Publishers Inc., San Francisco (2005). ISBN: 012620862X
34. Gallager, R.G.: Variations on a theme by Huffman. IEEE Trans. Inf. Theory **24**, 668–674 (1978). doi:10.1109/TIT.1978.1055959
35. Taubman, D.S., Marcellin, M.W.: JPEG2000: Image Compression Fundamentals, Standards and Practice. Kluwer, Boston (2002)
36. Ugur, K., et al.: Description of video coding technology proposal by Tandberg, Nokia, Ericsson. Doc. JCTVC-A119. 1st Meeting: Joint Collaborative Team on Video Coding (JCT-VC) of ITU-T SG 16 WP 3 and ISO/IEC JTC 1/SC 29/WG 11, Dresden, Germany (2010)
37. Golomb, S.W.: Run-length encodings. IEEE Trans. Inf. Theory **12**(3), 399–401 (1996). doi:10.1109/TIT.1966.1053907
38. Weinberger, M.J., et al.: The LOCO-I lossless image compression algorithm: principles and standardization into JPEG-LS. IEEE Trans. Image Process. **9**(8), 1309–1324 (2000). doi:10.1109/83.855427
39. Teuhola, J.: A compression method for clustered bit-vectors. Inform. Process. Lett. **7**(6), 308–311 (1978). doi:10.1016/0020-0190(78)90024-8
40. Video coding for low bit rate communication. ITU-T Rec. H.263. http://www.itu.int/rec/T-REC-H.263/en (2005). Accessed 14 Apr 2014
41. Marpe, D., et al.: Context-based adaptive binary arithmetic coding in the H.264/AVC video compression standard. IEEE Trans. Circ. Syst. Video Technol. **13**(7), 620–637 (2003). doi:10.1109/TCSVT.2003.815173
42. Sullivan, G.J., Wiegand, T.: Rate-distortion optimization for video compression. IEEE Signal Process. Mag. **15**(6), 74–90 (1998). doi:10.1109/79.733497
43. Oppenheim, A.V., et al.: Signals and Systems, 2nd edn. Prentice-Hall, Englewood Cliffs, NJ (1997)

44. Orchard, M.T., Sullivan, G.J.: Overlapped block motion compensation: an estimation-theoretic approach. IEEE Trans. Image Process. **3**(5), 693–699 (1994). doi:10.1109/83.334974
45. Malvar, H.S.: Biorthogonal and nonuniform lapped transforms for transform coding with reduced blocking and ringing artifacts. IEEE Trans. Signal Process. **46**(4), 1043–1053 (1998). doi:10.1109/78.668555

Chapter 3
Design and Specification

During the design and drafting phase of a video coding specification, the involved experts develop a common understanding of the technical tools and features which are adopted into the draft. For the final specification, an unambiguous representation of this common understanding is required which further must be completely independent of anything that was commonly understood in the development phase. In this chapter, scope of and requirements on specification text are presented. While there are general rules applicable to all standards and recommendations specified by the standardization bodies, some additional constraints on the video coding specification design as applied during the development of HEVC are presented as well.

Both ITU-T and ISO/IEC have normative requirements and recommendations for their recommendations and standards. Formal rules for structure and drafting of international standards in ISO/IEC are specified in the ISO/IEC Directives Part 2 [1]. The corresponding recommendations for the ITU-T are provided in the "Author's guide for drafting ITU-T Recommendations" [2]. Both organizations have approved the formal structure and rules for presentation of ITU-T | ISO/IEC common text in Appendix II of their "Guide for ITU-T and ISO/IEC JTC 1 cooperation" [3].[1]

In the light of the aforementioned guidelines, this chapter summarizes the general concepts and principles behind the development of a video coding specification. It further includes definitions and explanations of terminology and methodology used for specification and introduces the structure of the specification and how to assess it. Furthermore, aspects of determination and selection of tools appropriate for inclusion in the specification are discussed and the methodology for the evaluation of tools is presented.

[1] In the final development phase of H.264 | AVC in the Joint Video Team (JVT) after the July 2002 JVT meeting in Klagenfurt, Aharon Gill held a tutorial for the JVT editing team how to write high-quality interoperability standards. The guidelines established by this tutorial were the fundamentals of the editing work related to H.264 | AVC and its extensions. These fundamentals were reinforced for continued use in the development of HEVC [4].

M. Wien, *High Efficiency Video Coding*, Signals and Communication Technology,
DOI 10.1007/978-3-662-44276-0_3

3.1 Specification Fundamentals

In this section, the main goal of standardization, the scope of the specification, and the main classification of specification text are outlined. When editing specification text, a neutral perspective has to be adopted in order to make the specification as concise and unambiguous as possible.

3.1.1 Interoperability

The starting point of a standardization activity is the desire to enable devices and applications from different manufacturers and sources to interoperate. Such *interoperability* is of utmost importance for enabling information exchange and communication.

In order to enable the specification to fulfil this task, a clear definition of the *requirements* which are imposed on the specification is needed. These requirements must reflect the application needs as envisioned by the parties who plan the usage of the specification in their application scenario. At the same time, the specification must allow for sufficient extensibility in order to adapt the specification to use cases which have not been regarded in the original development process. The fulfilment of the imposed requirements must be monitored throughout the standardization process, and the lack of the fulfilment of requirements may trigger new technical development in order to accomplish the task.

While within one application scenario interoperability is elementary, the application space of the specification may be too wide and too heterogeneous to make interoperability between all application scenarios feasible. In order to cope with this flexibility challenge, the concept of *profiles* is introduced. A profile specifies a set of tools which are suitable to fulfil the requirements of a respective application scenario. This set of tools usually is a subset of the specified tools, tailored to the application needs. While all profiles share common parts of the specification, not all profiles need to support all tools that are specified. The extension of the application space for a specification usually implies the definition of one or more corresponding profiles that define the specification tool set which should be applied in the respective applications.

Besides the tools that are activated in the given application scenario, the computational power required for processing according to the specification plays an important role. In order to discriminate between applications with diverging computational load, the concept of *levels* is introduced. A level sets limits on the required buffer size, picture size, and other parameters which impact the required computational load to decode. In HEVC, levels are further organized in *tiers* which enable the distinction between application spaces (e.g. professional and consumer). Within a tier, a lower level indicates lower computational requirements compared to a higher level. The profiles, tiers, and levels as specified in HEVC are presented in Chap. 11.

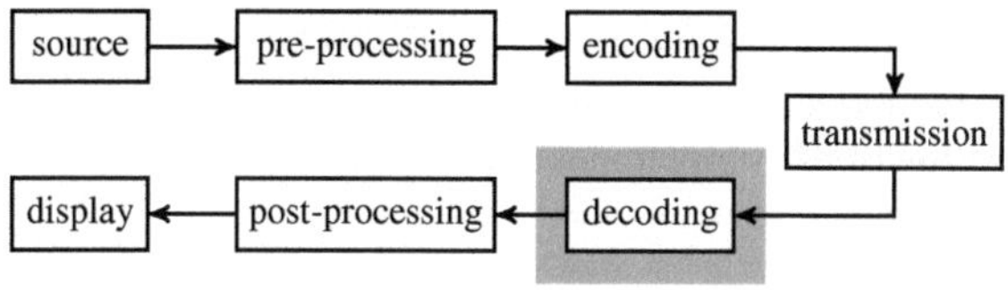

Fig. 3.1 The normative part of a video coding system (marked by a *gray box*)

3.1.2 Specification Scope

In Fig. 3.1, the general block diagram of a video coding system is presented, including the indication of the scope of a video coding specification (marked gray). The parts that need to be normatively specified include the syntax and semantics of the bitstream which is fed into the decoding process, the decoding process itself, and the resulting decoding output. This restriction provides full freedom to the developers of encoders which produce bitstreams conforming to the specification. Future optimization of the usage and exploitation of specification features is enabled. Thereby, the reconstruction quality of the decoded output can improve over time without the need to change the specification.

On the other hand, since this concept does not require a defined minimum encoder performance there is no guarantee of a minimum reconstruction quality which might be reached by following the specification. The capability of the scheme and the achievable reconstruction quality is commonly demonstrated by a reference encoder implementation. Furthermore, the method of transport for the bitstream to the decoder is not specified. This aspect is left to dedicated transport standards. The specification should however provide sufficient 'hooks' that enable the integration into appropriate transport formats.

3.1.3 Text Classification

Different parts of a specification text serve different purposes. The normative impact of the text must be clear in order to avoid confusion. In the HEVC specification, like in H.264 | AVC, a clear distinction between normative and informative text is established. Informative text is indicated by the term "NOTE" and a special formatting including text width and font size. Informative clauses and figures are clearly marked to be informative in order to distinguish them from the normative specification text.

The normative text can be categorized as follows:

- Information required for the normative decoding process to achieve the bit-exact reconstruction of the coded video sequence.
- Video usability information, which does not affect the process to reconstruct the pictures but specifies how the decoded pictures shall be presented (i.e. picture timing, color format, cutout of the region of the decoded picture to be displayed).

- Supplemental information that can be useful to the decoder but is not required to reconstruct the pictures of the video sequence. It should be noted that the specification of the supplemental information including syntax and semantics is normative nonetheless.

As indicated, some of the normative text does not directly affect the decoding process. However, if the decoder makes use of information related to such text (e.g. like supplemental enhancement information) it must follow the specified normative behaviour. If interpretation of coded information is open to the decoder this must be clearly indicated in the specification text.

3.1.4 Editing Perspective

Once the editing process for a specification has been finalized, the actual text of the specification forms the only basis for interpretation of the specification intent. Discussions and common understanding from the development phase are only known to the participants in that activity but not to the public. The editing process therefore takes the results and decisions of the development phase and generates the draft specification which is to be adopted and set in force by the standardization body. Elements of the development process and potential deviations that this process took in preparation of the final result are not relevant any more. In order to make the specification as reliable and as clear as possible, some elementary rules should be obeyed in the editing process at this stage [4]:

- The specification must not rely on assumed background information. Everything that is required to achieve the intended result must be specified. If something has been specified elsewhere and shall be used, an unambiguous reference to that specification, or the part of the specification which is concerned, is required.
- The specification text must be unambiguous by itself. To prevent misinterpretation, constraints must be well-defined and processes must be specified such that no other than the intended decoding result is possible.
- "The specification is not a tutorial." It must be written from a neutral perspective and does not require motivations or simplifying summaries. It must not advertise properties or features and it does not need to describe intentions behind design decisions.
- Informative text must not contain normative parts. Normative text does not need to and should not include informative parts. Informative text may be included but must be clearly marked as this to be distinguishable from normative text.
- All specific terms must be precisely defined. The definition is set once and the defined terms shall be used accordingly in the text. In general, the same things should be expressed by the same phrases in the same way. All acronyms used in the text must be defined and listed.
- Words, variable names, functions, and syntax elements must be clearly distinguishable. Existing conventions which are recognized and have proved relevance

in previous specifications should be reused. Variable and syntax element names should be unique and clear. To ease search within the specification document, no names should be prefix of other names.
- When extending an existing specification, renumbering should be avoided. For example, if values of a syntax element are prohibited, the values should be marked as prohibited. An annex that has been deleted should be marked as deleted and the number of the annex should not be reused.

Formatting rules and naming conventions play an important role in the editing process. The more consistent the layout and the phrasing of the specification, the more useful do these conventions become when interpreting the specification text. Both, ISO/IEC and ITU, provide comprehensive layout rules and examples to aid authors in the editing process [1, 2].

3.2 Specification Elements

The concepts and elements for defining a video coding specification are abstract and do occur similarly in other contexts like philosophy, linguistics, or computer programming languages.

The coded bitstream can be seen as a written text in a specific language which is read and understood by the decoder. The *syntax* provides the available set of 'words' and their order, the *semantics* specify the meaning of these words and constraints on the usage and combination of words. Together, syntax and semantics specify the grammar of the language. The 'meaning' of the received text is revealed by the *decoding process*. In order to read the words of the text that has been written, the decoder needs to extract the words from the bitstream, i.e. scan the received stream of characters (from a defined alphabet) and identify the words contained therein. This process is called the *parsing process*.

These elements of a specification are described in the following according to the conventions used in HEVC.

3.2.1 Syntax and Semantics

The syntax is specified in syntax structures, which represent a logical entity of the information coded in the bitstream, such as parameter sets, slices, or coding tree units. Each syntax structure is specified in a syntax table, which comprises the complete specification of all possible syntax variations.

The syntax element names are unique such that e.g. activation flags for tools on the picture or the slice level are differentiated. In the syntax tables, the syntax elements are written in bold font [3]. They may occur as flags, values, one-dimensional lists, or multi-dimensional arrays. In the latter two cases, indices are used to refer to a specific element. If the value of a syntax element is referenced in a syntax table or in the text, it is written in non-bold font. The occurrence of syntax elements in the syntax structure may further be conditional on the values of other syntax elements or variables as derived during the decoding process. For each syntax element, the applicable parsing process is specified by a dedicated descriptor in the syntax table.

The semantics specify meaning of the values of each syntax element. They further impose constraints on the values of the syntax elements. If the semantics specify allowed values of a syntax element in one or multiple tables, only the values explicitly defined in the tables are allowed to be present in a bitstream. Depending on the values of syntax elements, the semantics further include the definition of derived variables which are further used in the decoding process.

3.2.2 Decoding Process

The decoding process takes the syntax elements of the bitstream as input and reconstructs the video sequence according to the semantics of the coded syntax elements.

The specification of the decoding process comprises all possible normative operations that are allowed. If the decoder comes into an unspecified state, e.g. by losses in the received bitstream, it does not have a normative option how to act. In this case, the decision how to proceed and how to react on the unspecified state is up to the application. In most cases, the application will try to catch the undefined condition and conceal the detected issue in some way.

The decoding process includes a precise specification on all possible prediction modes, including e.g. filtering, interpolation, inverse transform, or buffer management operations. The specification of the decoding process is complete in the sense that no further information but the information and constraints available from and for the bitstream is required to decode and reconstructed the incoming bitstream to the intended video sequence.

3.2.3 Parsing Process

As indicated above, the syntax tables include a descriptor for each syntax element, which indicates the associated parsing process. Each available parsing process specifies how bits are read from the input bitstream and how the read bits are processed to decode the value of the corresponding syntax element.

The parsing process is only specified for error-free incoming bitstreams. In case of transmission errors and losses, external means as determined by the application have to be taken. In many cases, the erroneous information cannot be recovered and therefore has to be discarded. The syntax specification includes means to allow for resynchronization and re-access to the bitstream in case of transmission problems.

Conceptually, three types of descriptors are used to specify the parsing mode for the syntax elements in HEVC:

- Indication of fixed-length codes with one or more bits,
- Indication of variable length codes where the length of the code word and the corresponding value is derived by the parsing process,
- Indication of arithmetic coding, where no direct association between an integer number of bits from the bitstream and the values of the parsed syntax elements is established. In HEVC, this type is used on the coding tree unit level.

For each parsing mode, the specification ensures a prefix-free bitstream structure. Thereby, it is asserted that the syntax elements in the bitstream are uniquely decodable.

3.3 Specification Principles

As a fundamental principle, a specification must be complete. This means that all intended states and all intended processing operations must be addressed and specified in the specification text. Normative references to other standards and recommendations may be included as appropriate in order to avoid duplication. The complete specification of the possible states of the decoder specifies its normative behaviour. If a decoder for some reason enters an undefined state the normative range is left and the decoder may do anything.[2] Such cases may occur e.g. in the case of transmission losses, when only partial information is received at the decoder side.

Note that the specification of HEVC (as well as H.264 | AVC) does not include any provisions for recovery after bit errors or resilience against such errors. By concept, the specification assumes packet-based transmission, which implies either complete loss or perfect reception of the coded video data. Single bit errors in the coded video data may imply undefined states and may easily lead to a non-decodable packet or even a non-decodable bitstream.

In the following, a set of design principles is presented which has been applied during the development of the HEVC specification.

3.3.1 *Loss Robustness*

Transmission of the coded video sequence may be subject to errors and losses such that parts of the bitstream are not available to the decoding process. In order to improve the loss robustness of the scheme, the organization of the coded information

---

[2] "...including the option to catch fire." (Illustrative JVT/JCTVC saying to emphasize the freedom of choice that opens for the decoder in this case)

must follow hierarchical concepts where the syntax elements are strictly separated according to their scope and impact on the overall stream. Based on this separation, the protection of the information can be applied in a dedicated fashion. In H.264 | AVC and HEVC, the key design aspect in this regard is definition of a video coding layer (VCL) and a network abstraction layer (NAL) in the structure of the bitstream. The video coding layer includes all syntax elements on the slice level which are required for the actual decoding of the pictures. The network abstraction layer provides the encapsulation and identification of the coded video data in the VCL as well as non-VCL data for transport. The non-VCL data includes the high-level syntax such as parameter sets or supplemental enhancement information.

In H.264 | AVC and HEVC, parameter sets can be transmitted 'in-band' or 'out-of-band'. This means that they may be part of the transmitted bitstream, or may be made available to the decoder by other means. The same applies to the video usability information. This concept helps to assert that decoders have the required parameter set available regardless of the transmission conditions. In a dedicated application scenario, the parameter sets could e.g. be hard-coded and fixed at the decoder such that it cannot be affected by transmission.

Loss robustness is further supported by other design aspects such as the principle of independent parsing and the concept of slices, which are detailed below.

3.3.2 Independent Parsing

A fundamental rule in the design of video coding specifications is the rule of independent parsing. The parsing process must not depend on the decoding process. This property is crucial to sustain the ability to parse a bitstream under the condition of transmission losses. For example, the occurrence of a syntax element in the bitstream must not depend on a decision derived by analyzing a decoded picture.

Slices support independent parsing in order to enable independent processing of the different parts of the picture represented by the slices even under the condition of transmission loss. Regarding the aspect of computational complexity and implementation cost, parsing should conceptually be completely decoupled from all other processing [5]. Thereby, greatest flexibility and independence in the implementation of the parsing processes is guaranteed.

As a further design principle, parsing dependencies between parameter sets should be excluded in order to avoid issues with future changes and extensions of the specification [6]. For example, the parsing of parameter sets should not directly depend on the selected profile as future extensions may define profiles which were originally not envisaged. Such dependencies might induce an unnecessary burden especially on hardware decoders, which may not be modified easily. Provisions for extensions to the parameter sets are consequently encapsulated by dedicated flags whose semantics are constrained by the respective profiles. Future extensions may also make use of reserved syntax elements in the parameter set syntax structures. The values of these syntax elements are ignored by the current version of the specification such that a future modification of their value does not interact with older implementations.[3]

3.3.3 Bit-Exact Specification

Previous standards including H.261, H.262 | MPEG-2, H.263, and MPEG-4 Visual use a floating point specification of the inverse transform. Since different floating point implementations may yield slightly different results depending on the implementation strategy or the bit depth of the floating point representation in the encoders and decoders, different decoders may reconstruct slightly different residual signals from the transform coefficients. This may induce drift between the encoder and one or more decoder implementations even though the decoders conform to the specification. In order to assert that different implementations on different platforms produced sufficiently similar inverse transform results, separate standards were developed to measure the deviations of a floating point inverse transform implementation, e.g. IEEE 1180 [7] or MPEG-C Part 1 [8].[4]

In order to avoid these drift issues, a specification of the decoding process yielding an exact reconstruction of the pictures in the coded video sequence is desirable. With such an unambiguous decoding result, all conforming decoders produce the identical output video from a given bitstream. This property helps specifically when testing the conformance of a decoder. It further asserts that no drift occurs when decoding a video sequence with different decoder implementations.

To achieve the goal of bit-exact specification, all normative mathematical operations of the decoding process in H.264 | AVC and HEVC are specified in integer arithmetic. Thereby, deviations in the implementation of floating point operations do not affect the decoding result.

3.3.4 Parallelization

With the increasing picture resolution of video sequences, and with the prevalence of multi-thread and multi-core processors, concepts of parallelization become more and more important to achieve real-time processing capability [9]. The concept of independent parsing contributes to the parallelization capabilities of the specification as thereby, the parsing process can be generally performed in parallel to the decoding process.

---

[3] An example is the `vps_reserved_0xffff_16bits` syntax element in the video parameter set. The 16 bits of this syntax element are specified to have a fixed value for the first version of HEVC, and the value of this syntax element is ignored in the decoding process. The syntax element is foreseen for use in the 3D extension of HEVC.

[4] IEEE 1180 was withdrawn in 2003 and has been replaced by MPEG-C Part 1 in 2006.

At the sequence level, e.g. the decoding of non-reference pictures can be performed in parallel to reference pictures or other non-reference pictures. If decoding delay is not a critical parameter and if sufficient memory is available for decoded pictures, parallel decoding may start at each random access point of a coded video sequence.

At the picture level, the introduction of slices further allows for parallel parsing and parallel decoding of multiple slices. In HEVC, the newly introduced concept of tiles further allows for partitioning of the picture into rectangular regions (see Sect. 4.2.3). Since prediction dependencies as well as entropy coding dependencies are reset at tile boundaries, tiles enable parallel processing for both parsing and decoding on a sub-picture level. HEVC further includes the concept of wavefront parallel processing, which allows for parallel processing on a CTB row basis, with a predefined CTB offset from row to row (see Sect. 4.2.2.4).

At the block level, inter prediction for multiple blocks can be processed in parallel. Note that intra prediction of neighboring blocks cannot be performed in parallel due to local prediction dependencies from block to block. If constrained intra prediction is used, i.e. only prediction from intra coded neighboring blocks is allowed, intra prediction can be performed in parallel to inter prediction. The HEVC loop filter design has been specified such that independent operation on an 8×8 grid is achieved, see Sect. 9.1.

3.3.5 Dynamic Range

The dequantization and inverse transform operations require a large dynamic range, i.e. a high bit depth for the registers where intermediate values are stored may be required. Furthermore for access to the buffer memory, the bit depth of stored values plays an important role in the computational cost and effort that is required to implement a video coding scheme.

Therefore, the consideration of the required dynamic range is an important aspect in terms of complexity for the overall video codec design.[5] Regarding dequantization in connection with the inverse transform, three options are available:

(a) Normative restrictions on the encoding process such that the decoder will never receive values which exceed the dynamic range limits,
(b) Specification of the decoder side processes at a dynamic range which does not exceed some maximum limits,
(c) Truncation of dequantized and intermediate values in order not to exceed a predefined value range.

[5] In the development phase of HEVC, a concept called Internal Bit-Depth Increase (IBDI) was evaluated which allows for adaptation of the precision of the arithmetic operations in the coding scheme [10].

Option (a) has been used in the specification of H.264 | AVC. It may be non-trivial for an encoder to determine if these constraints are met. Option (b) may lead to issues with the available value range for coefficients and the inverse transform matrices. Option (c) is non-complex to implement and has been selected for the specification of HEVC [11, 12].

3.3.6 *Timing*

As part of the effort to separate different types of information to different and independent parts of the bitstream, the concept of time is an important aspect. Generally, information on the precise timing of pictures is not needed in the decoding process for the pictures, as the parameters for prediction do not necessarily relate to time. As an example, the distance between pictures is important for appropriate scaling of motion vector predictors. However, the expression of the distance does not need to relate to the actual picture rate. The distance can be expressed by the picture order count which indicates the play-out order of the decoded pictures, see Sect. 5.4.1.

While not required in the decoding process itself, timing information is of utmost importance for correct play-out of the pictures for display. This includes the play-out at a specified picture rate or synchronization with other information, specifically audio which is associated with the video signal. Furthermore, timing information is required for control of the coded and decoded picture buffers, see Sect. 5.6. Therefore, the timing information is not ignored but rather separated from the coded video data.

While some applications require a constant picture rate for display, other applications may treat the timing of pictures differently. For example, conversational applications may focus on a low delay between capture and display while a constant picture rate may be less important. Other special applications may code pictures at variable time distances, depending on the application needs. As a consequence, timing is classified as video usability information. If not present, the pictures can still be decoded conforming to the normative decoding process. If timing information is present, a normative interpretation is specified to enable correct play-out timing and buffer operation as indicated above.

3.4 Conformance

As stated before, the specification is complete in the sense that all information needed to precisely reconstruct the coded video from a given bitstream is provided. For decoder manufacturers and for content providers it is very important to be able to test and prove their products to be conforming to the specification. In addition to conformity of the decoding process itself, additional constraints and requirements are imposed by the full video system. These are e.g. imposed by the available decoder buffer for the incoming bitstream on the one hand and the buffer for the outgoing reconstructed pictures of the video sequence on the other hand. If a bitstream is to be sent at a given bitrate in a scenario with given input and output buffers, the encoder must ensure that the buffer constraints at the decoder side are observed.

For the purpose of testing conformance, a hypothetical reference decoder is specified. The hypothetical reference decoder is a model of an ideal decoder which instantaneously performs all decoding processing steps. Thereby, timing constraints imposed by information in the decoded bitstream are regarded in conformance testing, but the decoder capability of decoding the bitstream within the time constraints is not regarded. This decoder capability is expressed by the indicated decoder conformity to a given profile, tier, and level. A decoder which claims to be conforming must be able to successfully decode all bitstreams that are conforming to that profile, tier, and level of the specification.

Two types of conformance can be claimed by a decoder: output timing conformance and output order conformance. For output timing conformance the decoder must be able to deliver the decoded pictures according to the timing constraints set by the video usability information. For output order conformance, the decoder just needs to deliver the decoded pictures in the right output order but is not required to be able to decode the video sequence within these timing constraints. For bitstream conformance, the hypothetical reference decoder is used to test if the bitstream is constructed such that the buffer for the incoming bitstream is never overfilled, or unintentionally emptied completely. Bitstream conformance can be specified on different levels, depending if only the video layer part of the bitstream is considered or the full bitstream including all additional side information is included. Furthermore, it checks if all decoded pictures which are used as reference pictures for inter prediction are available at the time they are to be used in the prediction process. An overview of the hypothetical reference decoder of HEVC is presented in Sect. 5.6.

3.5 How to Read the Specification Text

The specification is a comprehensive and condensed piece of text, which may not be easy to access in a first step. According to the specification principles listed earlier, this specification is not illustrative but as compact and unambiguous as possible. This makes the specification text hard to read for the start but also a reliable and clear subject of interpretation when used in practice.

The outline of the HEVC specification is given in Table 3.1. For reading the specification, one should start with Clause 0, which actually does not form an integral part of the specification. It includes a brief prologue on the genesis of the specification and the involved standardization bodies, explanations on the purpose, envisioned applications, as well as the employed profiles, tiers, and levels concept. It further includes a recommendation on how to read the specification. The recommendations given there are paraphrased here.

Table 3.1 Clauses of the HEVC specification

Clause	Title
Clause 0	Introduction
Clause 1	Scope
Clause 2	Normative references
Clause 3	Definitions
Clause 4	Abbreviations
Clause 5	Conventions
Clause 6	Source, coded, decoded, and output data formats, scanning processes, and neighboring relationships
Clause 7	Syntax and Semantics
Clause 8	Decoding process
Clause 9	Parsing process
Clause 10	Specification of bitstream subsets
Annex A	Profiles, tiers, and levels
Annex B	Byte stream format
Annex C	Hypothetical reference decoder
Annex D	Supplemental enhancement information
Annex E	Video usability information

Clause 1 defines the scope of the specification (which is High Efficiency Video Coding) and in Clause 2, normative references are provided. These should be referred to as needed. As a next reading step, one should proceed with a review of Clauses 3 to 6. Clause 3 provides the definition of all specific terms used throughout the text. The definitions in this clause form the 'vocabulary' to read the specification text. In Clause 4, all abbreviations used in the text are decrypted. In Clause 5, the notation of all arithmetic and logical operators, mathematical functions and relations is specified. Furthermore, the invoking of processes and the notation and formatting of variables, syntax elements, and tables is described. Clause 6 contains the collection of all geometrical structures of sample arrays, blocks, and slices. It further includes the definition of the various scanning processes used in the specification, and also the definition of availability processes which are used in the text to test for the existence of neighboring blocks (which is needed e.g. in the context of slices or tiles). With these Clauses the terminology and the notation used in the specification are set. While working on the other Clauses the reader will frequently come back and reiterate on the precise understanding of the terminology and relations, especially in Clauses 3 and 6.

Clauses 7.1 and 7.2 settle the methodology used in the syntax tables that specify the bitstream syntax and employed syntax functions and descriptors. These are used in Clauses 7.3 and 7.4, which specify the bitstream syntax and the semantics of the syntax elements. The syntax tables determine the structure and composition of a conforming bitstream and the naming for all syntax elements is provided. The semantics further specify relation and constraints on the syntax elements.

The decoding process itself is specified in Clause 8. It acquires the parsed syntax elements and specifies the process to reconstruct the pictures of the video sequence based on these. It can be considered as the 'main method' of the specification and controls the progression of the sub-steps for reconstructing the pictures. All parts of the decoding process are successively specified as processes with defined input and output variables. The syntax elements specified in Clause 7 are considered to be available to the decoding process as required.

The process of bitstream parsing is specified in Clause 9. Depending on the applicable parsing method as indicated in the syntax tables, the syntax elements are read from the bitstream. In a first step, Clause 9 may be studied separately from the decoding process and the values of the syntax elements can be considered as given. The sub-bitstream extraction process described in Clause 10 is needed to specify how to operate the decoding process in the case that parts of the bitstream have been previously removed (e.g. to decode the bitstream at a lower temporal resolution). This process may be reviewed upon need.

The Annexes are integral parts of the specification and concern normative aspects around the decoding process itself. Annex A specifies profiles, tiers, and levels, that is, the applicable subset of tools that may be used in a bitstream and the maximum expected buffer sizes and computational load to decode the bitstream. In Annex B a byte stream format is specified which can be used for transportation (but other methods may be used as well). Annex C specifies the hypothetical reference decoder which determines the decoder timing and is used for conformance testing. The supplemental enhancement information specified in Annex D can carry useful information about the bitstream, or e.g. for recovery of lost information. However, this part of the bitstream may be completely removed from the bitstream without affecting the normative decoding process. Annex E includes the video usability information which comprises all parameters about the video that affect the way of displaying the video. Like the supplemental enhancement information, the video usability information does not directly affect the decoding process.

3.5.1 Terminology

The specification text uses a precise terminology to name objects and processes, as collected in Clause 3 of the specification [13]. This book follows the definitions and terminology of the specification text. The terms related to the different parts of the decoding process are explained in the respective chapters. Some more general aspects are summarized here.

3.5.1.1 Specification

The term *specification* is used synonymously for the terms *standard*, which denotes the norms which are published by the ISO/IEC, and *recommendation*, which denotes the norms published by the ITU. Since HEVC has been jointly developed and is a technically-aligned twin text published by both organizations, the term specification is used in the text to embrace both terms.

In the specification text, the *processes* which are specified in the normative text can be operated in different *modes*. Processes operate on *objects* that are input or output (or used only within a process). Objects can be of identical or different *type*.

3.5.1.2 Coded Representation of the Video Sequence

In this subsection, terminology related to the coded representation of a video sequence are presented. The terms are defined in Clause 3 of the specification [13]. Here, the definitions are paraphrased.

The term *bitstream* primarily denotes a sequence of bits. This sequence of bits can be a so-called *NAL unit stream*, i.e. a sequence of NAL units (see Sect. 5.2), or a *byte stream* in which the NAL unit stream is encapsulated by preceding each NAL unit with a start code, see Sect. 5.1. The term bitstream is used in the text when a differentiation between the two formats is not required.

A *coded video sequence* (CVS) denotes the coded representation of a sequence of pictures which can be decoded without reference to pictures not belonging to the same coded video sequence. A bitstream can consist of one or more coded video sequences. A *coded picture* is the coded representation of a single picture, consisting of one or more NAL units. In the bitstream, the coded picture is embedded in an *access unit* which may contain additional NAL units besides the ones of the coded picture.

If it is possible to remove access units in a systematic way from the bitstream to achieve the video sequence at a different temporal resolution and the bitstream is still decodable, *temporal scalability* is supported. Each achievable temporal resolution is denoted as a *temporal sub-layer*.

A bitstream can include further—non-temporal—*layers* which are denoted by a layer identifier. Such layers can e.g. represent different spatial resolutions of a video sequence in a spatial scalability scenario or different views in a multiview sequence. Each layer may have temporal sub-layers but temporal sub-layers cannot have lower (e.g. spatial) layers. The term layer is also used to indicate hierarchical dependencies in terms of coding layers, such as network abstraction layer, video layer, sequence layer, or picture layer.

3.5.1.3 Verbs

In the specification text, a predefined set of verbs is used to express the level of "normativeness" of a sentence. The verbs are exclusively used in normative or in informative text and indicate the strictness in which a certain behaviour or requirement has to be followed. The verbs and their definitions from Clause 3 of the HEVC specification text are given below [13]:

can A term used to refer to behaviour that is allowed, but not necessarily required. (Definition 3.20)

may A term used to refer to behaviour that is allowed, but not necessarily required. In some places where the optional nature of the described behaviour is intended to be emphasized, the phrase "may or may not" is used to provide emphasis. (Definition 3.79)

must A term that is used in expressing an observation about a requirement or an implication of a requirement that is specified elsewhere in the specification. This term is used exclusively in an informative context. (Definition 3.81)

shall A term used to express mandatory requirements for conformance to the specification. When used to express a mandatory constraint on the values of syntax elements or on the results obtained by operation of the specified decoding process, it is the responsibility of the encoder to ensure that the constraint is fulfilled. When used in reference to operations performed by the decoding process, any decoding process that produces identical cropped decoded pictures to those output from the decoding process described in the specification conforms to the decoding process requirements of the specification. (Definition 3.127)

should A term used to refer to behaviour of an implementation that is encouraged to be followed under anticipated ordinary circumstances, but is not a mandatory requirement for conformance to this Specification. (Definition 3.130)

3.5.2 *Conventions and Geometric Relations*

The formatting of the specification text follows a set of conventions which are e.g. specified in [3]. The aim of such conventions is to increase the readability, the conciseness, and the clarity of the text. Some relevant conventions are summarized in the following.

nal_unit(NumBytesInNalUnit) {	**Descriptor**
nal_unit_header( )	
NumBytesInRbsp = 0	
for(i = 2; i < NumBytesInNalUnit; i++)	
if(i + 2 < NumBytesInNalUnit && next_bits(24) = = 0x000003) {	
rbsp_byte[NumBytesInRbsp++]	b(8)
rbsp_byte[NumBytesInRbsp++]	b(8)
i += 2	
emulation_prevention_three_byte /* equal to 0x03 */	f(8)
} else	
rbsp_byte[NumBytesInRbsp++]	b(8)
}	

Fig. 3.2 Example syntax table: The NAL unit syntax structure [13]

3.5.2.1 Syntax Elements, Variables, and Functions

According to [3] syntax element names are indicated by a bold font in the syntax tables. In the text and in equations, they are written in regular font. The syntax elements are written in all lower case letters with underscore characters between the parts of the name. Syntax elements which have a fixed length and a fixed value indicate these properties in their name. For example, the syntax element forbidden_zero_bit has as code word length of one bit and the fixed value "0". The syntax element emulation_prevention_three_byte has a code word length of 8 bits (=1 byte) and the fixed value of "3".

Variables are used in the syntax tables and in the specification of processes. They are written using a mixture of lower case and upper case letters ("CamelCase notation"). If a variable is local to a syntax table or to the process where it is used, the first letter of the variable name is written in lower case. If the variable is global and may be referenced in other syntax tables or processes, it is written with an upper case first letter.

In order to make the meaning of the values of some syntax elements more readable, a set of mnemonic names are introduced for those values in the semantics of the syntax elements. For example, the syntax element pred_mode_flag can take the values 0 = MODE_INTER or 1 = MODE_INTRA.

The names of syntax functions are written using the same convention with lower case letters and underscore characters as for syntax element names and a set of parentheses at the end. The parentheses may include one or more parameters that are used by the function. Other functions, e.g. mathematical functions, are written using the convention of global variables, using an upper case first letter and a mixture of lower case and upper case letters in the following. The parentheses at the end of the function name may include one or more parameters that are used by the function.

An example syntax table is provided in Fig. 3.2 showing the syntax structure of a NAL unit. This syntax function only reads all bytes of the NAL unit payload. The bytes are parsed in the other syntax structures which operate on the read bytes. The structure is defined as a syntax function which is passed a (global) variable

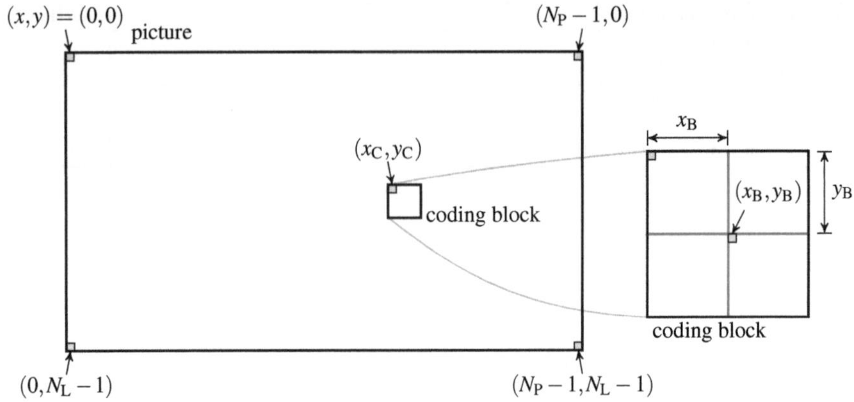

Fig. 3.3 Denoting the location of blocks relative to the picture origin and relative to the coding block

as parameter, nal_unit(NumBytesInNalUnit). In the syntax structure another syntax structure, nal_unit_header(), is called first. The global variable NumBytesInRbsp is initialized to zero, and the bytes that belong to the payload of the NAL unit are read from the bitstream. For detection and prevention of start code emulation within the payload of a NAL unit, the syntax function next_bits() is called which returns the value of the next 24 bits of the payload without shifting the bitstream pointer. The bitstream pointer marks the bitstream position to be read next. Depending on the return value of the next_bits() syntax function the next payload bytes are read from the bitstream (note that the byte of the syntax element emulation_prevention_three_byte is not counted by the variable NumBytesInRbsp), see also Sect. 5.2.1.

3.5.2.2 Locations and Positions

The *location* of a sample (= pixel) in a picture is indicated by the sample number in horizontal direction and by the line number in vertical direction. If an element of an array shall be addressed, the *position* in terms of rows and columns is denoted. Columns and samples in a line are usually addressed by the variable x while rows and lines are addressed by the variable y. Array elements are referenced in the specification either by the array name followed by the column and row indicators in squared brackets, or by a subscript index in reversed (mathematical) notation,

$$s[x][y] \quad \text{or} \quad s_{yx}.$$

For referencing the location of a block in a picture, the samples and lines are given relative to the origin of the picture coordinate system in the top left corner. The conventions are illustrated in Fig. 3.3. The variable pair (x_C, y_C) denotes the location of the top left sample of a coding block relative to the top left corner sample of the picture. Inside the coding block, the location of a block is indicated relative to the current coding block location by (x_B, y_B).

At various places of the specification, blocks in the temporal or spatial vicinity of another block must be referenced. These blocks are identified by any of the samples within the block. Often, the spatially neighboring blocks are reference by the directly adjacent samples.

3.6 Drafting Methodology

When drafting the specification, the adoption of a basic structure as well as the adoption of many dedicated tools have to be decided.

The fundamental criteria for evaluating a video coding specification are defined by the requirements that have been imposed for its development. While general technical requirements like the support of certain color formats or a certain form of scalability are supported by implementing the respective features, the selection of specific tools must be based on measurable criteria which are applied when assessing competing proposals.

In video coding standardization, a two-stage process is used for the development of the final specification. In the first, *competitive* phase, independent full proposals are evaluated. If a new specification is to be developed from scratch, this evaluation includes a formal subjective assessment of the proposals on a predefined test set with defined coding conditions. Among the proposals, the best suited one is selected to form the test model for the new specification. The test model can potentially also be constructed from merging elements from multiple proposals. The decision which proposal to choose as this starting point is not alone based on the result of the subjective tests and technical requirements but also on aspects like simplicity, expandability, and software quality. In the second, *collaborative* phase, all tools are individually investigated in terms of benefit and drawbacks and are further developed based on these findings. The evaluation is based on the implementation of the tools into the reference software of the test model, which is further developed within each meeting cycle.

In both phases, three major criteria are used for analysis:

- Fulfilment of technological requirements. At least most of the given requirements must be answered by a proposal. For full proposals, the completeness of the addressed requirements is assessed. Proposals which are responding to a subset of the requirements are analyzed with respect to these, and in respect to integration with other proposals responding to other requirements.
- Compression performance (complete proposals and single tools). The compression performance is measured by evaluating the observed reconstruction quality of decoded test sequences relative to the bitrate spent for encoding. Subjective visual assessment by experts or non-experts is applied to determine the perceived

reconstruction quality. The application of objective distortion metrics is used as a means to measure the objective reconstruction quality. Objective measures have the advantage of uncomplicated determination and easy reproducibility. Sufficient congruity of subjective assessment and objective measurement has to be asserted.
- Tool complexity and soft- and hardware implementation friendliness. The complexity impact of each proposed tool regarding the computational effort and required (uniform and irregular) memory access is evaluated. The description complexity and the required optimization effort for tool parameter derivation are taken into account.

Competing proposals for the tools are compared and the appropriate solution is adopted into the test model upon consensus of the group. While in the first phase, formal subjective assessment is mostly applied, the compression performance is mostly evaluated using objective measures in the second phase. Although the objective measurements cannot reproduce the outcome of a subjective assessment, this approach can still very well produce reliable results as the variation between the compared versions of the coding scheme are relatively small and the visual impact of the tools is distributed over the coded video. For tools which directly affect the visual impression of the reconstructed video sequence, such as interpolation or deblocking filters, subjective assessment is required nonetheless. Throughout the development process, the validity of the objective measurements must be assessed and confirmed in order to avoid decisions based on unsuitable or even misleading figures on experimental results.

3.6.1 Measuring the Compression Performance

For assessment of the compression performance of the video coding scheme under development, the reconstruction quality of the video is evaluated subject to the bitrate that is required for encoding. In order to enable systematic comparison of coding tools throughout development process, sustained common testing conditions [14] are defined, which aim to sufficiently reflect the target application areas.

3.6.1.1 Common Testing Conditions

The common testing conditions include a set of test sequences of various content and various resolutions and predefined encoding configurations for typical applications. In the development phase of HEVC, three types of encoding configuration settings were used. These are defined in the JCT-VC common testing conditions (CTC) which were to be used for core experiments between the meetings [14]:

- All Intra (AI): All pictures are coded independently with intra prediction only. This configuration reflects applications like editing where picture-wise access to the sequence must be enabled.

- Random Access (RA): Coding at very high compression performance, where the bitstreams allow to enter the decoding of the sequence roughly every second (varying depending on the picture rate). Coding order and output order of the pictures differ and thereby induce a structural coding delay. This configuration represents broadcasting and streaming applications.
- Low Delay (LD): Coding with no structural delay (coding order equal to output order). This configuration represents conversational applications with low-delay requirements but no necessity of random access. This configuration is evaluated using P pictures with uni-prediction only or B pictures with uni- and bi-prediction.

These three configuration settings are evaluated in a configuration corresponding to the Main profile of HEVC and in a configuration for highest coding efficiency with a higher internal processing bit depth. For test evaluation, video sequences of 10 s length are commonly used. This duration is regarded as being sufficient for a qualified evaluation of the visual quality of the reconstructed sequence, while keeping the effort for producing the coded bitstream for the evaluation at an acceptable level at the same time.

The test sequences of the JCT-VC common testing conditions that have been used during the development of HEVC are listed in Table 3.2. A screen shot for a representative sequence from each class is shown in Fig. 3.4. All sequences but two of the sequences in class A have a bit depth of 8 bit. The class A sequences are cropped regions of 2560×1600 samples from ultra high resolution sequences (Traffic: 3840×2160 8 bit, PeopleOnStreet: 3840×2048 8 bit, Nebuta and Steam Locomotive: 7680×4320 10 bit). As the class A test sequences represent application areas like editing or ultra high definition broadcast they are not used for low-delay coding tests. Similarly, since the sequences of class E are considered to represent conversational applications, they are not included in the random access test cases.

For evaluation, all test configurations are applied to all respective sequences at four different quantizer configurations, which are kept constant over the whole test sequence, making a total of more than 500 measurement points for comparison in a complete assessment. In many experiments, multiple different configurations and settings have been evaluated on this basis.

It should be noted that 'real-world' applications usually apply rate control mechanism and optimization strategies according to the application needs. The encoder control is designed to fulfill the application needs rather than responding to experimental coding conditions. The defined constant test settings are utile for tool evaluation in the standardization activity in order to prevent an overlay of effects of encoder optimization control and the tool itself. This separation (encoder optimization vs. tool impact) has always to be taken into account when comparing video coding schemes. When evaluating coding schemes based on different software implementations it is often not easy to conclude on the performance of either of the competing schemes as usually the encoder implementations and optimization strategies differ.

Table 3.2 Test sequences from the JCT-VC common testing conditions [14]

Class	Sequence Name	N_F	Resolution	r_F	AI	RA	LD
A	Traffic	150	2560 × 1600	30	×	×	
	PeopleOnStreet	150	2560 × 1600	30	×	×	
	Nebuta[a]	300	2560 × 1600	60	×	×	
	SteamLocomotive[a]	300	2560 × 1600	60	×	×	
B	Kimono	240	1920 × 1080	24	×	×	×
	ParkScene	240	1920 × 1080	24	×	×	×
	BasketballDrive	500	1920 × 1080	50	×	×	×
	Cactus	500	1920 × 1080	50	×	×	×
	BQTerrace	600	1920 × 1080	60	×	×	×
C	RaceHorses	300	832 × 480	30	×	×	×
	PartyScene	500	832 × 480	50	×	×	×
	BasketballDrill	500	832 × 480	50	×	×	×
	BQMall	600	832 × 480	60	×	×	×
D	RaceHorses	300	416 × 240	30	×	×	×
	BQSquare	600	416 × 240	60	×	×	×
	BlowingBubbles	500	416 × 240	50	×	×	×
	BasketballPass	500	416 × 240	50	×	×	×
E	FourPeople	600	1280 × 720	60	×		×
	Johnny	600	1280 × 720	60	×		×
	KristenAndSara	600	1280 × 720	60	×		×
F	BaskeballDrillText	500	832 × 480	50	×	×	×
	ChinaSpeed	500	1024 × 768	30	×	×	×
	SlideEditing	300	1280 × 720	30	×	×	×
	SlideShow	500	1280 × 720	20	×	×	×

N_F number of frames, r_F frames per second
AI All intra, RA Random access, LD Low delay
[a] 10 bit sequences

3.6.1.2 Subjective Quality Assessment

The rating of the perceived visual quality is a highly subjective task, with expectedly different results depending on the personal background and preferences of the viewer. Therefore, in order to get reliable results, precise methodologies for formal visual tests have been developed and standardized which are used in the standardization process for video quality assessment, see e.g. [15–17].

Formal visual testing requires a high number of test subjects who represent the "naive viewer" and who shall therefore not be familiar with the tested video coding technology. In the context of evaluation of video coding standardization, such subjective testing campaigns are run at a high logistic effort. For evaluation of the responses to the call for proposals for HEVC, more than 130 formal test sessions with more than 850 test subjects were performed [18].

Fig. 3.4 Example test sequences for each test sequence class printed in with appropriate relative size

At standardization meetings, informal subjective experts viewing sessions are used if subjective assessment is indicated and the effort of a formal subjective test is not considered practical. In these viewing sessions, experts from different organizations are presented anonymized sequences for evaluation. This assessment method is described e.g. in [19]. While experts in this informal assessment may recognize single proposals which may induce a bias, the method still seems to provide reliable results such that it can be used as a basis for further decisions.

3.6.1.3 Bjøntegaard Delta Measurements

When comparing two rate-distortion curves, e.g. the performance measurement for two variations of the coding scheme under consideration, a number representing the overall bitrate savings or the overall quality difference is very useful. Gisle Bjøntegaard proposed a delta calculation method which has become the state-of-the-art evaluation metric in the context of video coding standardization [20, 21]. Two types of Bjøntegaard Delta (BD) measurements are available: The BD-rate provides a number for the overall rate savings in percent, and the BD-PSNR provides the overall PSNR difference.

The original proposal assumes the measurement of the rate-distortion performance of two coder configurations at four different quantizer settings. The basic concept is to interpolate the curve between the rate-distortion points which have been measured in the experiment and to integrate over the difference. The calculation is performed over the logarithmic rate in order to derive the relative rate difference. Since the PSNR already is a logarithmic measure, it can directly be applied.

Example plots for illustration of the BD measurements are shown in Fig. 3.5. In the figure, the curves are shown with a linear rate axis as commonly used for presentation of rate-distortion curves. The curves in the figure are further drawn by linear interpolation. For the BD calculations, the rate axis is logarithmized and the rate-distortion curves are interpolated by an advanced interpolation method as described below. The overlapping portions of the rate-distortion curves are used for calculation of the BD values. For calculation of the BD-PSNR, the difference in vertical direction is evaluated. For calculation of the BD-rate, the difference in

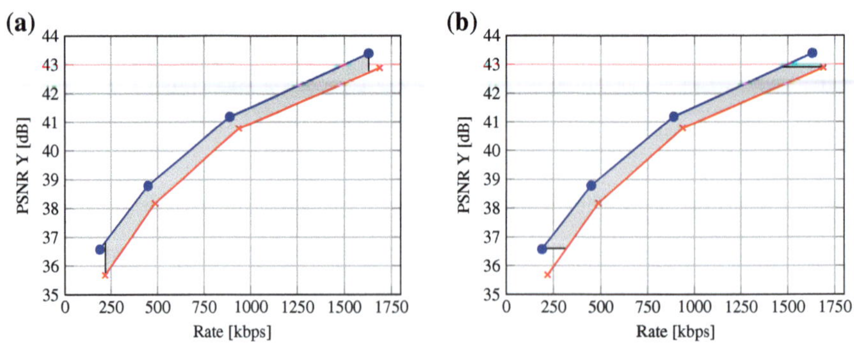

Fig. 3.5 Schematic presentation of example rate-distortion plots with indication of the calculation range used for **a** BD-PSNR and **b** BD-rate

horizontal direction is used. Depending on the range of overlap of the two curves, the resulting BD measurements get more or less meaningful. Therefore, careful interpretation of the resulting numbers is required.

Originally, the BD measurements were calculated by interpolating the rate-distortion curves using a 3rd-order polynomial which was fitted through the four measurement points [20]. This method was applied with relatively small quantizer step sizes (e.g. Δ_{QP} of 3 between the four rate points). Over the logarithmic rate axis, the rate-distortion plots evolved almost linearly within the given rate range and could be well approximated by the polynomial. With increasing quantizer step sizes of 4 or 5, the original interpolation method using the 3rd-order polynomial becomes less reliable for such large rate ranges as the polynomial interpolation could strongly deviate from the intended close-to-linear curve over the logarithmic rate axis [22]. In order to limit this effect, it was proposed to modify the calculation by reporting overall results for the lower and upper parts of the rate-distortion curve separately [21]. The JCT-VC selected a pice-wise cubic interpolation method for calculation the BD measures which shows a smooth behaviour [23].

In the JCT-VC, prearranged reporting sheets are provided with the definition of the common testing conditions which include the calculation of the BD measures [14]. It is required to use these reporting sheets when responding to core experiments. It is strongly recommended for other input contributions to also follow the common testing conditions and report the results based on the provided sheets. This formalized reporting method helps for fast access and judgement on proposals.

3.6.2 Proposal Assessment

In this section, three relevant categories of proposal assessment are summarized. These comprise the evaluation of complete coding schemes and tools, as well as the core experiment, which marks a stage in the process for potential adoption of a proposal into the working draft of a specification.

3.6.2.1 Evaluation of Full Proposals (Call for Proposals)

If full coding schemes, which are implemented in different software packages, shall be investigated, a strict separation of encoder control and tool impact is impossible. This is typically the case for the evaluation of full proposals in the context of a Call for Proposals for a new standard. In such a call, a precise formulation of the testing conditions and optimization constraints is required. Based on these requirements and the fulfillment of the testing conditions, the responses to the call are evaluated and the way to proceed is determined. Usually, responses to a call are required to provide decodable bitstreams according to the specified test points, a decoder executable to produce the reconstructed video sequences from the bitstreams, and a comprehensive description of the proposed coding scheme. Upon selection for further evaluation, the proponents are required to reveal a software implementation of their proposal which is capable of reproducing the results delivered as response to the call.

As mentioned in the beginning of Sect. 3.6, the comparison of different responses to a call for proposals constitutes the competitive phase of the standardization process. The major task in the evaluation of the results is the determination of the software package which shall become the test model software for the collaborative part of the process. While the performance in the subjective or objective assessment of the proposals plays the premier role in narrowing the set of selectable packages, different aspects have to be assessed when the test model software is determined. These include parameters such as readability, extensibility, or encoding and decoding speed.

Once a software package has been selected for the test model, the further development is organized by the appointed software coordinator, who is responsible for a clean and consistent implementation of adopted proposals into this software.

3.6.2.2 Tool Evaluation

The performance evaluation of full proposals as well as the evaluation of single tools within the testing environment of the reference software aims at using the best possible assessment metrics to determine on the tools that are to be included in the final specification. In the context of such decisions, it is of utmost importance to separate the performance impact of the tool implementation from the impact of the tool itself as well as to identify the specific performance contribution of a tool without side effects of other tools. The continuous development and maintenance of common testing conditions serves exactly this purpose: the definition of clear and stable test setups which allow for distinct assessment of the tool under investigation.[6]

For a comparison which shall reveal the performance impact of a tool, it is important to keep all other parts of the reference software but the tool under consideration unmodified. Specifically, the encoder control strategy must be identical for the comparison points, which is one of the reasons to require the integration of all tools in a common reference software and to perform all analysis on that software basis. Thereby, the impact of the tool shall be isolated from other effects. Sometimes, it is not possible to achieve this goal of complete "tool isolation" e.g. if a tool requires a dedicated encoder side decision algorithm. In such cases, the impact of the tool on both, encoder and decoder, has to be taken into account.[7]

[6] In the context of the definition of the first working draft and first official test model for HEVC, multiple *tool experiments* were conducted, which served the purpose of identifying the performance impact and performance contribution of isolated tools which were implemented in the so-called *test model under consideration* (TMuC). These experiments were used to decide on the inclusion or removal of the tools in the then adopted working draft and test model.

[7] For example, Sample Adaptive Offset requires a dedicated encoder side implementation for demonstration of its benefits, see Sect. 9.2.

3.6.2.3 Core Experiments

As discussed in Sect. 1.4.4, core experiments are a formal tool in the process of JCT-VC, which is used to gather all required information on a tool before the decision to adopt or reject inclusion in the working draft. A precise formulation of the tool under investigation and a strict formulation of the testing conditions aim at the production of a most stable and clear result. It is important that during a core experiment, no modifications to the proposed design are included. The performance of the scheme as presented at the meeting is assessed. If improvements or fixes are found during the core experiment process, these changes may be input to the definition of a continued core experiment from the next meeting on.

Through the assignment of proponents and cross-checking participants, a core experiment provides as much verification of the claimed performance as possible. By cross-checking activities from (potentially competing) independent proponents, the performance and tool impact are studied to the most thorough extent.

References

1. ISO/IEC: ISO/IEC directives, Part 2: Rules for the structure and drafting of international standards. Technicla report. Sixth edition. ISO/IEC, 2011. http://www.iso.org/iso/home/standards_development/resources-for-technical-work/iso_iec_directives_and_iso_supplement.htm (2014). Accessed 14 Apr 2014
2. Author's guide for drafting ITU-T Recommendations. ITU-T Rec. http://www.itu.int/oth/T0A0F000004 (2011). Accessed 14 Apr 2014
3. Guide for ITU-T and ISO/IEC JTC 1 cooperation. ITU-T and ISO/IEC JTC 1 Rec. A.23 Annex A. http://www.itu.int/rec/TREC-A.23/en (2010). Accessed 14 Apr 2014
4. Sullivan, G.J.: The art of writing standards: some "shalls" and "shoulds" for better Quality Interop Specs. Doc. JCTVC-E390. 5th meeting: Joint Collaborative Team on Video Coding (JCT-VC) of ITU-T SG 16 WP 3 and ISO/IEC JTC 1/SC 29/WG 11, Geneva (2011)
5. Sze, V., Budagavi, M.: High throughput CABAC entropy coding in HEVC. IEEE Trans. Circ. Syst. Video Technol. **22**(12), 1778–1791 (2012). doi:10.1109/TCSVT.2221526
6. Sullivan, G.J., et al.: Overview of the high efficiency video coding (HEVC) standard. IEEE Trans. Circ. Syst. Video Technol. **22**(12), 1649–1668 (2012). doi:10.1109/TCSVT.2221191
7. IEEE Standard Specifications for the Implementations of 8 $\times$ 8 Inverse Discrete Cosine Transform. IEEE 1180 (1991). doi:10.1109/IEEESTD.101047
8. Information technology—MPEG video technologies—Part 1: Accuracy requirements for implementation of integer-output 8 $\times$ 8 inverse discrete cosine transform. ISO/IEC 23002-1:2006 (MPEG-C). http://www.iso.org/iso/home/store/catalogue_tc/catalogue_detail.htm?csnumber=42030 (2006). Accessed 14 Apr 2014
9. Bossen, F., et al.: HEVC complexity and implementation analysis. IEEE Trans. Circ. Syst. Video Technol. **22**(12), 1685–1696 (2012). doi:10.1109/TCSVT.2012.2221255
10. Chujoh, T.: Tool experiment 2 on IBDI and memory compression. Doc. JCTVC-A302. 1st meeting: Joint Collaborative Team on Video Coding (JCT-VC) of ITU-T SG 16 WP 3 and ISO/IEC JTC 1/SC 29/WG 11, Dresden, Germany (2010)
11. Budagavi, M., Kreofsky, L.: BoG report on quantization. Doc. JCTVC-G1044. 7th meeting: Joint Collaborative Team on Video Coding (JCT-VC) of ITU-T SG 16 WP 3 and ISO/IEC JTC 1/SC 29/WG 11, Geneva (2011)

12. Kerofsky, L., Segall A.: Non CE4: Limiting dynamic range when using a quantization weighting matrix. Doc. JCTVC-H0541. 8th meeting: Joint Collaborative Team on Video Coding (JCT-VC) of ITU-T SG 16 WP 3 and ISO/IEC JTC 1/SC 29/WG 11, San Jose, USA (2012)
13. High efficiency video coding. ITU-T Rec. H.265 (HEVC). http://www.itu.int/rec/T-REC-H.265/en (2013). Accessed 14 Apr 2014
14. Bossen F.: Common test conditions and software reference configurations. Doc. JCTVC-K1100. 11th meeting: Joint Collaborative Team on Video Coding (JCT-VC) of ITU-T SG 16 WP 3 and ISO/IEC JTC 1/SC 29/WG 11, Shanghai (2012)
15. Methodology for the subjective assessment of the quality of television pictures. ITU-R Rec, BT.500-13. http://www.itu.int/rec/R-REC-BT.500/en (2012). Accessed 14 Apr 2014
16. Subjective assessment methods for image quality in high-definition television. ITU-R Rec, BT.710-4. http://www.itu.int/rec/R-REC-BT.710/en (1998). Accessed 14 Apr 2014
17. Subjective video quality assessment methods for multimedia applications. ITU-T Rec. P.910. http://www.itu.int/rec/TREC-P.910-200804-I/en (2008). Accessed 14 Apr 2014
18. JCT-VC. Report of subjective testing of responses to Joint Call for Proposals (CfP) on video coding technology for high efficiency video coding (HEVC). Doc. JCTVC-A204. 1st meeting: Joint Collaborative Team on Video Coding (JCT-VC) of ITU-T SG 16 WP 3 and ISO/IEC JTC 1/SC 29/WG 11, Dresden, Germany (2010)
19. MPEG Video Group: Conditions for visual comparison of VCB, IVC and WVC codecs. Doc. N13943. CH: ISO/IEC JTC1/SC29/WG11 MPEG, Geneva. http://mpeg.chiariglione.org/standards/mpeg-4-exploration/web-video-codinginternet-video-coding-video-coding-browsers/conditions (2013). Accessed 14 Apr 2014
20. Bjontegaard, G.: Calculation of average PSNR differences between RD curves. Doc. VCEG-M33. ITU-T SG16/Q6 VCEG, Austin, USA (2001)
21. Bjøntegaard, G.: Improvements of the BD-PSNR model. Doc. VCEG-AI11. 35th meeting: ITU-T SG16/Q6 VCEG, Berlin (2008)
22. Zhao, J., et al.: On the calculation of PSNR and bit-rate differences for the SVT test data. Doc. SG16-C404. ITU-T, Geneva, CH (2008)
23. Wang, J., et al.: On BD-rate calculation. Doc. JCTVC-F270. 6th meeting: Joint Collaborative Team on Video Coding (JCT-VC) of ITU-T SG 16 WP 3 and ISO/IEC JTC 1/SC 29/WG 11, Torino, IT (2011)

Chapter 4
Coding Structures

In this chapter and the following Chap. 5, the general organization of HEVC bitstreams and decoded HEVC video sequences is explained. Picture types and the partitioning of the coded pictures into different types of units and blocks specify the spatial and temporal structures which are used to represent the coded video sequence. While this chapter focuses on the relation of pictures and the spatial structures that are used to represent the picture for coding, Chap. 5 deals with the high-level syntax and organization of the coded data in the bitstream. Due to the strong interrelation between the spatio-temporal structure and the coded representation, some aspects of the bitstream representation are mentioned in this chapter nonetheless. Terms that are used here include the NAL (Network Abstraction Layer) unit, the video parameter set (VPS), the sequence parameter set (SPS), and the picture parameter set (PPS). These are further detailed in Chap. 5.

To represent a video sequence in a coded form in the bitstream, flexible temporal coding structures may be used as detailed in Sect. 4.1. These coding structures describe the *coding order* of the pictures to be encoded relative to the *output order* in which the pictures finally are arranged for output e.g. to a screen.[1] The chosen coding structure impacts the *structural delay* of the coded video sequence, i.e. the maximum number of pictures that precede a picture in output order but follow the picture in coding order. This number is also referred to as the *maximum latency increase* and is indicated in the VPS and the SPS as a characterization of the coded video sequence.

The organization of a single picture into units and blocks is detailed in Sect. 4.2. A picture is partitioned into one or more slices where each slice can consist of one or more slice segments. A slice segment is organized into coding units, prediction units, and transform units, which all together contain the coded representation for blocks of samples within the slice. Three different slice types are specified in HEVC, allowing for intra-only coding (I slice), intra coding and predictive coding with one reference picture list (P slice), and intra coding and predictive coding with two reference picture lists (B slice). The picture may be partitioned into tiles which change the coding order

[1] The output order is sometimes also referred to as the display order.

M. Wien, *High Efficiency Video Coding*, Signals and Communication Technology,
DOI 10.1007/978-3-662-44276-0_4

of the blocks and units the picture consists of. The interaction of slices and tiles is discussed in Sect. 4.2.3.

4.1 Temporal Coding Structures

A video of a given number of pictures may be partitioned into one or multiple coded video sequences (CVS, see Sect. 3.5.1.2). Each CVS can be decoded independently from other coded video sequences that may be contained in the same bitstream. In this section, the focus is on a single coded video sequence and the organization of the pictures in a coded video sequence in terms of coding order and in terms of output order. The *output order* of the pictures is the order in which the pictures have been generated (e.g. recorded or rendered), and in which the pictures should be output for display to resemble the picture order of the original input sequence for the intended visual impression. In the specification, the output order is represented by the *picture order count* (POC), which uniquely identifies a picture within the coded video sequence. Note that POC distance between two successive pictures does not need to be constant throughout the CVS. It is only required to be strictly increasing along the output order, see Sect. 5.4.1 for a more comprehensive discussion.

The *coding order* specifies the order in which pictures are reconstructed at the decoder, and thereby defines which pictures may be used for reference. Obviously, only previously decoded picture can be referred to for prediction. The dependencies between the pictures determine the coding structure of the coded video sequence. The *coding structure* comprises a consecutive set of pictures in the sequence with a specific coding order and defined dependencies between the included pictures. The full video sequence may be represented by periodic repetition of this coding structure.[2] The set of pictures that is comprised in the coding structure is often called a *Group Of Pictures* (GOP).[3] In the context of the HEVC specification, this set is called a *Structure of Pictures* (SOP). In the text, we will use the previously established term GOP.

Sometimes, the first picture of a GOP in coding order is referred to as the *anchor picture* or *key picture* of the GOP. Note that these terms as well as the term Group of Pictures are not used in the specification. They are however commonly used in literature for description of such coding structures.

In Fig. 4.1, two example coding structures with different coding order and output order are provided. Both coding structures have a GOP size of eight pictures,

---

[2] Encoders do not need in any way to stick to a single coding structure when encoding a video sequence, but may adapt the coding structure e.g. according to local features of the sequence, or application requirements as needed. However for some features, e.g. temporal scalability, it may be required that the adapted coding structure fulfills certain specific constraints.

[3] Note that in MPEG-1 and MPEG-2, a group of pictures denotes a set of pictures which starts with an intra picture. The decoding of the bitstream can be started with a GOP [3]. Hence, the definition of a GOP in MPEG-1/2 has some similarity with the definition of a coded video sequence in HEVC.

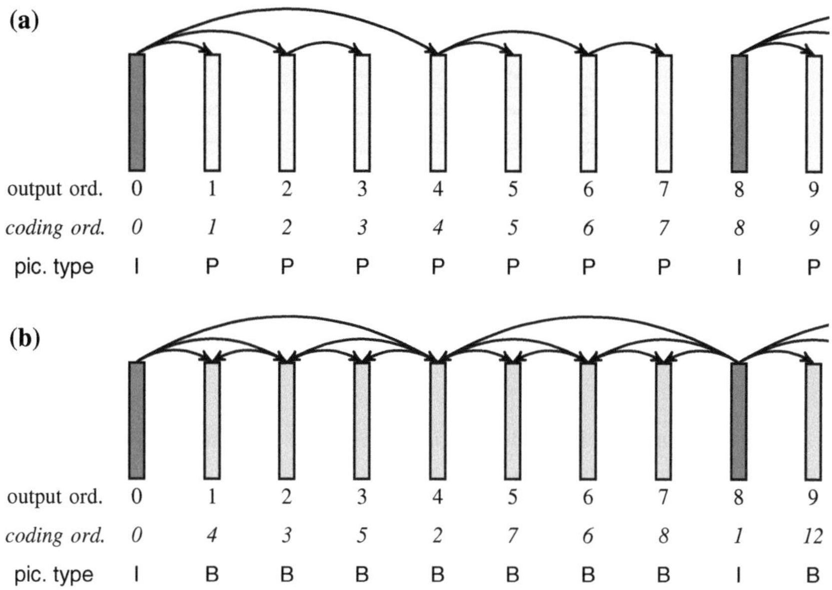

Fig. 4.1 Examples for coding structures with different output order and coding order. **a** Hierarchical-P coding structure (coding order = output order). **b** Hierarchical-B coding structure (coding order ≠ output order)

with the key pictures of the GOP marked in dark grey.[4] Figure 4.1a shows a coding structure where all pictures of the GOP are predicted from past encoded pictures (all arrows from prediction reference to prediction target are directed towards future pictures). Since there is no dependency from any coded picture to a future picture in output order, the coding order of the pictures in fact corresponds to the output order, i.e. the order in which the pictures should be displayed. In the given example, each picture uses a single reference picture for prediction. Prediction from a single reference is called *uni-prediction*. Historically, such pictures are called *P pictures*.[5] The coding structure further exhibits a certain hierarchy of the pictures. For example, the pictures with odd output numbers may be dropped without impacting the decodability of the other pictures. The coding structure of the other pictures is independent of these pictures. Similarly, every fourth picture can be decoded without impact from the intermediate pictures with even output numbers.Such a structure is called a *hierarchical* coding structure. In the given example using only P pictures,

[4] The key pictures in Fig. 4.1 are random access points. In the following figures, random access points are depicted in dark gray. Pictures which are output before the associated random access point are marked gray. Pictures which follow the random access point (trailing pictures) are marked light gray.

[5] P picture = predicted picture. Uni-prediction of a picture may use a single reference block selected from multiple available reference pictures for prediction of each block. The reference pictures further do not need to be restricted to be in the past of the current picture.

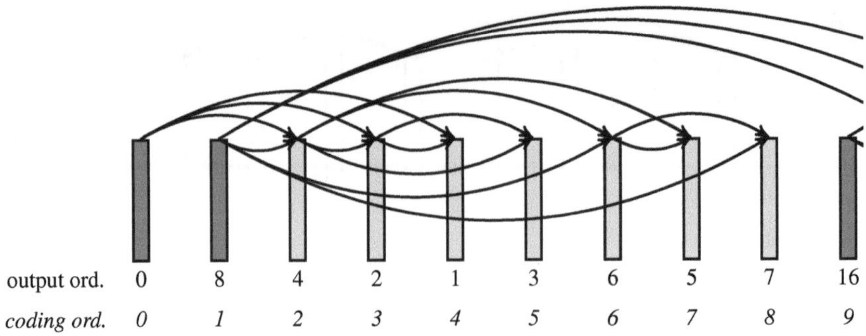

Fig. 4.2 The hierarchical-B coding structure of Fig. 4.1b in coding order

the coding structure is referred to as a *hierarchical-P* structure. It should be noted that uni-prediction of course may refer to a reference picture list including more than one reference picture for prediction. Still, for each block, only one reference picture can be used.

Figure 4.1b provides an example for a coding structure with prediction arrows from past pictures as well as from future pictures. For this coding structure, the coding order is different from the output order, i.e. some pictures need to be stored until they are to be output. This induces the structural delay described earlier. Coding structures with references from the past and the future are typically used with *bi-prediction*, where the prediction for a picture is formed by the combination of two references. Such pictures are referred to as 'B pictures'.[6] Like the P picture example, the presented structure exhibits a prediction hierarchy where the prediction dependencies are nested such that certain pictures may be dropped without impacting the decodability of the other pictures. Such a structure is called a 'hierarchical-B' coding structure.[7] Figure 4.2 shows the coding structure of Fig. 4.1b in coding order.

There may be prediction relations between consecutive GOPs of a coding structure. An example for such prediction dependencies is the structure shown in Fig. 4.1b. Such a structure is called a *Open-GOP* coding structure: Pictures of one GOP reference pictures from a different GOP. Structures that omit inter-dependencies between consecutive GOPs are called *Closed-GOP* coding structures. An example for such a coding structure is given in Fig. 4.3. The structure shown here comprises I pictures (intra-only coded), P pictures, and B pictures and has been similarly used for encoding in H.262 | MPEG-2 or MPEG-4 Visual.

The HEVC specification includes an SEI message (supplemental enhancement information, see Sect. 5.5) on the 'structure of pictures' which can be used by the encoder to describe the used GOP structure, including the corresponding picture

[6] B picture = bi-predicted picture. For bi-prediction, the two references may be temporarily arbitrarily located in the future or past relative to the current picture.

[7] The coding structures in Fig. 4.1 are examples for *dyadic* hierarchical structures as the dependencies are dyadic. Hierarchical structures do not need to be dyadic but can also comprise other dependencies, see e.g. Fig. 4.4b.

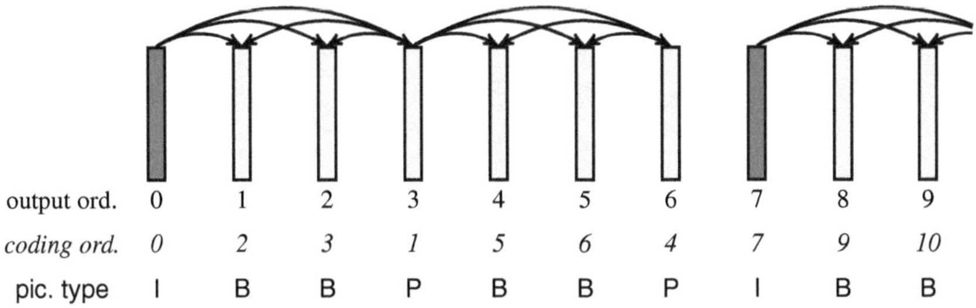

Fig. 4.3 Closed-GOP structure. Here, an I picture is followed by two P pictures with two associated B pictures each (IBBPBBP structure)

types, the applicable reference picture sets, and the temporal relation between the pictures in the GOP.

4.1.1 Temporal Layers

In HEVC, a temporal layer concept very similar to the temporal layer concept for temporal scalability in H.264 | AVCAnnex G [4, 5] is available in the base specification. Each picture has an associated temporal level identifier t_{id}. Random access pictures are specified to always have $t_{id} = 0$. With the concept of temporal layers, extracting a coded video sequence of lower temporal resolution from a given video sequence can be simply achieved by discarding all NAL units with t_{id} larger than a selected value.

Furthermore, the feature of temporal nesting can be indicated in the VPS and SPS. If temporal nesting is indicated, pictures with lower t_{id} must not refer to a picture with higher or equal t_{id} for prediction within a GOP. Further, with temporal nesting activated it is always possible to switch the decoding from a lower temporal level to a higher temporal level (i.e. switching to a higher temporal resolution, starting from any decoded picture at the lower temporal resolution).

Examples for coding structures with temporal nesting are shown in Fig. 4.4. In this presentation, the pictures are vertically arranged according to their temporal layer to ease the identification of the prediction relations. Since there is no dependency of pictures with lower t_{id} to pictures with higher t_{id}, decoding can be switched between the temporal layers at any picture. e.g., the coding structure in Fig. 4.4a enables switching between 60, 30, 15, and 7.5 Hz, while the structure in Fig. 4.4b enables 60, 20, and 10 Hz output rate.

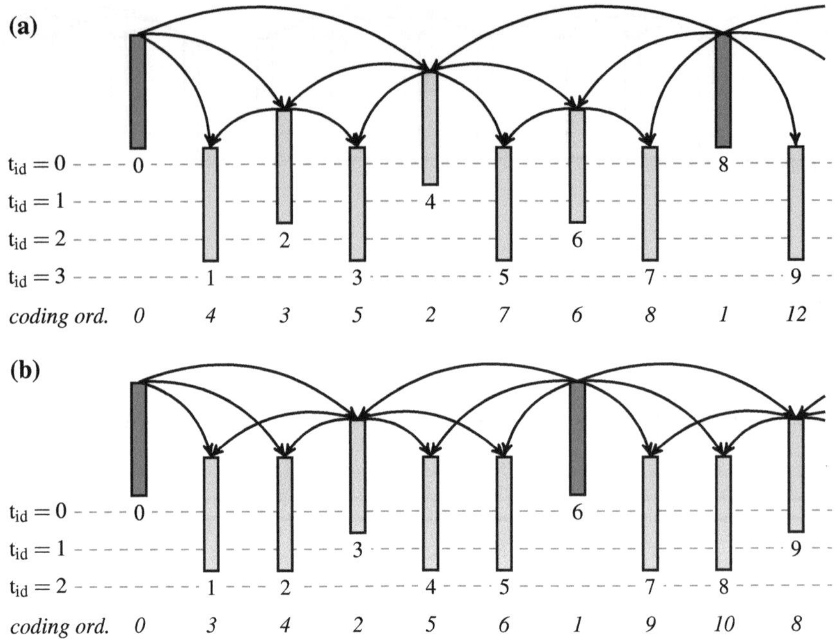

Fig. 4.4 **a** Hierarchical-B coding structure with dyadic temporal nesting and four temporal layers. **b** Hierarchical-B coding structure with non-dyadic temporal nesting and three temporal layers

4.1.2 Picture Types

The picture type is an indication of the role of the given picture within the coded video sequence. The picture type is indicated by the NAL unit type of the coded video data.

HEVC specifies four classes of picture types:

- Random Access Point Pictures
- Leading Pictures
- Temporal Sub-Level Access Pictures
- Trailing Pictures

For each of the classes, a set of variants can be indicated. This variety provides explicit indications of the organization and coding structure of the bitstream at the highest possible level (i.e. the NAL unit header). It includes information on the current picture, but may further provide information on other pictures that are associated with the current picture. This type of information can be useful for external processing and parsing of the bitstream (e.g. media aware network elements, MANEs).

The picture type imposes constraints on the available prediction methods by allowing or disallowing certain slice types. The picture types, their constraints, and their usage in different scenarios are detailed in the following.

4.1.2.1 Random Access Points

The term *Random Access Point* (RAP) indicates the main feature of this picture type: The decoding may be started at such pictures (specifically, at the access unit containing the slices of the picture). RAP pictures can only contain I slices, i.e. only intra prediction is applied for coding the picture. Therefore, random access point pictures are denoted by the acronym IRAP (for *intra random access point*).

It is required that the corresponding VPS, SPS, and PPS are available to the decoder at the time when the decoding of the RAP picture starts. The picture types that provide the random access point functionality are detailed below.

Instantaneous Decoder Refresh Picture: IDR

An IDR picture resets the decoding process and always starts a new coded video sequence. If an IDR picture is received at the decoder, the decoded picture buffer is reset before the decoding of the IDR picture starts. Therefore, an IDR can only be referenced from pictures that are succeeding the IDR picture in coding order. Pictures preceding the IDR in coding order are deleted and are therefore not available for prediction. The concept of Instantaneous Decoder Refresh (IDR) has been used in H.264 | AVC before [4, 5].

Since the IDR starts a new coded video sequence, the POC value is reset to zero. It should be noted that in applications with intra-only coding, it is advisable not to mark all pictures as IDR since the ability to use the POC for detection of the boundaries of access units in the NAL unit stream would get lost.[8]

Clean Random Access Picture: CRA

In contrast to the IDR, the Clean Random Access picture (CRA) picture does not reset the reference picture buffer and does not necessarily start a new coded video sequence. Thereby, a CRA may occur within a coded video sequence when decoding it.The CRA type indicates that the decoding process *may* be started with the current picture. When parsing the bitstream, detected CRA pictures indicate potential entry points for decoding the video sequence. The decoding result after such a new start is asserted to be conforming to the specification for pictures that follow the CRA picture in output order. This property is achieved by restricting pictures that follow the CRA in both, coding and output order, to not use pictures before the CRA for reference.

The NAL unit stream may contain pictures that follow the CRA in coding order but are to be output *before* the CRA. These pictures may contain further references to pictures before the current CRA in coding order, and hence may not be decodable (at least not without serious degradation of the reconstruction quality due to missing

[8] The access unit includes all NAL units assigned to one picture, see Sect. 5.2.3.

or wrong references). The corresponding picture types and the methods how to deal with these issues are discussed in the following subsection.

If a CRA picture does not occur at the start but within a coded video sequence, it is regularly decoded as part of the coding structure encoded in the NAL unit stream. CRA pictures are especially useful for application in 'open-GOP' coding structures. Examples for such structures are given in Fig. 4.5.

Broken Link Access Picture: BLA

Broken Link Access (BLA) pictures are useful for bitstream splicing operations, see Sect. 4.1.3.[9] The BLA indication is not originally intended to be coded into the bitstream but can be applied to a CRA in order to indicate the concatenation of two sequences at this point. For this purpose, only the NAL unit type has to be modified to indicate the switch from CRA to BLA. This change does not affect the picture itself but might affect the subsequent pictures in coding order.

As stated before, decoding of a video sequence may be started at a CRA picture. If however a sequence is spliced to another sequence it might happen that pictures following the CRA in decoding order reference to pictures of the original sequence which are not further available after splicing to the new sequence. Now, pictures from the spliced sequence might reside in the decoded picture buffer at the same reference picture index as pictures in the previous stream before the splicing operation. The BLA indicates that pictures with such references shall not be displayed. These pictures might otherwise refer to different pictures than originally available at the encoder, such that the reconstructed pictures might not be appropriate for display.

4.1.2.2 Leading Pictures

A *leading picture* precedes the associated random access point picture in output order but follows it in coding order. Depending on the reference pictures used for inter prediction, the leading picture may or may not depend on pictures that precede the associated IRAP in decoding order. The existence of such a dependency characterizes the two picture types specified for leading pictures.

Leading pictures may contain I, P, or B slices or combinations thereof. The dependency property is only relevant if at least one non-I slice is present in the picture.

Random Access Decodable Leading Pictures: RADL

A *decodable leading picture* precedes the associated IRAP picture in output order and can be correctly decoded if decoding of the video sequence starts at the associated IRAP picture. Decodable leading pictures are denoted as RADL (*Random Access*

[9] The concept of Broken Link Access has been present in MPEG-2 [3, 6].

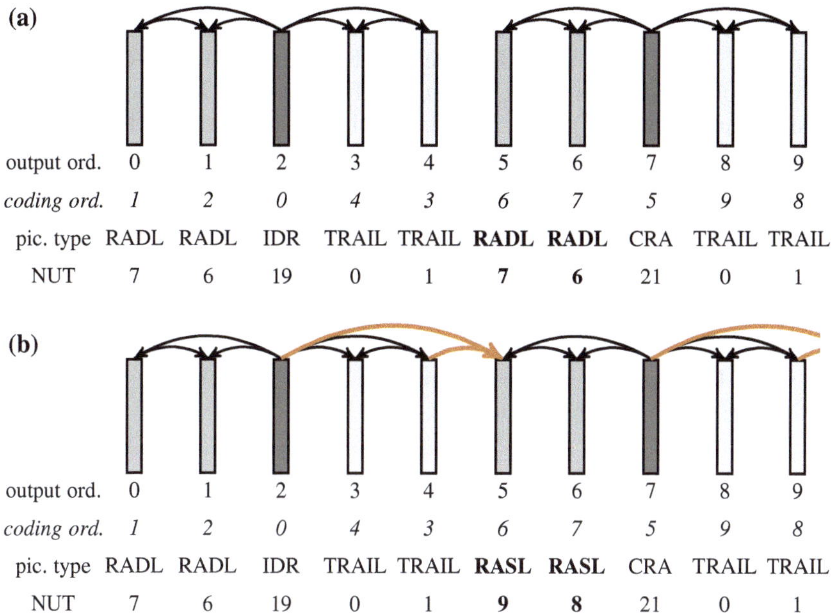

Fig. 4.5 Example for the usage of leading pictures. **a** Coding structure with RADL pictures. **b** The same coding structure with additional prediction from picture 2 to picture 5, turning RADL pictures into RASL pictures

Decodable Leading) pictures. A RADL picture contains no references to pictures prior to the IRAP picture in decoding order. An example for a coding structure with RADLs is given in Fig. 4.5a.[10]

Random Access Skipped Leading Pictures: RASL

If a leading picture is coded as a Random Access Skipped Leading (RASL) picture, it is marked to potentially contain references to pictures prior to the associated IRAP picture in decoding order. This means that it may contain dependencies to pictures which may not be available if decoding of the video sequence is started at the associated IRAP. Therefore, a RASL picture shall not be output for display if the associated IRAP picture is marked as a broken link access picture (BLA), as the IRAP picture being marked as a BLA indicates that the pictures belong to a coded video sequence that was spliced into the current NAL unit stream. RASL pictures must also not be output for display if the associated IRAP picture is a CRA picture

[10] This example is given as an illustration only and should not be regarded as a recommended GOP structure. In the following figures, NUT indicates the NAL unit type, see Sect. 5.2.2.

and is the first IRAP picture in the coded video sequence. In this case, reference pictures which the RASL may point to would be missing.[11]

In Fig. 4.5b, an example for the difference between RADL and RASL pictures is given. Compared to the RADL structure in Fig. 4.5a, this sequence contains an additional reference from picture 2 to picture 5 in output order. This dependency on a previous IRAP picture (or on any picture preceding the associated IRAP) induces the picture type of pictures 5 and 6 to change from RADL to RASL. If the decoding would start at picture 7 in output order, pictures 5 and 6 would need to be discarded, and output would start at picture 7. In the case of Fig. 4.5a, pictures 5 and 6 do not have this feature. They would hence be decoded and displayed.

According to the specification, any picture that is a leading picture must be marked as either being a RADL or a RASL picture. Furthermore, any RASL picture associated with a CRA or BLA picture shall precede any RADL picture associated with the CRA or BLA picture in output order.

4.1.2.3 Temporal Sub-Layer Access

The temporal coding structure of a coded video sequence may be organized in a way that allows for decoding of subsets of pictures in the sequence representing different temporal resolutions. Each temporal resolution is associated to a temporal layer as described in Sect. 4.1.1. With temporal nesting, only hierarchical prediction dependencies exist from lower t_{id} to higher or equal t_{id}. The availability of this property is indicated in the SPS.

The signaling of the use of temporal nesting indicates that temporal switching is permitted at all pictures. Since temporally nested configurations are not the only possibility to define a temporally scalable coding structure, the indication of alternative switching points for $t_{id} > 0$ may be useful. Two special picture types are defined for this purpose.

- The indication of a picture to be a *Temporal Sub-layer Access* picture (TSA) marks the possibility for the decoder to switch to decoding of a higher temporal layer of the coded video sequence at this picture. The decoder can decide which temporal layer to switch to as the TSA property indicates that any higher temporal layer may be accessed, starting from the TSA picture on. To achieve this property, no pictures preceding the TSA in coding order, which have t_{id} equal to or higher than the TSA, shall be referenced by the TSA picture or any picture following the TSA picture in coding order.
- The indication of a picture to be a *Stepwise Temporal Sub-layer Access* picture (STSA) indicates that from the current STSA picture on, the decoder may switch decoding to the temporal layer of this STSA picture. In contrast to TSA pictures, switching to a higher temporal level than the one of the STSA picture is not indicated, e.g. due to prediction dependencies.

[11] RASL pictures have also been referred to as *tagged for discard* pictures in draft versions of the specification.

Examples for a coding structure with TSA and STSA pictures are given in Fig. 4.6.[12] In Fig. 4.6a, a coding structure including TSA pictures is shown. At the TSA pictures, the decoder can switch to decode the sequence at any higher temporal layer. In Fig. 4.6b the same coding structure has been modified to include additional prediction dependencies on the highest temporal layer ($t_{id} = 3$). Due to these dependencies, only stepwise temporal sub-layer access is available for some pictures at $t_{id} = 2$.

As stated before, leading pictures are required to be marked either as RADL or as RASL. Therefore, the integration of temporal sub-layer access indication into the picture type is only available for trailing pictures. However, it should be noted that in many applications the frequency of IRAP pictures may not be high, or that

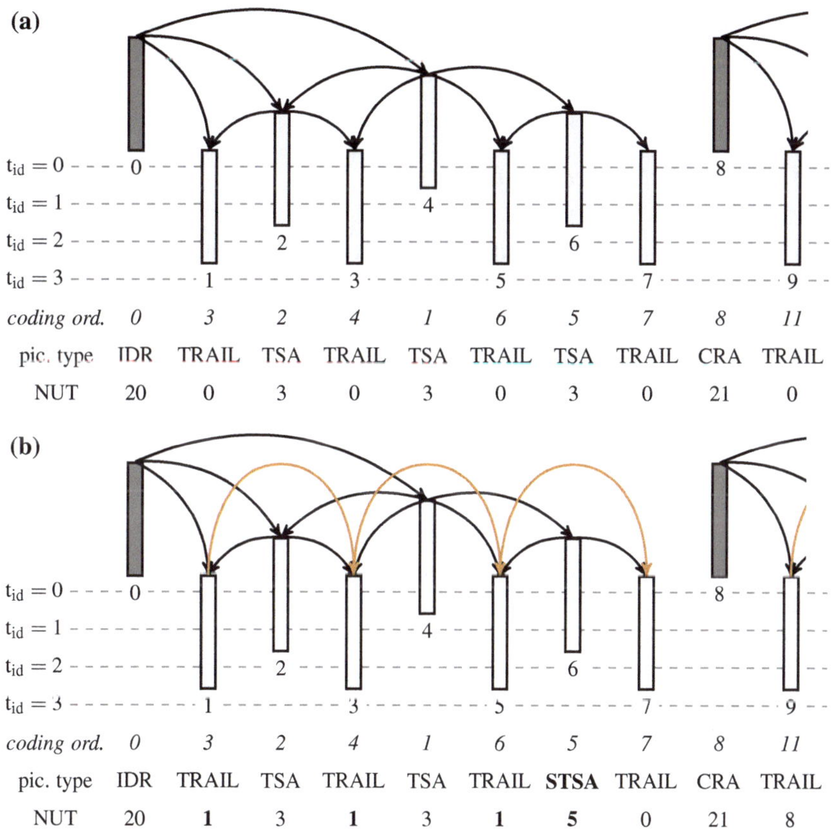

Fig. 4.6 Example for a coding structure with temporal sub-layer access indication. **a** Coding structure with TSA pictures. **b** A similar structure with additional prediction dependencies and stepwise temporal sub-layer access at picture 6 (marked bold)

[12] This example is given as an illustration only and should not be regarded as a recommended GOP structure.

leading pictures are not used due to delay constraints (e.g. visual communication). The missing provision of temporal sub-layer access indication for leading pictures can therefore be considered a balanced compromise.

4.1.2.4 Trailing Pictures

Trailing pictures follow the associated IRAP picture in output order as well as in coding order. This picture types collects all pictures that do not fall in any of the previously mentioned categories.

According to the specification, any picture that is a trailing picture shall not be marked as a RADL or a RASL picture. All leading pictures associated to an IRAP picture shall precede the trailing pictures associated to the same IRAP in coding order. As all picture types except the random access points, trailing pictures may contain I, P, or B slices or combinations thereof.

4.1.3 Splicing of Video Sequences

In various applications, coded video sequences from different sources need to be concatenated or spliced together. With the available picture types, HEVC provides flexible and low-complex means to support this task. With a few operations, multiple coded video sequences can be spliced into a joint NAL unit stream. It should be noted that for a successful splicing operation, it has to be asserted that all parameter sets referred to by of the appended sequences are available for decoding of the spliced sequence.

As discussed above, the decoding process for a NAL unit stream can be started at any random access point picture. Each random access point may therefore also serve as a splicing point. Special care has to be taken if the chosen IRAP is a CRA picture as a CRA picture may have associated leading pictures. If the CRA has leading pictures of type RASL, the picture type for the IRAP at the splicing point has to be changed to a Broken Link Access (BLA) picture accordingly.

An example for a splicing operation is given in Fig. 4.7. In the example, a part of the coded video sequence of Fig. 4.5b is appended to a different short sequence. It is assumed that all required parameter set are available. In the figure the pictures are arranged in coding order rather than in output order to better illustrate the splicing operation. It can be seen that the RASL pictures of the spliced sequence must be discarded as they contain references to pictures that are not available in the new NAL unit stream.

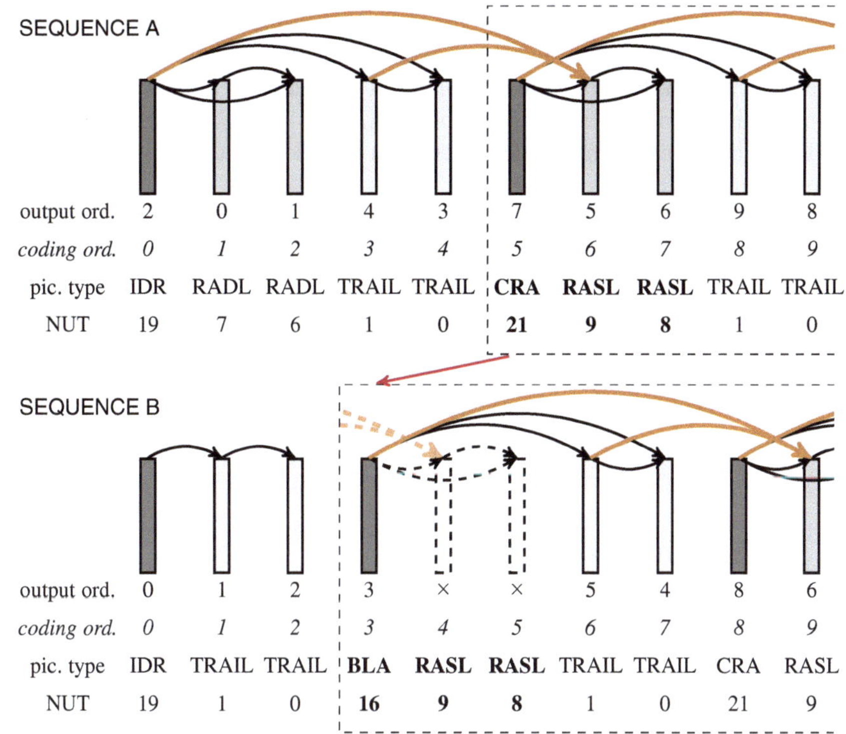

Fig. 4.7 Example for splicing video sequences of two coded video sequences (pictures shown in coding order)

4.1.4 Comparison to H.264 | AVC

The presentation in the previous subsections reveals a rich set of picture types, which carry high-level information on the picture itself or on subsequently coded pictures in HEVC. The characterization of pictures in H.264 | AVC is much less elaborate. Here, three categories of picture classification are provided: Instantaneous decoder refresh, non-IDR pictures which are used for reference, and non-IDR pictures which are not further used for reference [4, 5]. In H.264 | AVC, some of the functionality provided by the new HEVC picture types has been enabled by insertion of supplemental enhancement information (SEI messages). The broken link access (BLA) or the clean random access (CRA) functionalities are provided by the recovery point SEI message. In contrast to the new HEVC picture types, the handling of pictures which should not be displayed is not normative but only recommended in H.264 | AVC. This is due to the fact that normative suppression of picture output has not been specified in H.264 | AVC.

In H.264 | AVC, a clean concept of temporal sub-layers has only been introduced in the scalable extension specified in Annex G where a temporal sub-layer (called a *temporal level* in H.264 | AVC) is assigned to each picture in the coded video sequence. The concept of stepwise temporal sub-layer access has been enabled in this extension as well through the definition of the temporal level switching point SEI message. Like in the case of BLA and CRA, these features have been promoted to normative features of the pictures in contrast to indication through supplemental information.

While the new HEVC picture types are very helpful with providing information on sequence characteristics local to a picture and its vicinity, the encoder has to spend more logic on the selection and determination of the right picture type for a picture at the time of encoding the NAL units of the picture into the bitstream. The effort is limited if fixed coding structures are employed where the picture types are pre-determined and stable but may be larger if the coding structure is dynamically adapted and modified. In H.264 | AVC, the only effort in this regard is to determine if a picture is used as reference or not.

4.2 Spatial Coding Structures

In this section, the available partitioning methods for the pictures of a video sequence are presented, and the terminology to address specific areas of a picture is introduced.

For coding, a picture is partitioned into coding tree blocks (CTB) of square shape. A consecutive set of coding tree blocks is grouped into a slice. The CTBs have a configurable size per coded video sequence and replace the 'macroblock', which was used in the previous standards and recommendations of ISO/IEC and ITU-T. The basic concept of slices is also inherited from the previous standards. In HEVC however, a slice can be partitioned into multiple segments which are coded in separate NAL units. A CTB is the root for a tree of coding blocks, which are in turn further partitioned for prediction and for transform coding of the residual. The organization of the blocks and slices within a picture can further be influenced by the new concept of tiles, which modify the otherwise fixed raster-scan order of the CTBs.

In the following, after introducing the relation of blocks and units, the spatial coding structures are detailed starting from the picture level and going down to the prediction and transform blocks.

4.2.1 *Blocks and Units*

The HEVC specification text distinguishes between *blocks* and *units*. While the former address a specific area in a sample array (e.g. luma, Y), the latter comprise the collocated blocks of all encoded color components (Y, C_b, C_r, or monochrome) as well as all syntax elements and prediction data that is attached to the blocks (e.g. motion vectors).

The base entities are the Coding Tree Block (CTB) and the corresponding Coding Tree Unit (CTU). The CTU contains the CTBs of the encoded color components and forms a complete entity in the bitstream syntax. A CTB is the root of a quadtree partitioning into Coding Blocks (CB). A Coding Block is partitioned into one or more Prediction Blocks (PB) and forms the root of a quadtree partitioning into Transform Blocks (TB). A corresponding set of units is specified which comprise the block and the respective syntax structure, each. Accordingly, a Coding Unit (CU) contains the Prediction Units (PU) and the tree-structured set of Transform Units (TU). While a PU contains the joint prediction information for all color components, a TU contains a separate residual coding syntax structure for each color component. The location and size of the CBs, PBs, and TBs of the luma component applies to the corresponding Coding Unit (CU), Prediction Unit (PU), and Transform Unit (TU). Accordingly, the locations and sizes for the chroma blocks are derived from the corresponding luma blocks. Some deviations that result from the sub-sampling of the chroma components in 4:2:0 YCbCr video are detailed in the following.

The presentation here focuses on the blocks and their relations. If not otherwise mentioned the luma component is considered.

4.2.2 Slices and Slice Segments

A picture may be constructed from one ore more *slices*. A slice is constructed of one or more *slice segments*. Slice segments are non-overlapping and for each slice segment, the entropy coding process is reset. Thereby, each CTB of a picture is included in exactly one slice segment, and the whole picture area is covered by the constituting slices. As an important feature, each slice with its slice segments can be decoded independently from other slices in the same picture. This means that no prediction (i.e. intra prediction or motion vector prediction) is performed from one slice to the other. It further means that each slice of a picture can be decoded regardless of the fact that other slices in the picture may have been lost. Among the slice segments which belong to the same slice, prediction is allowed.

Each slice segment contains information on the location of the first CTB in the slice such that the area in the picture represented by the slice segment is specified. Thereby, slices and slice segments constitute very useful tools for recovery and resynchronization in case of data loss. Furthermore, all slices in a picture use the same reference picture set, i.e. the construction of the reference picture set only needs to be processed for the first decoded slice in a picture. Slices within a picture may be further forced to use identical reference picture *lists* for inter picture prediction.

4.2.2.1 Slice Segment Header

The coded CTUs of a slice are preceded by a slice segment header indicating the slice type and an identifier of the picture parameter set to be activated. Besides the

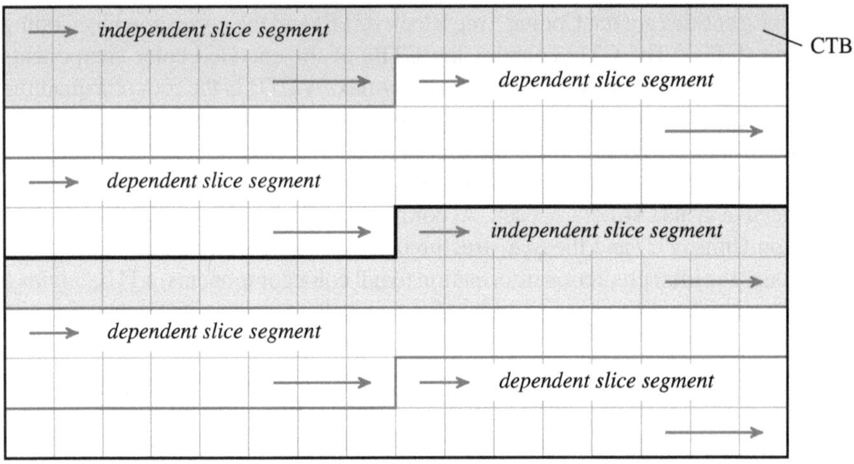

Fig. 4.8 Illustration of a picture with 16 × 9 CTBs partitioned into slice segments with two independent segments and two associated dependent slice segments each

information on the start address, i.e. the start of the area represented by the slice in the picture, it further provides information on the construction of the applicable reference picture set, the configuration of inter prediction tools, loop filtering, and quantization parameter offsets. If tiles or wavefront parallel processing are enabled, the slice segment header further provides information on entry points into the coded bitstream. These entry points indicate byte positions where a decoder may start (parallel) decoding.

4.2.2.2 Independent and Dependent Slice Segments

The option to split a slice into multiple slice segments is indicated in the picture parameter set. The slice segments of a slice share the information of the slice segment header sent with the first segment.[13] The first slice segment which includes the slice segment header is called an *independent slice segment*. It can be decoded independently from any other slice or slice segment in the same picture. Slice segments that follow the independent slice segment are denoted as *dependent slice segments*. They contain a widely reduced slice segment header which indicates the active PPS and the slice segment address, i.e. the location of the first CTB in the slice segment. A dependent slice cannot be decoded without the header information from the associated independent slice segment. Furthermore, the slice segment concept can be used in very-low delay applications to start transmission of segments of the current slice at the encoder side while the dependent slice segments are still being encoded.

[13] In the specification, slices are completely replaced by slice segments. The notion of a slice is used here nonetheless for sake of clarity.

The relation between independent and dependent slice segments is illustrated in Fig. 4.8. A slice segment contains an integer number of consecutive CTBs in raster scan order and is the entity to be packed into a NAL unit for transmission (see Sect. 5.2). In transmission scenarios with strict packet size limits, the coded size of a slice segment (i.e. the number of bytes that represent the slice segment) may determine the number of CTBs in that segment.

4.2.2.3 Slice Types

Three slice types are defined that indicate what type of prediction methods are used for coding of the contained CTBs, where the slice type of independent slice segment determines the slice type of the dependent slice segments. The available slice types are listed below:

I Slice Only intra prediction is applied for coding of the CTBs.

P Slice Intra prediction, or motion compensated prediction from a single reference picture from a reference picture list with a single motion vector per prediction partition can be applied.

B Slice Intra prediction, or motion compensated prediction from one or two reference pictures from two reference picture lists with a single or with two motion vectors per prediction partition can be applied.

As mentioned before, random access point pictures can only contain I slices as inter picture prediction is precluded by the random access functionality. All other picture types may contain I, P, or B slices and combinations thereof.

4.2.2.4 Wavefront Parallel Processing

The HEVC specification includes several tools to enable parallel processing at the decoder side. *Wavefront parallel processing* (WPP) allows the decoder to synchronously perform entropy decoding for multiple CTU rows in a slice. The concept is illustrated in Fig. 4.9. Each row is considered to have its own entropy decoding process. The entropy decoding of the second row can start with two CTUs delay compared to the first row, the third row can start with two CTUs delay to the second row, and so on. Thereby, a diagonal 'wavefront'-like simultaneous decoding of the picture is possible.

To enable the WPP functionality, the slice segment header contains a list of byte-aligned entry points to the bitstream that mark the beginning of the first CTU in each row. The scheme is illustrated in Fig. 4.10. The byte alignment is ensured by the encoder by a byte alignment syntax structure, which is included in the bitstream before each CTU that marks an entry point, see also Chap. 10. The decoding process for each row starts from the respective entry point in the bitstream, subject to the

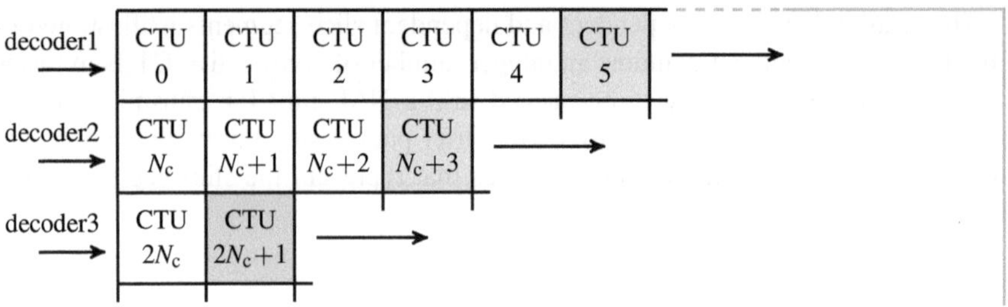

Fig. 4.9 Wavefront parallel processing of N_c CTUs per row. The decoders can start decoding each row with a two-CTU-delay relative to the previous row

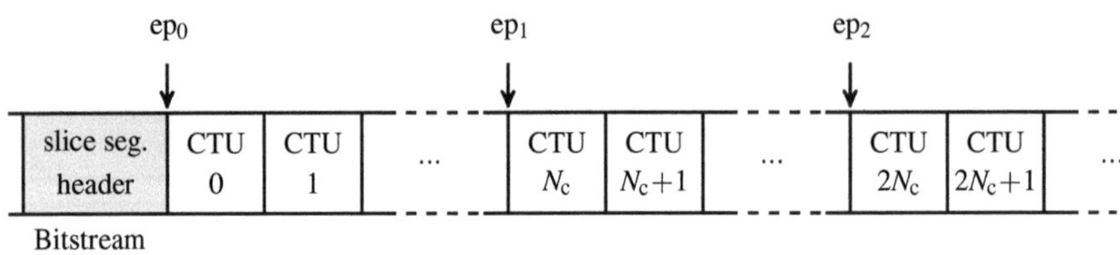

Fig. 4.10 Entry points ep_i for starting decoding at byte positions indicated in the slice header

aforementioned synchronization which confirms availability of all required context information from the neighborhood.

For the first CTU in each row, the arithmetic coding engine is initialized with the context state of the second CTU in the previous row, i.e. CTU $nN_c + 1$, $n = 0, 1, \ldots, N_r - 1$, where N_r is the number of CTU rows in the picture. Thereby, context update and adaptation across the CTU rows are achieved.

4.2.3 *Tiles*

A slice is the entity in which CTUs are encapsulated for transmission purposes. Traditionally, slices follow the raster scan order of the macroblocks or CTUs in the picture. In HEVC, the *tiles* concept is used to organize the coding arrangement of CTUs in a picture in a more flexible way. Tiles allow for the definition of additional horizontal and vertical scan boundaries within a picture. Each area that is framed by such boundaries is called a tile. The boundaries modify the scan order of the CTUs in the picture. They further break prediction dependencies as intra prediction as well as motion vector prediction are disabled across the tile boundaries. Tiles are processed in tile raster scan order. Accordingly, the slices in a picture with tiles are organized along the tile scan. If tiles are used and there is more than one slice in the picture, a slice must not contain incomplete portions of two or more tiles. I.e., slices must include complete tiles or tiles must include complete slices. It is possible to have a combination of both settings within the same picture.

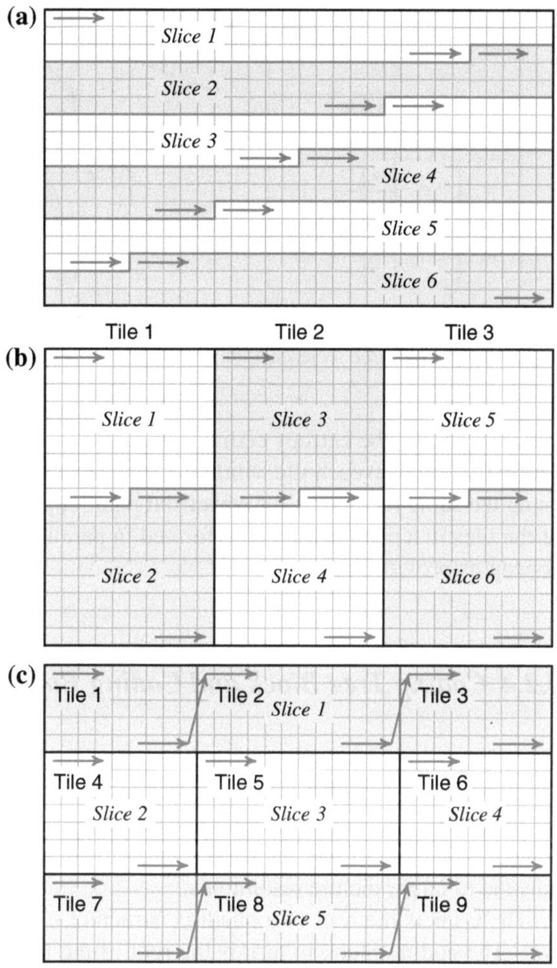

Fig. 4.11 Examples for partitioning a picture with 30 × 17 coding tree blocks into slices and tiles. Tiles must include complete slices or slices must include complete tiles. The slices can be composed of multiple slice segments. The slice segments must not cross tile boundaries. **a** Slices only. **b** Slices in tiles. **c** Tiles in slices

An example for slices and tiles in a picture is given in Fig. 4.11. Besides a configuration only with slices in Fig. 4.11a, configurations with slices in tiles and tiles within slices are shown in Fig. 4.11b and c. In the example, the configuration with slices in tiles in Fig. 4.11b has the same number of CTUs per slice as the slice-only configuration in Fig. 4.11a.

The applicable tile configuration is encoded in the picture parameter set. The tile boundaries can only be located at CTU boundaries. The tile boundary location can be either signaled explicitly, or the application of a uniform spacing is indicated. In case of uniform spacing, the size of the tiles is derived from the size of the picture in CTB units.

Note that within one slice, the start of a new tile induces byte alignment of the CTU coding and re-initialization of the entropy coding stage CABAC for context

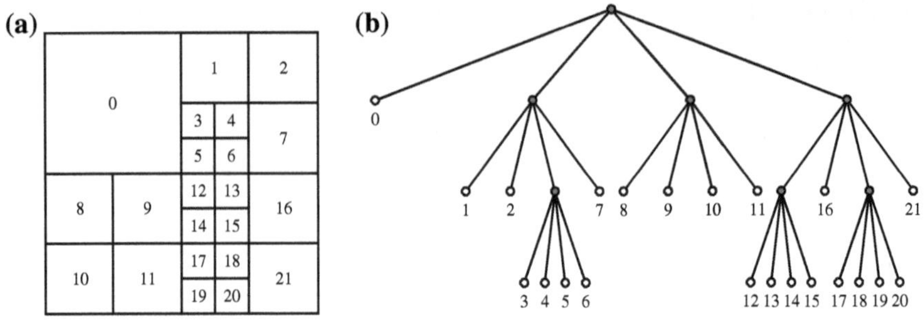

Fig. 4.12 Quadtree partitioning of a coding tree block into coding blocks. **a** Spatial partitioning. **b** Corresponding quadtree representation

variables.[14] I.e., the entropy coding states are reset for each tile, even within a slice. The entry points to each tile within the slice are encoded in the slice segment header using the same method as for wavefront parallel processing. Thereby, the decoding of the tiles in a picture can be performed in parallel. An overview article on the tile concept is provided in [7].

4.2.4 Coding Tree Block and Coding Block

The picture is partitioned into squared *Coding Tree Blocks* (CTBs) of $2^n \times 2^n$ samples, where $n \in \{4, 5, 6\}$ is specified in the Main Profile. The smallest CTB size of 16×16 corresponds to the macroblock size as used in previous video coding standards. By modifying the CTB size up to 64×64, the coding structure can be adapted to larger picture resolutions.[15]

The CTB is the root of a quadtree structure of *Coding Blocks* (CB) of squared size. The coding block in turn is the root of the prediction block structure and the residual quadtree. A CTB can be represented by a single CB, or it can be split into four partitions of half horizontal and half vertical size. Each of these partitions may be a CB itself or may be further partitioned to half horizontal and half vertical size, and so forth. In the Main Profile of HEVC, a coding block may have a minimum size of 8×8 samples. The resulting quadtree partitioning is scanned using a z-scan to form the sequence of CBs for encoding. In Fig. 4.12, an example CTB partitioning into CBs with a quadtree depth of three is shown. The figure also indicates the CB order along the z-scan.

The potentially large size of the basic coding block may lead to the issue that the picture size of the encoded sequence can not be partitioned into an integer number

[14] See Sect. 10.2 for a description of CABAC.

[15] A sequence with pictures of $1,920 \times 1,088$ samples comprises 510 CTBs of 64×64 samples. The number of CTBs here is comparable to a CIF sequence with pictures of 352×288 samples with 398 macroblocks of size 16×16.

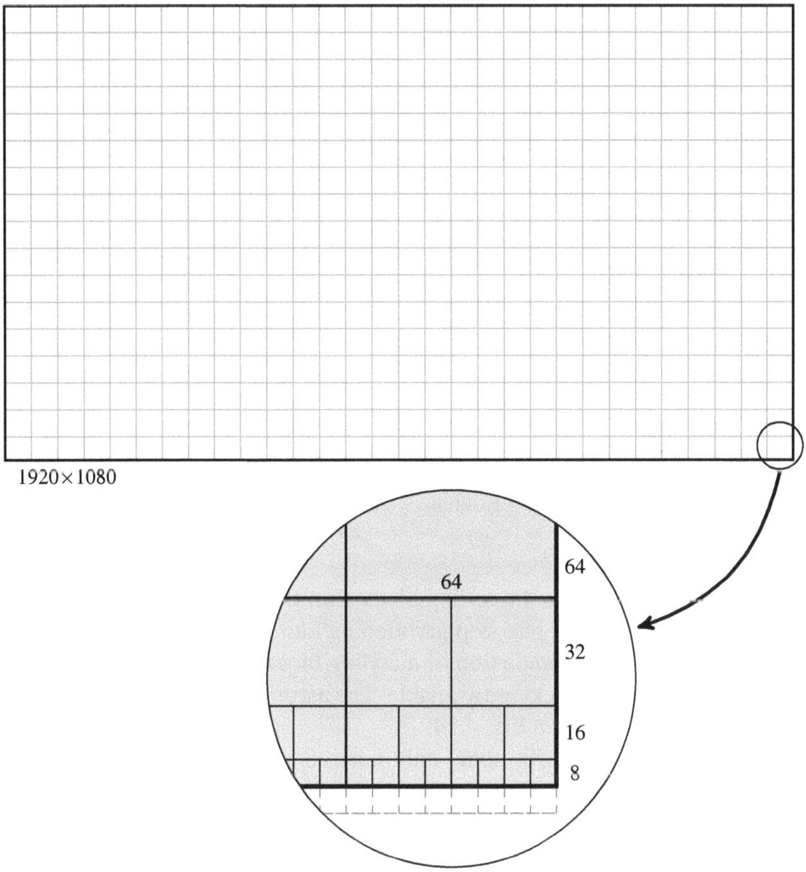

Fig. 4.13 Example picture format 1080p with CTB size 64 × 64 using implicit coding block partitioning for picture formats with non-integer number of CTB lines

of CTBs. The partitioning may cause the CTBs in the last CTB column or last CTB row to be incomplete. To solve this issue, these CTBs are implicitly partitioned such that the picture area is covered by complete CBs. In Fig. 4.13, this partitioning is shown for a picture of $1,920 \times 1,080$ samples and a CTB size of 64. The picture is partitioned into 30×17 CTBs. Since 1080 is not an integer multiple of 64, the last CTB row is truncated by implicit partitioning.

4.2.5 Prediction Block

For motion compensated inter prediction and for intra prediction, the coding block is partitioned into *Prediction Blocks* (PBs). For intra prediction, the prediction block

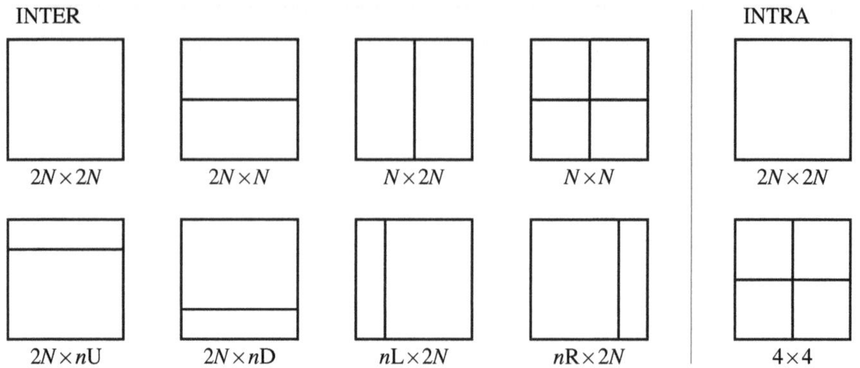

Fig. 4.14 Partitioning of a coding block of size $2N \times 2N$ into prediction blocks, with $n = N/2$. For asymmetric partitioning of inter coding blocks, the indices U, D, L, and R denote Up, Down, Left, and Right, respectively. For intra coding, partitioning of the coding block is only allowed for 8×8 blocks resulting in four 4×4 partitions

size is generally identical to the coding block size. Only for the minimum possible CB size of 8×8, a further one-step partitioning into four 4×4 prediction blocks may be applied. For inter prediction of a coding block, eight different partitioning schemes into prediction blocks are available. The partitioning for both, inter and intra prediction blocks is shown in Fig. 4.14.

Besides the structures with none, only horizontal, only vertical, or symmetric horizontal and vertical partitioning as used in H.264 | AVC, another four asymmetric partitioning schemes are specified. These partitioning schemes include a rectangular partition of quarter CB height or width. This partition can be horizontally oriented at the top or bottom of the block (Up, Down), or vertically oriented at the left or right side of the block (Left, Right), as shown in Fig. 4.14.

In order to limit the complexity of the coding scheme, the usage of 4×4 inter prediction blocks is prohibited in HEVC. Furthermore, for 4×8 and 8×4 prediction blocks, only uni-directional prediction from List 0 or List 1 is enabled.

4.2.6 Transform Tree and Transform Block

Besides the CB partitioning into prediction blocks, the (square) CB is the root of another quadtree partitioning called the *transform tree*. The leafs of this tree are square *transform blocks* (TBs). The transform tree is also referred to as the *residual quadtree* (RQT). The size of the transform block defines the size of the applicable residual transform, see Chap. 8. The minimum TB size is 4×4, the maximum TB size is 64×64. A TB size of 64×64 induces an implicit splitting into four 32×32 transform blocks as the maximum available transform is of that size.

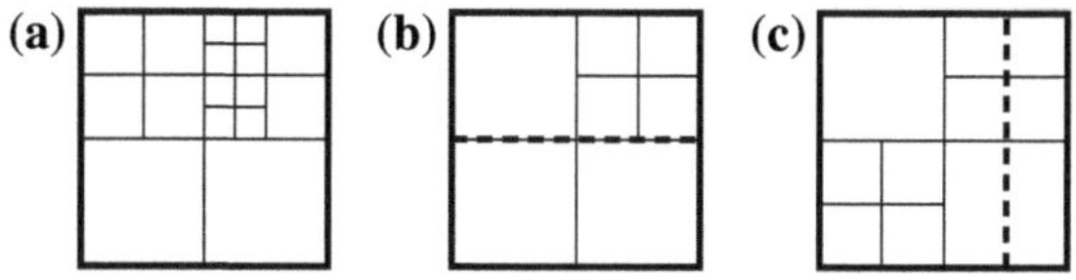

Fig. 4.15 Example CB partitionings into prediction blocks (*dashed lines*) and transform blocks (*solid lines*). **a** PB=CB. **b** Two horizontal PBs. **c** TB across PB boundary

Three examples for TB partitionings are shown in Fig. 4.15. It should be noted that the transform tree is constructed aside from the prediction block partitioning. For inter predicted blocks, the chosen transform block size may be larger than the sub-block size of the prediction block partitioning. Thereby, the encoder may choose to apply a residual transform across the boundaries of the prediction blocks. I.e., the residual of multiple prediction blocks including the prediction block boundary is jointly transformed, quantized and coded. Care has to be taken regarding potential visual artifacts arising from this option. If the coefficients in the overlapping transform block are all quantized to zero however, the selection of such overlapping transform blocks may help to save bits by reducing the signaling overhead of encoding an empty tree structure and the indication of no encoded coefficients. The transform block size can never be larger than the CB size as this constitutes the root of the transform tree.

The minimum and maximum applicable transform block sizes as well as the maximum applicable transform tree depth are configured in the sequence parameter set. If the transform tree depth is set to zero the transform tree degenerates to a single transform block. In this case, no transform tree signaling is included in the TU syntax. Within the range of the minimum and maximum applicable transform block size, the TB size is derived to be equal to the CB size. This configuration results in a scheme similar to adaptive block-size transforms which had been proposed during the development of H.264 | AVC [8], with the transform shape being set to square-sized only.

4.2.7 Comparison to H.264 | AVC

In H.264 | AVC, the picture is partitioned into macroblocks of size 16×16 samples [4, 5]. The size of the macroblocks cannot be modified. H.264 | AVC further does not include the concept of dependent and independent slice segments. It also does not include the concept of tiles for changing the scan order of the macroblocks in the picture. Instead, the Baseline and Extended profiles include the concept of slice groups, slice group maps, and arbitrary slice ordering. With arbitrary slice ordering, the slices of a picture do not need to be encoded in the order of appearance in the picture. With the definition of slice group map types, the assignment of macroblocks and their coding order within the slice can be flexibly configured. Besides the conventional raster scan order, macroblocks can e.g. be assigned to a checker board structure, or the scan can be transposed to vertical, or slice groups for the foreground and slice groups for the background can be defined, see e.g. [9, 10]. Note that these special modes are not included in other profiles such as the High profile, the Main profile, or the Constrained Baseline profile.

An inter coded macroblock can be partitioned according to the symmetric partitioning structure shown in the top row of Fig. 4.14 leading to prediction blocks of size 16×16 down to size 8×8. The 8×8 sub-macroblocks can be further partitioned according to the same scheme to achieve a prediction block size down to 4×4 samples. The partitioning of macroblocks into sub-macroblocks can be interpreted as a simplified pre-version of the concept of coding tree blocks which are partitioned by a quadtree into coding blocks. In HEVC, the symmetric partitioning of prediction blocks available in H.264 | AVC has been further extended by the asymmetric partitioning shown in Fig. 4.14. For intra macroblocks, either a fixed 4×4 or 8×8 block partitioning, or a 16×16 prediction block size can be selected. The partitioning information for inter and intra macroblocks is jointly encoded in the macroblock type syntax element. In HEVC, such information is separated in dedicated syntax elements and flags.

For support of coding of interlaced video sequences, the H.264 | AVC specification includes the concept of macroblock pairs. A macroblock pair consists of two macroblocks in adjacent macroblock rows in the same macroblock column. With macroblock adaptive frame/field coding, H.264 | AVC allows for the selection of coding two interleaved or two stacked macroblocks within each macroblock pairs. On the picture level, switching between an interlaced and a progressive scan can also be indicated (picture-adaptive frame/field coding). It should be noted that HEVC does not include any dedicated coding tools for interlaced video on the slice level.

Besides the I, P, and B slices which are used in both specifications, H.264 | AVC includes two additional slice types referred to SI and SP slices. These slice types allow for the reconstruction of an identical picture from otherwise differently encoded references and thereby enable seamless switching from one bitstream to another. This concept is only available in the Extended profile of H.264 | AVC and has not been transferred to HEVC.

It should be noted that HEVC includes multiple tools to enable parallel processing at the encoder and/or decoder-side. On the picture level, these include the concept of tiles as well as the concept of wavefront parallel processing. Provisions for parallel processing have not been incorporated in the design of H.264 | AVC. Besides slices which are present in both specifications, no further parallelization tools are available here.

4.3 Reference Pictures

Decoded pictures are stored in the *Decoded Picture Buffer* (DPB). The pictures in the DPB may be used as reference pictures or be stored until they shall be output.

The HEVC specification distinguishes between three types of pictures: pictures that are used for short-term reference, pictures that are used for long-term reference, and pictures that are unused for reference. Pictures that are to be used for prediction in the current or a future picture of the coded video sequence are collected in the *Reference Picture Set* (RPS). All pictures that are not included in the RPS are marked to be unused for reference. These pictures are removed from the DPB once their output time has been reached.

The HEVC specification distinguishes between reference picture sets and reference pictures lists. While the reference picture set comprises all pictures that are held for potential prediction in the current or future pictures, the reference picture list contain the pictures that are actually used for prediction in the current picture. The reference picture list is constructed as a subset of the reference picture set. The number of pictures to be actually used in the reference picture lists can be controlled on a slice basis [11].

4.3.1 Reference Picture Sets

A reference picture set is a set of previously decoded pictures which may be used for inter prediction of a coded picture. The reference pictures in a reference picture set are explicitly identified by their POC value. Before starting to decode a picture, the applicable reference picture set is constructed once for all slices in the picture. The pictures in the decoded picture buffer may be included in the reference picture set and may be used for reference in the current picture or in a following picture in decoding order. If a picture is not included in the reference picture set, it is marked to be "unused for reference" and can be deleted from the DPB after being output.

The HEVC specification allows for predefined reference picture sets which are to be sent in the SPS. The applicable set for the current slice is then specified by an index into the list of predefined reference picture sets in the slice segment header. Additionally, an extra reference picture set can be explicitly signaled in the slice segment header. Since the reference picture set is indicated in each slice segment header, the explicit signaling of the reference picture set may need to be repeated in each corresponding header.

Conceptually, two types of reference pictures are differentiated in the reference picture set, *Short-Term Reference Pictures* and *Long-Term Reference Pictures*. *Short-Term Reference Picture Sets* are designed to provide reference to pictures in the temporal proximity of the current picture, e.g. within the same structure of pictures.

Long-Term Reference Picture Sets can be used to address selected pictures for longer term reference purpose.[16] These long-term reference picture sets are constructed independently from the short-term picture sets. The syntax allows for up to 32 pictures which can be specified for long-term prediction in the SPS. In the

[16] For example content that occurs frequently in the coded video sequence (without interruption by IDR pictures).

Main profiles of HEVC, the maximum number of long-term reference pictures is constrained to 15. Like for the short-term picture sets, a long-term reference picture set can also be signaled in the slice header for specific use in the current slice. The possible number of long-term reference pictures signaled in the slice header is determined by the remaining size of the decoded picture buffer.

For construction of the applicable reference picture set, five subsets are derived. The assignment of a reference picture to one of the subsets is determined by the relation of their POC value in relation to the POC value poc_c of the current picture. The short term pictures are classified by POC value smaller than poc_c ('Current Short-Term Before'), larger than poc_c ('Current Short-Term After'), and not used in the current picture but held in the DPB for possible use as reference in a future picture ('Short-Term Following'). The long term pictures are classified into pictures used in the current picture ('Current Long-Term') and not used in the current but used in a following picture ('Long-Term Following'). These subsets are used in the process of reference picture list construction.

As stated before, pictures in the DPB that are not in one of these lists are marked as unused for reference and can be deleted from the buffer. Each picture in the DPB can be listed in exactly one or none of the five subsets.

Since the applicable reference picture set is indicated in each slice segment header and the pictures from the DPB which belong to the reference picture set are identified by their unique POC values, the construction of the current reference picture set does not depend on the construction of the set in previous pictures. Furthermore, if no picture with a requested value of POC is available in the DPB for construction of the current reference picture subsets, the decoder has a clear indication that a transmission loss has occurred. Both, the explicit RPS signaling and the absolute POC indication thereby improve the error robustness of the coding scheme.

4.3.1.1 Reference Picture Set Example

In Fig. 4.16, an example for the applicable reference picture sets used in coding structure with a GOP size of 8 is given.[17] The reference picture sets as coded in the SPS are shown according to their RPS index for each POC position in the group of pictures. Reference picture sets 8–10 only refer to one previous picture and to pictures within the GOP otherwise. These sets are used if an IDR is coded at POC position 0. This is the case e.g. at the beginning of the sequence, or if an IDR would be inserted at this position into the sequence later on. If a CRA or a trailing picture is coded at this position, the reference picture sets 0–2 are used. The reference picture sets 0–7 are applicable for GOPs following the first GOP in the coded video sequence. In the figure, POC positions −8, 0, 8 constitute possible IRAP positions. In a Random

[17] The shown configuration corresponds to the Main profile Random Access encoder configuration setting of the JCT-VC common testing conditions, see [12].

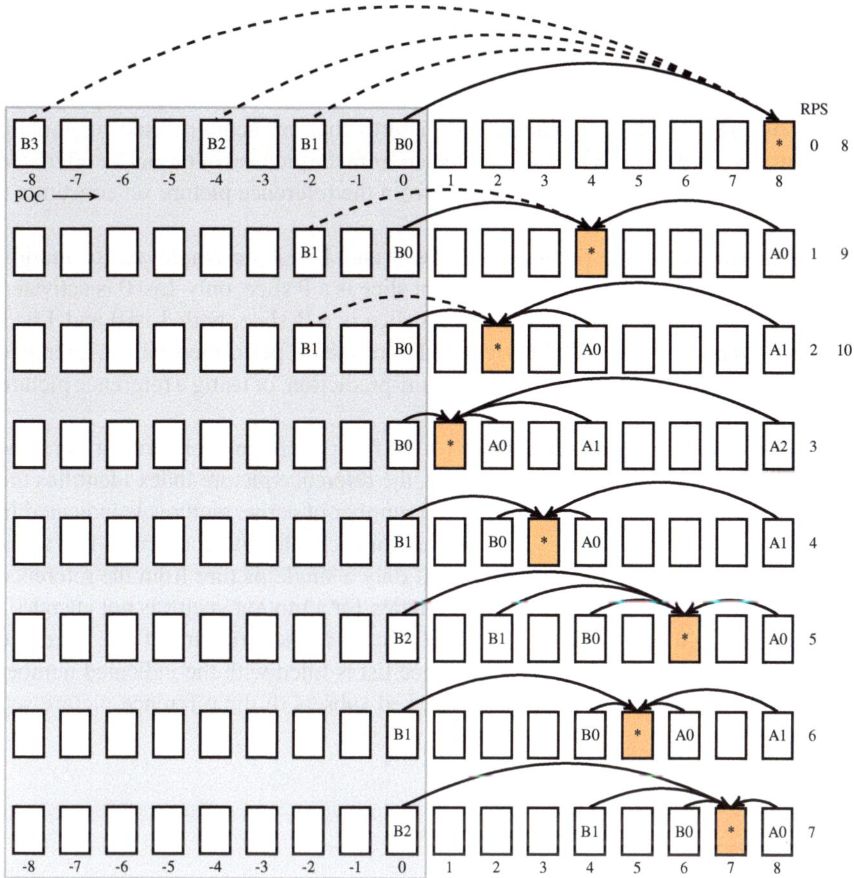

Fig. 4.16 Example for the short-term reference picture sets (RPS) of a GOP structure with 8 pictures. The pictures are shown in output order with short-term reference pictures before the current picture (*) denoted by Bi and short-term reference pictures after the current picture denoted by Aj. In each row, one reference picture set is shown. RPS 8, 9, 10 do not include the pictures with the dashed arrows

Access scenario, CRA pictures can be frequently inserted at these positions.[18] In GOPs without a CRA picture, the pictures are all trailing pictures, with some of them being further use for reference. In a GOP with a CRA, these pictures are leading pictures and are therefore coded as RASL pictures, accordingly.

[18] For example, an IRAP is inserted every six GOPs, i.e. every 48th picture, to achieve approx. 1 s random access in a 50 Hz sequence.

4.3.2 Reference Picture Lists

The reference picture sets contain the set of pictures that are available for prediction of the current and future pictures. The pictures that are actually used for motion compensated prediction in a slice are selected from the pictures of the active reference picture set. The set of selectable pictures from the reference picture set constitute a *reference picture list*.

One or two reference picture lists, List 0 and List 1 are employed for motion compensated inter prediction. If the current slice is a P slice, only List 0 is activated and used for uni-prediction. If the current slice is a B slice, both List 0 and List 1 are activated and motion compensated prediction can be performed using a reference picture from only List 0 or only List 1 for uni-prediction, or using a reference picture from each list for bi-prediction.

The number of pictures to be actually used from a reference picture list has to be indicated. Within the reference picture list, the reference picture index identifies the picture to be used for prediction. A default number of active pictures is indicated in the picture parameter set. It can be modified for each slice by an override-flag and a new number in the slice segment header. If only a single picture from the reference picture list is active, the reference picture index for a motion vector is not encoded.

The number of active pictures is used to control the structure of the reference picture list. For initialization, each reference list is filled with the indicated number of active pictures from the previously marked subsets of the reference picture set. For List 0, the order is

- Current Short-Term Before,
- Current Short-Term After,
- Current Long-Term.

For List 1, the order is

- Current Short-Term After,
- Current Short-Term Before,
- Current Long-Term.

Since the order of 'Current Short-Term Before' and 'Current Short-Term After' is exchanged between List 0 and List 1, the support of hierarchical coding structures is straight-forward, as can be seen in the example given below.

Modifying the order of the pictures in the reference picture lists can be enabled to allow the encoder to optimize the order according to the coding cost for the current picture. The modification of the reference picture list is signaled in the slice header. Note that changing the order of the reference pictures in the reference picture list is only possible if all slices in the picture use identical reference picture lists. This feature is indicated in the sequence parameter set.

Table 4.1 Constructed reference picture lists for the first 17 frames of a sequence coded with the reference picture sets from Fig. 4.16

POC	Pic.type (slice type)	RPS	#L0	#L1	List 0 POCs	List 1 POCs
0	IDR (I)	–	–	–	–	–
8	TRAIL (B)	8	1[a]	1[a]	0	0
4	TRAIL (B)	9	2	2	0, 8	8, 0
2	TRAIL (B)	10	2	2	0, 4	4, 8
1	TRAIL (B)	3	2	2	0, 2	2, 4
3	TRAIL (B)	4	2	2	2, 0	4, 8
6	TRAIL (B)	5	2	2	4, 2	8, 4
5	TRAIL (B)	6	2	2	4, 0	6, 8
7	TRAIL (B)	7	2	2	6, 4	8, 6
16	TRAIL (B)	0	4[a]	4[a]	8, 6, 4, 0	8, 6, 4, 0
12	TRAIL (B)	1	2	2	8, 6	16, 8
10	TRAIL (B)	2	2	2	8, 6	12, 16
9	TRAIL (B)	3	2	2	8, 10	10, 12
11	TRAIL (B)	4	2	2	10, 8	12, 16
14	TRAIL (B)	5	2	2	12, 10	16, 12
13	TRAIL (B)	6	2	2	12, 8	14, 16
15	TRAIL (B)	7	2	2	14, 12	16, 14
⋮	⋮	⋮	⋮	⋮	⋮	⋮

[a]Number of active reference pictures indicated in the slice header

4.3.2.1 Reference Picture List Construction Example

For the reference picture set example in Fig. 4.16, the reference picture list construction of the first pictures of a coded sequence is given in Table 4.1 below. In the table, the pictures are listed in coding order.

Some observations can be made in the given example: Picture 16 has identical entries in List 0 and List 1. I.e., the lists are used as if bi-prediction would be applied to a single list.[19] For poc = 9, 14, 15, not all available pictures from the reference picture set are activated (missing poc = 16 or 8). For some pictures, List 0 contains pictures with higher POC ('Current Short-Term After'), or List 1 contains pictures with lower POC ('Current Short-Term Before'). In these cases, the ordering of the reference picture subsets for List 0 and List 1 as explained above has been applied.[20]

---

[19] Pictures with this reference picture list structure are sometimes referred to as PB pictures. The identical lists List 0 and List 1 are sometimes referred to as a 'combined list'.

[20] The example reference picture list construction corresponds to the Main Profile Random Access encoder configuration setting of the JCT-VC common testing conditions [12] as in Fig. 4.16. Conceptually, the configuration corresponds to a hierarchical-B coding structure. It should be noted that due to activated references to Picture B1 in RPS 0 and RPS 1, this structure does not allow for general temporal sub-layer access. Consequently, all pictures are coded with $t_{id} = 0$.

4.3.3 Short-Term Reference Picture Set Signaling

In the sequence parameter set, up to $N_{\text{ST,max}} = 64$ different Short-Term Reference Picture Sets can be specified. The applicable reference picture set is addressed by an index i_{ST}, with $i_{\text{ST}} = 0, \ldots, N_{\text{ST}} - 1$, in the slice segment header. As mentioned above, the slice segment header may include an extra reference picture set. If the extra reference picture set is present, it is referred to by the index $i_{\text{ST}} = N_{\text{ST}}$. As stated above, the reference picture set is constructed only once per picture. Hence, each slice header in the picture must indicate the same reference picture set. The repeated explicit indication for each slice improves error robustness in cases of slice losses.

Reference picture sets are constructed with absolute POC values in the reference picture set construction process. These absolute POC values are derived by adding POC delta values Δ_{poc} encoded in the reference picture set to the POC value poc_{c} of the current picture. The reference picture set construction process is invoked once per picture before decoding any coding units of the picture.

Two methods of signaling short-term reference picture sets are specified in HEVC. In the *explicit* coding mode, the number of positive and negative POC deltas (i.e., the number of pictures marked as 'Current Short-Term Before' or 'Current Short-Term After') is encoded, and for each entry of the two subsets, a POC delta and a usage flag is provided. The usage flag indicates whether the picture specified by Δ_{poc} is active (i.e. used for prediction) or kept in the reference picture set for later use ('Short-Term Following').

In the *differential* coding mode, a previously coded reference picture set is indicated to be used for the current picture. For each of the pictures in this set, the activation in the current reference picture set is indicated by a flag. If a picture from the previous set is not activated in the current set, another flag is sent to indicate if the picture shall be kept in the set but marked as 'unused' or if the picture shall be completely removed from the current reference picture set. The classification into 'Current Short-Term Before', 'Current Short-Term After', or 'Short-Term Following' is derived from the relative position of the reference pictures to the current picture, and from their respective usage flag.

4.3.4 Long-Term Reference Picture Set Signaling

The occurrence of long-term reference pictures in the coded video sequence is indicated in the sequence parameter set as well. If long-term reference pictures are indicated to be present, a set of poc_{lsb} values and usage flags for each picture can be encoded in the sequence parameter set. The poc_{lsb} values of further long-term reference pictures can be encoded in the slice segment header. Note that the number of long-term reference pictures signaled in the sequence parameter set may be zero, such that the long-term reference pictures to be used are only signaled in the slice segment header. It is up to the encoder to decide, which of the signaling methods is appropriate for a given application. The poc_{lsb} values represent the least significant bits of the POC value of the pictures (see Sect. 5.4.1).

The slice header includes an indication of the number of long-term reference pictures from the sequence parameter set to be used in the current slice. For the long-term reference pictures from the sequence parameter set and the long-term reference pictures signaled in the slice header, information on potential modifications of the respective most significant bits of the POC is encoded. Thereby, the respective pictures can accurately be identified.

4.3.5 Comparison to H.264 | AVC

In H.264 | AVC, multiple identifiers for pictures are specified. The picture order count (POC) is used to identify the output order of all pictures independent of their type. In both specifications, the POC value of IDR pictures must be equal to zero. In addition to POC, H.264 | AVC includes a syntax element for the frame number. This number is used to identify reference pictures and is encoded in the slice header [4, 5]. For pictures which are not used for reference, the frame number must be different from the frame number of the previously coded reference picture. In regular operation, the value of the frame number must be strictly increasing. This feature is used to enable the detection of reference picture losses in the sequence. In order to enable temporal scalability, gaps in the values of the frame number can be signaled to be permitted. In this case, the detection of lost pictures may be impossible. As an additional tool for error resilience, IDR pictures are marked with a separate dedicated IDR picture identifier. This identifier is used to uniquely characterize sub-sequences within the coded video sequence.

H.264 | AVC uses the concept of reference picture marking in the decoded picture buffer with an implicit or explicit marking process. The implicit marking is performed by a sliding window process which does not need further control commands. The explicit marking operation is performed by memory management control operation (MMCO) commands. Pictures in the DPB may have three different states: They can be used for short-term or long-term reference, or they can be marked as unused for reference. After the decoding of each picture, the marking process is performed to update the status of the DPB. Since the marking process is performed relative to the current state of the DPB, synchronization issues between encoder and decoder may occur in the case of lost pictures. In such cases, potential MMCO commands are lost too, which may lead to an incorrect buffer state at the decoder. Due to the incremental nature from picture to picture of this marking process, it may be difficult for a decoder to detect such losses.

The scheme of reference picture sets and derived reference picture lists in HEVC tries to circumvent the aforementioned issues. By using the POC value for identification of all pictures and by signaling the state for all pictures in the DPB the identification of potentially lost pictures is facilitated. The design further enables the decoder to precisely identify the POC of a lost picture and thereby supports appropriate actions for concealment or recovery. The reader is referred to Sjöberg et al. [11] for further comparison of the two schemes.

References

1. High efficiency video coding. ITU-T Rec. H.265 (HEVC). http://www.itu.int/rec/T-REC-H.265/en (2013). Accessed 14 Apr 2014
2. Information technology—High efficiency coding and media delivery in heterogeneous environments—Part 2: High efficiency video coding. ISO/IEC 23008-2:2013 (HEVC). http://www.iso.org/iso/home/store/catalogue_tc/catalogue_detail.htm?csnumber=35424 (2013). Accessed 14 Apr 2014
3. Mitchell, J.L., et al.: MPEG Video Compression Standard. Digital Multimedia Standards Series. Springer, US (copyright Kluwer Academic Publishers) (1996). doi:10.1007/b115884
4. Advanced video coding for generic audiovisual services. ITU-T Rec. H.264 (AVC). http://www.itu.int/rec/T-REC-H.264/en (2014). Accessed 14 Apr 2014
5. Information technology—Coding of audio-visual objects—Part 10: Advanced Video Coding. ISO/IEC 14496–10:2012 (AVC). http://www.iso.org/iso/home/store/catalogue_tc/catalogue_detail.htm?csnumber=61490 (2012). Accessed 14 Apr 2014
6. Information technology—Generic coding of moving pictures and associated audio information—Part 2: Video. ISO/IEC 13818-2:2013 (MPEG-2). http://www.iso.org/iso/home/store/catalogue_tc/catalogue_detail.htm?csnumber=61152 (2013). Accessed 14 Apr 2014
7. Misra, K., et al.: An overview of tiles in HEVC. IEEE J. Sel. Top. Sign. Process. **7**(6), 969–977 (2013). doi:10.1109/JSTSP.2013.2271451
8. Wien, M.: Variable block-size transforms for H.264/AVC. IEEE Trans. Circ. Syst. Video Technol. **13**(7), 604–613 (2003). doi:10.1109/TCSVT.2003.815380
9. Wiegand, T., et al.: Overview of the H.264/AVC video coding standard. IEEE Trans. Circ. Syst. Video Technol. **13**(7), 560–576 (2003). doi:10.1109/TCSVT.2003.815165
10. Wenger, S.: H.264/AVC over IP. IEEE Trans. Circ. Syst. Video Technol. **13**(7), 645–656 (2003). doi:10.1109/TCSVT.2003.814966
11. Sjöberg, R., et al.: Overview of HEVC high-level syntax and reference picture management. IEEE Trans. Circ. Syst. Video Technol. **22**(12), 1858–1870 (2012). doi:10.1109/TCSVT.2012.2223052
12. Bossen, F.: Common HM test conditions and software reference configurations. Doc. JCTVC-I1100. 9th meeting: Joint Collaborative Team on Video Coding (JCT-VC) of ITU-T SG 16 WP 3 and ISO/IEC JTC 1/SC 29/WG, Geneva, CH, 11 Apr 2012

Chapter 5
High-Level Syntax

In this chapter, the high-level syntax and the high-level organization of the bitstream containing the coded video sequence are detailed. While in Chap. 4 the spatio-temporal relation and organization of coding structures was described, this chapter deals with the representation at the level of syntax elements and their semantics.

The high-level syntax provides the encapsulation for the coded video data for further processing. It structures the information for transport and makes it accessible and also searchable. It further provides means to code all specified high-level parameters and additional side information. In order to separate between the encoded information and the form in which this information is represented, a *Video Coding Layer* (VCL) and a *Network Abstraction Layer* (NAL) are defined. All VCL and non-VCL data are encapsulated in NAL units for transmission. These concepts are further detailed below. The high-level syntax scheme is conceptually inherited from H.264 | AVC with some revisions and extensions to match modified requirements and enable improved functionality.

5.1 Byte Stream Format

Like H.264 | AVC, the HEVC specification provides a self-contained byte stream format for representation of the NAL unit stream. The byte stream format is specified in in Annex B of HEVC. This format can be used e.g. to stream the coded video data to or from a file, or over a network. Usage of the byte stream format is not required if the transport layer provides other means to encapsulate the raw NAL unit data. A different type of transport which does not make use of the byte stream format is e.g. the payload format for the Real-time Transport Protocol (RTP).[1] An overview of the system layer integration of HEVC is provided in [3].

[1] The RTP payload specification for HEVC has not yet been finalized by the time or writing [1]. The RTP payload format for H.264 | AVC is specified in RFC 6184 [2].

M. Wien, *High Efficiency Video Coding*, Signals and Communication Technology,
DOI 10.1007/978-3-662-44276-0_5

In the byte stream format, the NAL units are encoded into a stream of bytes. In order to enable the decoder to detect and extract the NAL units from this stream, the byte stream format specifies a start code prefix that precedes each NAL unit in the byte stream. The start code prefix consists of three bytes, two zero bytes and a '1' byte:

$$\text{start code prefix} = \text{0x00 0x00 0x01} = \text{0000 0000 \ 0000 0000 \ 0000 0001}$$

An additional 0x00 (zero) byte is inserted before the start code if the NAL unit is the first NAL unit of the first access unit in a coded video sequence or if the NAL unit contains a parameter set.[2]

As further detailed in Sect. 5.2.1, the encoder must prevent the occurrence of the byte sequence of the start code prefix on a byte aligned position of the byte stream. If the coded data would result in this byte sequence, an emulation prevention byte is inserted. By scanning the byte stream for the start code prefix, the decoder can hence determine the starting byte position of each NAL unit and, by detecting eventual trailing zero bytes, the total size N_{U} of the NAL unit in bytes.

5.2 Network Abstraction Layer

The *network abstraction layer* (NAL) defines the encapsulation of the coded video data for transportation and storage. The encoded information is organized in NAL units, which contain either data of the video coding layer (VCL) or non-VCL data. The video coding layer comprises all slice and CTU data. The non-VCL data includes the parameter sets and other information as further detailed below.

The ordered set of all VCL and non-VCL NAL units that represent the coded video sequence is referred to as the NAL unit stream. Depending on the context, the NAL unit stream or the byte stream according to the byte stream format are also referred to as the *bitstream*.

5.2.1 NAL Unit Structure

The payload of a NAL unit (NALU) is a sequence of N_{U} bytes. The first two bytes contain the NAL unit header, the remaining bytes are referred to as the *raw byte sequence payload* (RBSP). The last byte of the RBSP containing coded data includes the *RBSP stop bit* ('1') followed by '0' bits such that byte alignment is achieved.[3] The concatenation of the bits of the RBSP bytes before the RBSP stop bit constitutes the *string of data bits* (SODB). The SODB contains all coded data that is needed

[2] I.e., the NAL unit type is VPS_NUT, SPS_NUT, or PPS_NUT, see Sect. 5.2.2 below.

[3] The byte with the RBSP stop bit may be followed by bytes with all '0' bits for NAL units that contain slice data. Such zero bytes may be induced by CABAC entropy coding, see Sect. 10.2.

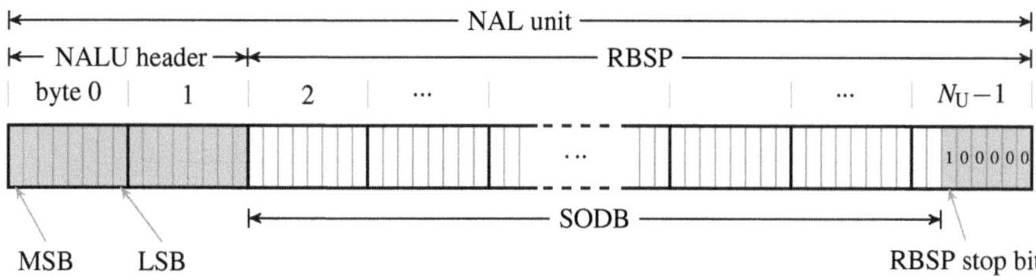

Fig. 5.1 Raw byte sequence payload of a NAL unit with N_U bytes with indication of the raw byte sequence payload (RBSP) and the string of data bits (SODB)

for the decoding process. Within the SODB, the bits of the bytes are concatenated from the MSB to the LSB. An illustration of the NAL unit bytes including RBSP and SODB is given in Fig. 5.1.

The specification requires the SODB to assert the ability for byte-aligned resynchronization. Thereby it is possible to scan the bitstream for a start code to find the byte-aligned start of the next NAL unit. To prevent the occurrence of start codes, the encoder must scan the RBSP for the following binary patterns at byte-aligned positions:

$$0000\,0000 \quad 0000\,0000 \quad 0000\,00\text{xx},$$

with xx $\in \{00, 01, 10, 11\}$. These binary patterns must then be modified using a start code emulation prevention byte with value '3':

$$0000\,0000 \quad 0000\,0000 \quad \mathbf{0000\,0011} \quad 0000\,00\text{xx}.$$

According to the specification, the earliest position in the NAL unit such an emulation prevention byte can occur is the fifth byte (i.e. three bytes after the NAL unit header). The encoder has to assert that no start code emulation occurs considering also the content of the NAL unit header.

The number of bytes in the NAL unit N_U includes the emulation prevention bytes. Considering the number of RBSP bytes, the emulation prevention bytes are *excluded* as these bytes are removed at the decoder side when the SODB is extracted from the RBSP. The number of bytes in the RBSP must be known to the decoder as it is used to stop entropy decoding on the slice level.

The NAL unit header identifies the NAL unit type and its relation to other NAL units in the NAL unit stream. The structure of the NAL unit header is shown in Fig. 5.2. It consists of two bytes. The first bit must be of value zero in a bitstream compliant to the specification.[4] It is followed by the six bit NAL unit type. The next six bits are reserved for the layer identifier l_{id}. This identifier will be used in the

[4] In external applications, such as the RTP payload format for H.264 | AVC [2], where the NAL unit header is also used as the transport payload header, a value of '1' may be used to indicate data loss which can be used to trigger error concealment at the receiver side.

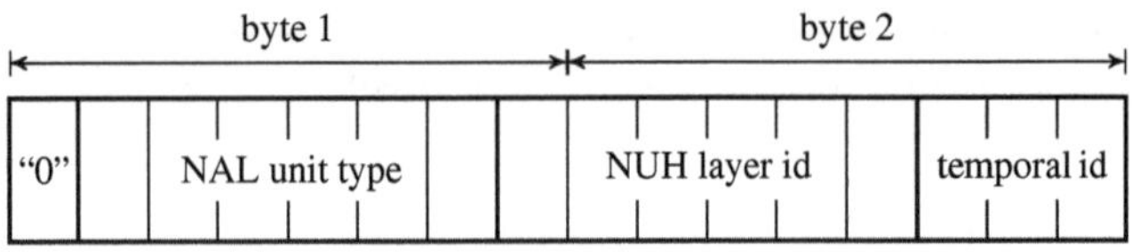

Fig. 5.2 Structure of the 2-byte NAL unit header

HEVC extensions for scalable video coding and multiview coding. In version 1 of the HEVC specification, $\mathrm{l_{id}} = 0$ is required. The last three bits of the header are assigned to the temporal identifier $\mathrm{t_{id}}$. The $\mathrm{t_{id}}$ indicates the assignment of the NAL unit to a specific temporal layer of the video sequence, see Sect. 4.1.1. For non-VCL NAL units, the $\mathrm{t_{id}}$ in the NAL unit header indicates the lowest temporal layer which the content of the NAL unit applies to.

In order to provide access to various high-level information within the NAL unit, the header as well as some other information for some NAL unit types is specified at fixed bit positions in the NAL unit, which further may be byte aligned. This concept enables media-aware network elements (MANEs) to easily parse and access this information for their application use.

5.2.2 NAL Unit Types

Compared to H.264 | AVC, the number of available NAL unit types has been extended from 32 to 64. This extension was introduced to prevent a lack of free NAL unit types for future extensions.[5]

A variety of NAL unit types is specified to classify the information contained in the NAL unit. The list of available NAL unit types is provided in Table 5.1. The NAL unit types are categorized into two classes: Video Coding Layer (VCL) NAL units and non-VCL NAL units. The VCL types comprise all NAL units that contain actual coded video data. The non-VCL types contain parameter sets, delimiters, supplemental enhancement information, or filler data. For both classes, reserved NAL unit types are specified which may be used in future extensions of the specification. The NAL unit types $48, \ldots, 63$ are not used and will remain unspecified also in future versions of the HEVC specification. This number range may be used for dedicated external needs such as the emerging HEVC0 RTP payload format [1].

A VCL NAL unit contains the coded video data of one slice segment, which corresponds to a set of CTUs in a defined area of the picture. All VCL NAL units

[5] In the course of the standard development of H.264 | AVC, the number of 32 NAL unit types turned out to be too low. The specified scalable and multiview extensions required the specification of additional NAL unit types while external applications made use of non-specified and non-reserved NAL unit types. For example in the RTP payload format for Scalable Video Coding, the complete H.264 | AVC NAL unit type range is covered and an extension mechanism has been specified to cover the needs of the application [4].

Table 5.1 NAL unit types

NUT[a]	Name	NAL unit content	Class
0	TRAIL_N	CSS[b] of a trailing picture (non-referenced)	VCL
1	TRAIL_R	CSS of a trailing picture (referenced)	VCL
2	TSA_N	CSS of a temporal sub-layer access pic. (non-ref.)	VCL
3	TSA_R	CSS of a TSA picture (referenced)	VCL
4	STSA_N	CSS of a step-wise TSA (non-referenced)	VCL
5	STSA_R	CSS of a STSA (referenced)	VCL
6	RADL_N	CSS of a random access decodable leading picture (non-referenced)	VCL
7	RADL_R	CSS of a RADL picture (referenced)	VCL
8	RASL_N	CSS of a random access skipped leading picture (non-referenced)	VCL
9	RASL_R	CSS of a RASL picture (referenced)	VCL
10	RSV_VCL_N10	Reserved non-IRAP sub-layer non-referenced	VCL
11	RSV_VCL_R11	Reserved non-IRAP sub-layer referenced	VCL
12	RSV_VCL_N12	Reserved non-IRAP sub-layer non-referenced	VCL
13	RSV_VCL_R13	Reserved non-IRAP sub-layer referenced	VCL
14	RSV_VCL_N14	Reserved non-IRAP sub-layer non-referenced	VCL
15	RSV_VCL_R15	Reserved non-IRAP sub-layer referenced	VCL
16	BLA_W_LP	CSS of a Broken Link Access (BLA) with leading pictures	VCL
17	BLA_W_RADL	CSS of a BLA with associated RADL pictures	VCL
18	BLA_N_LP	CSS of a BLA without leading pictures (LP)	VCL
19	IDR_W_RADL	CSS of an instantaneous decoder refresh picture with RADL	VCL
20	IDR_N_LP	CSS of an IDR picture without LP	VCL
21	CRA_NUT	CSS of a clean random access picture	VCL
22	RSV_IRAP_22	Reserved IRAP	VCL
23	RSV_IRAP_23	Reserved IRAP	VCL
24–31	RSV_VCL_24, … RSV_VCL_31	Reserved non-IRAP	VCL
32	VPS_NUT	Video Parameter Set (VPS)	non-VCL
33	SPS_NUT	Sequence Parameter Set (SPS)	non-VCL
34	PPS_NUT	Picture Parameter Set (PPS)	non-VCL
35	AUD_NUT	Access Unit Delimiter (AUD)	non-VCL
36	EOS_NUT	End of Sequence (EOS)	non-VCL
37	EOB_NUT	End of Bitstream (EOB)	non-VCL
38	FD_NUT	Filler Data (FD)	non-VCL

(continued)

Table 5.1 (continued)

NUT[a]	Name	NAL unit content	Class
39	PREFIX_SEI_NUT	Prefix Supplemental Enhancement Information (SEI)	non-VCL
40	SUFFIX_SEI_NUT	Suffix SEI	non-VCL
41–47	RSV_VCL_N41, ... RSV_VCL_N47	Reserved	non-VCL
48–63	UNSPEC48, ... UNSPEC63	Unspecified	non-VCL

[a] NUT: NAL unit type
[b] CSS: Coded slice segment

that belong to the same picture (and thereby to the same access unit, see below) are required to have the same NAL unit type. Thereby, the NAL unit type specifies the picture type. On top of the information regarding the picture type, the NAL unit type provides additional information on dependencies of other VCL NAL units on the current NAL unit. Some of these aspects are highlighted in the following.

For the NAL unit types 0, ..., 15, which do not provide random access functionality, the odd or even parity of the NAL unit type indicates the usage or non-usage of the current picture for reference in other pictures, respectively. NAL units which are thereby marked as non-reference correspond to pictures that are not included in the reference picture list. For the NAL unit types 16, ..., 21, which correspond to the specified random access point picture types, the specific NAL unit type indicates whether there are associated leading pictures or not. It is further indicated if these leading pictures are all decodable if decoding is started at the random access point, or if the set of leading pictures includes pictures that would need to be skipped if decoding is started here.[6]

The non-VCL NAL unit types include the video, sequence, and picture parameter sets which comprise profile/level information as well as all parameters for tool configuration as detailed below. Also, supplemental enhancement information (SEI) messages are carried in non-VCL NAL units. For the SEI messages, the parity of the NAL unit type indicates if the SEI message is required to be sent *before the last* VCL NAL unit in an access unit (prefix SEI message) or if it is required to be sent *after the first* VCL NAL unit (and potentially after the last VLC NAL unit, suffix SEI message). In general, SEI messages are not required by the decoding process but may be very helpful for the decoder to determine properties of the bitstream, especially in the case of data losses, or to convey additional side information.

Specific NAL unit types are provided that indicate the boundaries of access units, the end of a video sequence, or the end of the bit stream. While such information could be derived from the available NAL units and their ordering in the NAL unit stream, the availability of these NAL unit types is valuable in systems which are not capable

[6] The indication of existing skipped pictures is not needed for IDR pictures as by definition, those cannot be associated with skipped leading pictures.

of performing such an analysis. The end of sequence and end of bitstream NAL units do not convey further specific information besides their termination indication. The access unit delimiter signals the start of a new access unit and indicates if the access unit contains I slices only, or I and P slices, or if it may contain I, P, and B slices.

A special NAL unit type is defined for filler data. NAL units of this type may contain an arbitrary number of 0xFF bytes (all one-bits). Filler data NAL units may be used e.g. by systems that have strict requirements on avoidance buffer underflows or are required to provide video at a defined fixed bit rate.[7]

5.2.3 Access Units

In the previous sections, pictures and their properties were discussed. This approach is intuitive as the displayed sequence of pictures constitutes the video that is perceived. In terms of organization of NAL units and the bitstream, the concept of a picture is not directly applied. Instead, the concept of *Access Units* (AUs) is introduced. In contrast to a picture, which is constructed from VCL data, the access unit is a set of NAL units, including both, VCL and non-VCL NAL units.

Each coded picture in the sequence is contained in a set of one or more NAL units. An access unit contains all NAL units that are associated to one picture. All NAL units in the access unit (VCL and non-VCL) share the output time of the contained picture. Note that by specification, all NAL units in the bitstream are associated to access units, and an access unit must contain a coded picture. Accordingly, the complete NAL unit stream can be seen as a stream of access units.

The available set of picture types was discussed in the previous sections. These types are also used to address the access units. I.e., a BLA picture is contained in a BLA access unit, and so on. As mentioned above, all VCL NAL units within one access unit must have the same NAL unit type. This does not necessarily mean that they need to have identical slice types within one picture.

A visualization for the organization of NAL units in an access unit is given in Fig. 5.3. A similar figure is provided in the specification as well. An Access Unit Delimiter (AUD) may be used to indicate the start of a new access unit. If present, this NAL unit must be the first NAL unit in the access unit. Prefix SEI NAL units may contain one or more SEI messages. These NAL units may appear before and interleaved with coded slice NAL units, but must not follow the last VCL NAL unit. On the contrary, suffix NAL units follow the VCL NAL units. These must not precede the first VCL NAL unit. In addition the option to indicate the beginning of an access unit by an AUD, the end of a coded video sequence or the end of the complete bitstream can be indicated by EOS and EOB NAL units, respectively. Such

[7] Besides filler data NAL units, such data can also be conveyed in a dedicated SEI message (type 3). SEI messages are treated differently from Filler Data NAL units in the hypothetical reference decoder and therefore correspond to a different use case. See also Sect. 5.6.

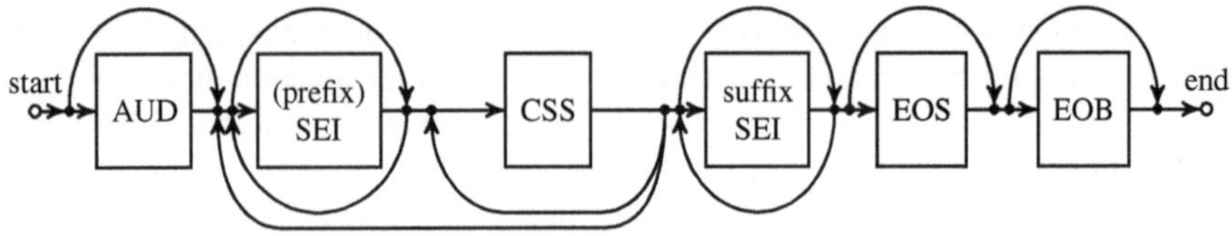

Fig. 5.3 Schematic diagram of the NAL unit organization in an access unit. The presentation does not include parameter sets, filler data, or NAL units of reserved or unspecified types

explicit high-level signalization can be useful to relieve the decoder from detection and identification of the respective information.

The VCL NAL units in an access unit are required to all have the same temporal identifier t_{id}, which at the same time is the t_{id} of the access unit. The t_{id} of non-VCL NAL units may be different (higher) compared to the t_{id} of the current access unit if the data persists and applies to later access units in the bitstream of a different temporal layer. This applies e.g. to picture parameter sets that are sent in the bitstream for later activation or to SEI messages.

5.2.4 *Decoding Units*

For very-low-delay applications, the concept of *Decoding Units* (DUs) has been introduced. Each decoding unit consists of a subset of the NAL units within an access unit. The ordered set of decoding units constitutes the access unit. While the access unit corresponds to the decoded picture, a decoding unit corresponds to a decoded sub-picture (e.g. an independent slice segment and its associated dependent slice segments, i.e. one or more NAL units).

If used, the operation of the hypothetical reference decoder is driven on a sub-picture basis. As the decoding process thereby also operates on a sub-picture basis it is possible to start output of sub-pictures before the full picture has been decoded. This very-low-delay operation can e.g. be useful in screen sharing applications [5].

The usage of DUs is indicated in the parameters for the HRD, see Sect. 5.6. If sub-picture operation is not indicated (because no HRD parameters are included in the SPS or sub picture operation is deactivated there), the access unit constitutes the decoding unit.

5.3 Parameter Sets

The concept of parameter sets helps to structure the high-level features of a coded video sequences and further high-level information on the sequence in a clear hierarchy. Information that applies to multiple slice segments, multiple pictures, or multiple layers of a coded video sequence is separated from each other as well as from the coded slice data itself. The strict separation of these different hierarchy levels has been designed to improve the error robustness of the coded video sequence representation under the condition of data loss. With this scheme, dedicated protection can be applied according to the needs of the respective hierarchy level. From the picture level on, this information is organized in dedicated parameter sets, which may be transmitted with the coded video data, sent over a separate (reliable) channel, or even be hard-coded and unchangeable for dedicated applications. Thereby, it is assured that the most important information necessary to decode the slice segments is made available to the decoder even under the condition of partial or severe data loss.

HEVC specifies three parameter sets. These are the *video, sequence, and picture parameter set* (VPS, SPS, and PPS). From these parameter sets, the sequence and picture parameter sets have already been present in H.264 | AVC.[8] The video parameter set is newly introduced in HEVC. Its purpose is to collectively provide information on different layers and sub-layers of the coded video sequence. While for the currently specified single layer profiles (Main, Main Intra, Main Still Image), the use of this parameter set may be limited, it will be very useful in foreseen and future extensions of the HEVC specification, e.g. for (spatial and quality) scalability, or for multiview coding [5].

In the decoding process, the three parameter sets are expected to be available at the decoder before the decoding of the first slice of the coded video sequence starts. The slice segment decoding process activates and processes the picture parameter set only at the beginning of a picture. The PPS then remains active for the whole picture. The VPS and the SPS are activated only at the start of a new coded video sequence. Therefore, the start of a new coded video sequence is required if parameters in these parameter sets shall be modified. As stated above, the parameter sets do not need to be transmitted with the video data if they are provided by other means ('out-of-band' transmission). If transmitted in the NAL unit stream, the occurrence of a parameter set NAL unit marks the start of a new access unit. Therefore, the parameter sets must be first in an access unit, after the access unit delimiter (if present).

It is possible to send multiple copies of the same parameter set in the bitstream for error resilience purposes. Parameter sets sharing the same parameter set ID in a NAL unit stream are required to contain identical values for all syntax elements. This restriction is necessary to avoid conflicts if for some reason different copies of the parameter set need to be activated at different times (or in different decoders). A NAL unit stream may further contain parameter sets that are never activated. However, all parameter sets that are encoded have to be valid parameter sets, obeying all constraints of the specification including profiles and levels.

For error and loss robustness reasons, each parameter set is self-contained, i.e., it can be decoded without reference to any other NAL unit in the NAL unit stream. For the VPS and the SPS, the most important information is fixed-length coded at

---

[8] In the course of the development of HEVC, the introduction of an Adaptation Parameter Set (APS) was discussed. This parameter set was intended to contain parameters e.g. for adaptive loop filters. The concept was later dropped from the draft specification [6].

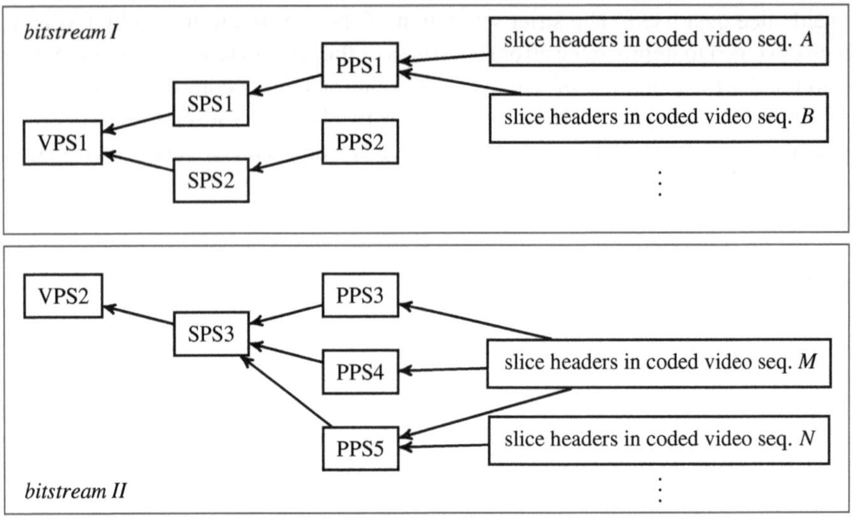

Fig. 5.4 Schematic illustration of two bitstreams with the activation of parameter sets through the slice headers of coded video sequences

the beginning to enable direct and low-complex access. All parameter sets include various hooks to enable future extensibility or modification.

The activation of the parameter sets follows a hierarchical approach as illustrated in Fig. 5.4. The picture parameter set is activated from the first slice segment header of a coded picture. As stated above, the subsequent slice segments of the picture must reference a PPS with identical content. It should be noted that it is not necessary for these PPS to have the same PPS identifier. Different pictures of the same coded video sequence however may well activate different PPSs with different parameter settings. The picture parameter set activates a sequence parameter set, which in turn activates the corresponding video parameter set. By allowing multiple PPSs referring to the same SPS, and multiple SPS pointing to a VPS, an efficient and flexible representation of the different types of high-level information on coded video sequences is achieved.

The content of the three parameter sets is briefly described in the following subsections. The description started from the highest level information in the VPS and moves on to the lowest level in the slice segment header. The presentation does not include a complete discussion of all available parameters. The main features are pointed out. For a full overview, the reader is referred to clauses 7.3 and 7.4 of the specification [7, 8].

5.3.1 Video Parameter Set

The video parameter set (VPS) has been introduced especially for information on multi-layer bitstreams. It provides general information on the coded video sequence, including the existence of one or multiple layers, and available operation points, which e.g. could be reached by extraction of sub-bitstreams from the given bitstream.

Since the VPS is the highest layer parameter set, all parameter sets that directly or indirectly refer to a specific VPS must fulfill the constraints set by this VPS. A VPS can only be activated at the beginning of a coded video sequence, i.e. an IDR or BLA access unit, or a CRA access unit which is the first access unit in the coded video sequence.

The VPS starts with a set of syntax elements which are fixed-length coded at defined positions in the first four bytes. Thereby, these syntax elements can be easily accessed by the decoder or external applications. The set of syntax elements includes the VPS identifier and information on available layers as well as on available temporal sub-layers. The VPS identifier is used by the SPS to activate the given VPS. The number of available layers is fixed to be set to one for single-layer sequences as described in this book. This syntax element will be e.g. used in the upcoming scalable or multiview extensions of HEVC to indicate the number of available (spatial or quality) layers or views, respectively. The number of temporal sub-layers corresponds to the available values for the temporal identifier t_{id} as detailed in Sect. 4.1.1. Furthermore, for the temporal sub-layers the property of temporal nesting is indicated.

The first four bytes are followed by a syntax structure containing information on the available operation points a decoder may work at for the given bitstream. An operation point is characterized by the applicable profile, tier, and level. It specifies constraints on coding tools that may be required for decoding the stream as well as constraints on bitstream size or buffer capabilities, see Chap. 11. The specification of the operation point for the whole bitstream can be followed by the specification of operation points for the temporal sub-layers of the bitstream, if available. Again, these syntax elements are specified using fixed-length code words such that the coded information continues byte aligned for easy access by the decoder or external applications.

The remaining information in the VPS has been considered not to require fixed-length byte-aligned access. The coding of these syntax elements is partly specified with variable length codes for the sake of an efficient representation. These syntax elements include information on buffering and latency for sub-layers, general timing information as well as parameters for the *Hypothetical Reference Decoder* (HRD) (see Sect. 5.6).

5.3.2 Sequence Parameter Set

While the VPS contains high-level parameters describing general features of the coded video sequence, the sequence parameter set (SPS) specifies the features and activated tools used in the coded video sequence. In contrast to the VPS, which concerns the whole bitstream, the SPS information applies only for the layer specified by the layer identifier l_{id}. In version 1 of the specification, only $l_{id} = 0$ is allowed, and hence, some global information on the bitstream in the VPS is simply doubled in this case.

Features specified in the SPS include the color format and bit depth as well as the sample resolution of the coded pictures. Like the VPS, the SPS is activated once at the beginning of a coded video sequence which may be an IDR, a BLA, or a CRA access unit.

The SPS starts with the SPS identifier, the number of temporal sub-layers and a flag indicating temporal nesting. This information is fixed-length coded in the first two bytes of the SPS. Then, the operation point syntax structure as described in the VPS section is included. The operation point information provided in the VPS comprised information on the full bitstream, which in the case of a single-layer bitstream would correspond to the complete operation point information. In the scalable and multiview extensions, a separate SPS may be activated for each layer or view. The operation point information in the SPS therefore specifies the parameters dedicated only to the respective layer or view.

Besides the picture and color format information mentioned above, the parameter set includes applicable tool configurations such as the minimum and maximum coding block sizes, i.e. the coding tree block parameters, the minimum and maximum transform block sizes and transform tree hierarchy depth for both, inter and intra blocks. Furthermore, the short-term and long-term reference picture sets, and various activation flags for coding tools are included in the SPS.

The SPS particularly includes the possibility to send *video usability information* (VUI). The VUI parameters provide detailed information on the coded video content, including color spaces, chroma positioning, sample aspect ratio etc., see Sect. 5.7. Furthermore, the VUI includes the parameters for the HRD. Again, those will be identical to the parameters in the VPS for single-layer coding, but may be different when being used in the envisaged HEVC extensions.

5.3.3 Picture Parameter Set

As the name indicates, the picture parameter set (PPS) contains parameters that apply for one picture. Consequently, a picture parameter set is activated on a picture level rather than a sequence level and the active PPS may change from picture to picture. Note again that all slices in a picture must refer to the same picture parameter set.

The PPS includes entropy coding configuration parameters for CABAC, the applicable quantizer configuration, offsets, and scaling lists, and the activation of dependent slices and wavefront parallel processing. The PPS further configures the tile partitioning for the current picture and the configuration of the loop filters. Due to this specification structure, the tile partitioning in a sequence may change from picture to picture. The application of a constant tile partitioning can be indicated in the VUI.

5.4 Slice Segment Header

The slice segment header contains the reference to the applicable PPS and thereby activates the chain of parameter sets. All slice segment headers in a picture must refer to a PPS with identical content. It is not necessary that these PPS have the same PPS identifier.

The slice segment header contains information that may change from slice. It should be noted that some slice segment header parameters are required to be identical for all slices in the picture. The parameters in the slice segment header of an independent slice segment apply to all associated dependent slice segments. Since the header is included in each independent slice segment, it is guaranteed that this slice segment and its associated dependent slice segments can be decoded independently from the other slice segments of the picture.

Note that the slice header contains a special flag indicating if the current picture shall be marked for output. If the flag is set to zero, the picture is not delivered for display but still stored in the decoded picture buffer. It can thereby be made available for later reference in displayed pictures.

Besides tool parameters e.g. for quantization, deblocking and loop-filtering, the slice segment header contains information on available entry points for wavefront parallel processing or for the application of tiles. It should be noted that the amount of bits needed for this information may be significant [5]. For non-IDR pictures, the slice segment header further contains the picture order count and the configuration of the short-term and long-term reference picture sets as well as the length of the applicable reference picture lists.

5.4.1 Picture Order Count

As in H.264 | AVC, the concept of time, i.e. the temporal relation between pictures in the video sequence, is not directly used in the coded representation. Instead, the relation of the pictures in terms of ordering and distance if used for prediction is expressed by the *picture order count* (POC).

The POC value is a index number that defines the output position of the current picture in the coded video sequence. As an important feature, the POC value is specifically used to identify the picture in the decoded picture buffer. For identification purpose, the value of POC is required to be strictly increasing with the output order of the coded pictures. The POC value of IDR pictures is specified to always be $\mathrm{poc}_{\mathrm{IDR}} = 0$.

For an efficient representation, the POC value is not directly coded in the syntax. Instead, it is represented by a least significant part $\mathrm{poc}_{\mathrm{lsb}}$ which is encoded in the slice segment header. This value must be identical for all slice segments in the picture. Additionally, a most significant part $\mathrm{poc}_{\mathrm{msb}}$ is derived during the decoding process. The available POC number range is specified by the value N_{poc} in the sequence parameter set which determines the maximum value of $\mathrm{poc}_{\mathrm{lsb}}$ as

$$\max\left(\mathrm{poc}_{\mathrm{lsb}}\right) = 2^{N_{\mathrm{poc}}} - 1, \text{ with } 4 \leq N_{\mathrm{poc}} \leq 16. \tag{5.1}$$

The value of $\mathrm{poc}_{\mathrm{msb}}$ is updated during the decoding process. For derivation, the $\mathrm{poc}_{\mathrm{lsb}}$ and $\mathrm{poc}_{\mathrm{msb}}$ values of the last picture with $\mathrm{t}_{\mathrm{id}} = 0$ are used. These are denoted by $\mathrm{poc}_{\mathrm{msb,prev}}$ and $\mathrm{poc}_{\mathrm{lsb,prev}}$. The values are used to determine a potential modification of $\mathrm{poc}_{\mathrm{msb}}$ for the current picture.

If $\mathrm{poc}_{\mathrm{lsb}} < \mathrm{poc}_{\mathrm{lsb,prev}}$ and $\left(\mathrm{poc}_{\mathrm{lsb,prev}} - \mathrm{poc}_{\mathrm{lsb}}\right) \geq 2^{N_{\mathrm{poc}}-1}$,

$$\mathrm{poc}_{\mathrm{msb}} = \mathrm{poc}_{\mathrm{msb}} + 2^{N_{\mathrm{poc}}}.$$

Otherwise, if $\mathrm{poc}_{\mathrm{lsb}} > \mathrm{poc}_{\mathrm{lsb,prev}}$ and $\left(\mathrm{poc}_{\mathrm{lsb}} - \mathrm{poc}_{\mathrm{lsb,prev}}\right) > 2^{N_{\mathrm{poc}}-1}$,

$$\mathrm{poc}_{\mathrm{msb}} = \mathrm{poc}_{\mathrm{msb}} - 2^{N_{\mathrm{poc}}}.$$

Otherwise, $\mathrm{poc}_{\mathrm{msb}}$ is not updated.

The POC value $\mathrm{poc}_{\mathrm{c}}$ of the current picture is finally derived as

$$\mathrm{poc}_{\mathrm{c}} = \mathrm{poc}_{\mathrm{msb}} + \mathrm{poc}_{\mathrm{lsb}}. \tag{5.2}$$

The number range for $\mathrm{poc}_{\mathrm{c}}$ is restricted to

$$-2^{31} \leq \mathrm{poc}_{\mathrm{c}} \leq 2^{31} - 1, \tag{5.3}$$

with a maximum POC difference between two pictures A and B within

$$-2^{15} \leq \left(\mathrm{poc}_{\mathrm{c,A}} - \mathrm{poc}_{\mathrm{c,B}}\right) \leq 2^{15} - 1. \tag{5.4}$$

Conceptually, the values of POC do not need to be changing with constant delta. The values $\mathrm{poc}_{\mathrm{c}}$ of the coded pictures can be chosen by the encoder. Since POC is used for scaling motion vectors, e.g. for derivation of appropriate candidates for merge mode, POC values should be chosen such that the distance between pictures can be represented appropriately by POC deltas.

The VPS contains a flag to indicate if a delta of POC values relates to the time difference between the corresponding coded pictures. In this case, POC is proportional to the time interval between successive pictures. This option can be used to e.g. synchronize the POC values of coded pictures to their capture time distance. This option can be useful e.g. to apply appropriate motion vector scaling in sequences with irregular picture capture times.

5.5 Supplemental Enhancement Information

Supplemental Enhancement Information (SEI) messages contain information which is not required for the decoding process and the reconstruction of the coded video sequence. SEI messages may be helpful though for the decoder to determine certain features of the coded video sequence that otherwise might be complicated to be derived. Further information e.g. regarding display of the sequence or additional side information that shall be attached and integrated into the coded bitstream for transport can be encoded into specific SEI messages. Dedicated SEI messages can further be of help for error recovery or for testing of the picture integrity. SEI messages are classified either as prefix SEI messages or as suffix SEI messages, indicating if the SEI must occur before the last VCL NAL unit or after the first VCL NAL unit in the access unit, respectively.

Besides the distinction between prefix and suffix SEI messages, the concept of SEI messages has been inherited from H.264 | AVC. A subset of the SEI messages specified in H.264 | AVC have also been specified for HEVC. The payload types indices of these identical SEI messages are synchronized between the two specifications in order to avoid confusion. Some H.264 | AVC SEI messages have not been considered useful in the context of HEVC and are therefore no longer provided in the specification.

An overview of the SEI messages specified in HEVC is provided in Table 5.2 for the prefix SEI messages and in Table 5.3 for the suffix SEI messages. Note that some SEI messages can be sent either as a prefix or as a suffix SEI message. The SEI messages have a specified persistence that indicates for which NAL units the information contained in the SEI message is valid. The persistence is indicated in Tables 5.2 and 5.3 as well. Many SEI messages persist for the access unit they are coded in. Others may persist for the whole coded video sequence or until an update is provided by a new SEI message of the same payload type. Multiple SEI messages can be contained in an SEI message NAL unit. Each SEI message is identified by an SEI payload type index and an SEI payload size, where an SEI message always consists of an integer number of bytes. Tables 5.2 and 5.3 include a brief characterization of the functionality provided by each SEI message.

Table 5.2 Prefix SEI messages

Plt	Name	Summary
0	Buffering period	Indication of the initial buffering duration for control of the HRD. Persists for the bitstream
1	Picture timing	Indication of coded and decoded picture buffer delay. Applies to the associated AU
2	Pan-scan rectangle	Sub-region of the picture to be displayed e.g. on small screens. Persists for one AU or until update
3	Filler payload	Filler data. Persists for the associated AU
4	User data ITU-T T-35	User data registered as specified by ITU-T T.35. Persistence not specified
5	User data unregistered	Unregistered user data. Persistence not specified
6	Recovery point	Indication of acceptable picture quality after random access at a non-IRAP picture. Applies to the associated AU or an AU in specified POC distance
9	Scene information	Identification of scenes, spec. of potential transitions. Persists until next scene information SEI
15	Picture snapshot	Use this picture for a still-image snapshot. Applies to the associated AU
16	Progressive refinement start	Refine reconstruction quality of the associated picture by following pictures. Applies to the associated AU or an AU in specified POC distance
17	Progressive refinement end	Last refinement picture (all pictures with identical POC). Applies to the associated AU
19	Film grain characteristics	Simulate film grain by a specified model for display. Persists for one AU or until update
22	Post filter hint	Proposed filtering for post processing prior to display. Applies to the associated AU
23	Tone mapping information	Re-mapping of the coded color sample values to a target color map. Persists until update
45	Frame packing arrangement	Indication of multiple pictures packed into one coded picture. Persists until update
47	Display orientation	Flip/rotate picture for display. Persists until update
128	Structure of pictures description	Describes the coding structure including used reference picture set for a SOP. Persists for the AUs associated to the SOP
129	Active parameter sets	Indication of the active VPS and possibly used SPS in the CVS. Persists for the coded video sequence
130	Decoding unit information	Indication of decoding unit boundaries and sub-picture operation of the HRD. Persists for the associated DU
131	Temporal sub-layer zero index	Detection of loss of pictures with $t_{id} = 0$, including unique identification of IRAP pictures. Applies to the associated AU
133	Scalable nesting	Wraps other SEI messages which are dedicated to a specified layer or temporal sub-layer. Persistence depends on included SEI messages
134	Region refresh information	Indication for a set of slice segments to be decoded (approximately) correct after a recovery point. For gradual decoder refresh. Persists until update

SEI messages inherited from H.264 | AVC share payload type index (plt)

Table 5.3 Suffix SEI messages

Plt	Name	Summary
3	Filler payload	Filler data that can be discarded. Persists for the associated AU
4	User data ITU-T T-35	User data registered as specified by ITU-T T.35. Persistence not specified
5	User data unregistered	Unregistered user data. Persistence not specified
17	Progressive refinement end	Last refinement picture (all pictures with identical POC). Applies to the associated AU
22	Post filter hint	Proposed filtering for post processing prior to display. Applies to the associated AU
132	Decoded picture hash	Hash to enable the decoder to check if the decoded picture matches the encoded picture. Applies to the associated AU

SEI messages inherited from H.264 | AVC share the original payload type (plt)

5.6 Hypothetical Reference Decoder

When encoding a video sequence it is crucial to have control over the decoder buffer state for many applications. This applies e.g. for communication applications or broadcast scenarios. In a simplified view, the encoder must guarantee to provide the transmitted data such that it is available at the decoder at decoding time of the corresponding picture. Further, the encoder guarantees that the bitstream does not overrun the input bitstream buffer of the decoder as well as the picture buffer in which the decoded pictures are stored.

The tool to provide and test this functionality is the *hypothetical reference decoder* (HRD). The HRD is a normative part of the specification. The parameters for configuration and operation of the hypothetical reference decoder can be provided in the VPS and in the SPS. An introduction and overview to the concept of HRD operation as it was used in H.264 | AVC is given by Ribas-Corbera et al. in [9]. The concept has been transferred to HEVC and is summarized in the following.

The HRD parameters can be given for multiple operation points for the bitstream as detailed below. This is important to provide information on the characteristics of the bitstream after further processing, e.g. like sub-bitstream extraction. It should be noted that the HRD cannot only be applied in encoders to control the produced bitstream, but also to verify the conformance of a given bitstream to the specification requirements. Further, the conformance of a given decoder implementation can be tested against the performance and timing requirements defined by the HRD. Thereby the HRD is the essential tool for conformance testing. An encoder can decide to omit signaling the HRD parameters for a bitstream. In this case, verification of bitstream conformance by the HRD is not possible.

A schematic diagram of the HDR is provided in Fig. 5.5. The properties and tasks of the building blocks are briefly described in the following. The scheduling of the transmission and the delivery of the coded bitstream to the decoder side is modelled by the hypothetical stream scheduler (HSS). The HSS provides the bitstream to the

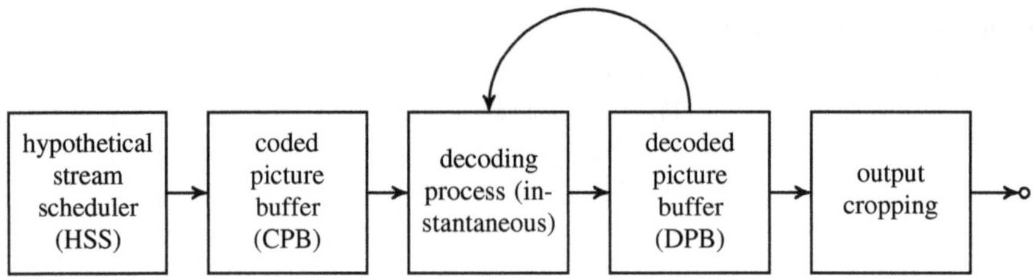

Fig. 5.5 Buffer model of the hypothetical reference decoder

coded picture buffer (CPB) in a timed manner. The CPB stores the coded data, i.e. the received NAL units, and provides them to the hypothetical decoding process which, in the idealized model, is assumed to instantaneously operate according to the control commands and coded video data provided in the bitstream. In terms of timing, the start of the decoding of a picture is defined to be initiated after processing the control operation for the reference picture set and reference picture lists as signaled in the slice header. The decoded pictures (or parts of them, if operating on sub-picture level, see below) are stored for output and used as reference pictures in the decoded picture buffer (DPB). As the final step, the pictures are passed to the output cropping stage where normative output cropping is applied to the output pictures according to conformance window parameters specified in the SPS. Note that the pictures from the DPB are not processed by the cropping stage before application as reference pictures in the decoding process. Within the normative cropping area specified in the SPS, an additional suggested cropping region may be signaled with the Video Usability Information, see below. Furthermore, the Pan-Scan Rectangle SEI message can be used to indicate a cropped sub-region of the picture e.g. for display on small screens.

5.6.1 Coded Picture Buffer

The configuration of the coded picture buffer (CPB) can be specified by three parameters: The *bitrate* (bits per time) at which the bitstream is streaming in, the *CPB size* which is required to hold the stored bits, and the *initial CPB removal delay* which specifies the time needed to fill the buffer such that the first picture can be decoded and no buffer underrun (i.e. unavailability of data) occurs.

As suggested in [9], the operation of the CPB can be illustrated by the filling level of the buffer under the assumption of constant input at the specified bitrate, see Fig. 5.6. After an initial CBP removal delay d, the decoding of the first picture starts at time instance t_0 when the initial filling level of the buffer B_{init} is reached. The bits b_0 required for decoding the picture are removed from the CPB and the decoding process is operated instantaneously. At time t_1, the bits b_1 for the next picture are removed from the buffer and the picture is decoded, and so forth for the following pictures. The parameters configuring the CPB operation warrant that the buffer size is never exceeded and also that the buffer is never emptied. The minimum buffer size for operating the illustrating example in Fig. 5.6 is indicated by $B_{\min}$.

The figure shows the decoder side buffer. In a similar fashion, the operation of the encoder can be modelled, where a hypothetical encoder instantaneously fills the bits required for encoding the picture into the encoder output buffer and the output buffer constantly drains the bits to the channel at the specified rate. Due to the similarity to a bucket with leaks which constantly drains the water that is filled into the bucket, the model for the buffer operation is called the *leaky bucket model* [9].

5.6.1.1 Constant Bitrate and Variable Bitrate Operation

In the above explanation of the CPB operation, a constant bitrate (CBR) scenario was assumed, where the video data is transmitted at a given constant bitrate to the decoder side. In an alternative mode of operation, variable bitrate (VBR) conditions are often observed. In a VBR scenario, the CPB operation can be described using a long-term average bitrate as the buffer parameter. If the scenario includes a peak bitrate which is larger than the long-term average and can be sustainably provided, a different CPB configuration may be beneficial in which this peak bitrate is indicated. The higher bitrate induces a smaller required CPB size as the removal of large pictures can be better compensated. This further implies that the buffer may run empty under some conditions.

5.6.1.2 Low-Delay Operation

In scenarios which require low-delay operation, the coded picture buffer can be operated in a dedicated mode which allows the decoder to decode a picture as soon as all data for the picture is available. In this mode of operation, the violation of buffer constraints (i.e. overrun or underrun of the buffer limits) is not strictly forbidden.

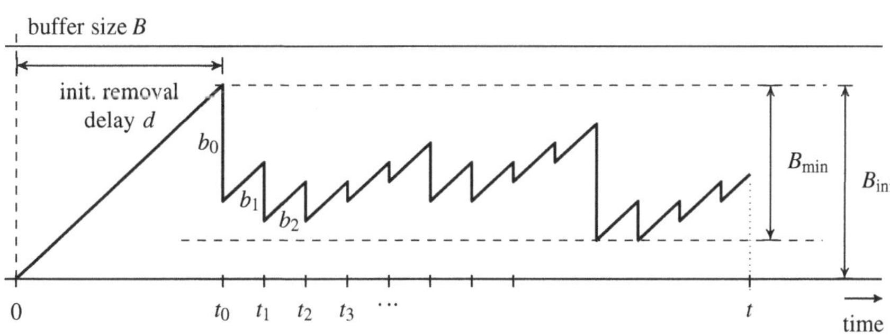

Fig. 5.6 Illustration of the filling level of the coded picture buffer which stores the received bitstream for processing at the decoder [9]

However, it is assumed that such violations only occasionally occur, e.g. if big pictures are coded from time to time. Since the buffer constraints are not enforced, the timing of the display of the pictures can also not be guaranteed in this mode of operation. While this may be acceptable or even intended in communication scenarios where the encoder and the decoder can communicate, e.g. to adapt the transmission bitrate if required, the relaxed play-out requirement may be undesirable in other application scenarios such as playback from file or broadcast live streaming over a network.

5.6.2 Decoded Picture Buffer

Besides the configuration of the coded picture buffer, the HRD parameters further specify the timing for output of pictures from the decoded picture buffer (DPB) for display. The DPB holds all reconstructed pictures which are used as reference pictures (long-term or short-term) and all pictures which have not been scheduled for display yet. Note that the syntax of HEVC allows for explicitly suppressing the output of a picture for display, which may be utile under specific application conditions.[9] Furthermore, the prevention of output of pictures for display may be indicated in cases where a BLA picture has been encountered and the content of the decoded pictures may not be suitable for display (e.g. because the referenced pictures in the DPB do not have the right content after splicing of sequences).

Pictures that are marked as unused for reference are removed from the DPB upon display. The output timing of pictures from the DPB for display is signaled as a delay relative to the time when the access unit comprising the picture has been removed from the CPB. The marking of pictures to be used or unused for reference is performed using the reference picture set syntax structure which applies for the current picture. By default, a picture is marked as used for short-term reference when being stored in the DPB.

5.6.3 Sub-picture Operation

While in previous standards the HRD operated on a picture basis, the HEVC HRD can be configured to operate on a sub-picture basis. In this mode, the operation is performed using decoding units which have been presented in Sect. 5.2.4. If sub-picture processing is activated, corresponding parameters for the CPB operation as well as DPB output delay information is provided on a DU basis.

Conceptually, the sub-picture operation provides the HDR functionality previously specified for full pictures for parts of pictures. It thereby enables early output of sub-pictures for very-low-delay.

[9] For example, pictures are collected in the DPB to be used for reference but not be used for display.

5.6.4 Operation Points

The HRD in HEVC follows the generalized HRD concept which has been introduced in H.264 | AVC: multiple HRD parameter configurations can be provided for the same bitstream. Thereby, adapted settings for different combinations of buffer sizes and bitrates can be provided, which allows for greater flexibility in the HRD operation.

Configurations are provided for the whole bitstream but can also be provided at a sub-bitstream level. Specifically, HRD parameters can be provided for different layers and temporal sub-layers. For example in the planned extensions for scalability and multiview coding, the HRD parameters can be specified to apply to a specific layer index which corresponds to a given spatial resolution or a given view, respectively.

An *operation point* is defined as a bitstream potentially extracted from another bitstream according to a specific t_{id} or layer id. The extracted bitstream must represent a coded video sequence. Note that different values of layer id only apply to the planned extensions but not for the single-layer profiles discussed in this book. The basic classification of the operation points is provided by the selected profile and level which the bitstream corresponds to (see Chap. 11).

5.6.5 Conformance Points

As detailed in Sect. 5.2 an HEVC bitstream consists of NAL units of various types. The NAL unit types are categorized into two main classes: NAL units containing syntax elements directly used for reconstructing the coded picture (VCL NAL units) and NAL units containing high-level information like parameter sets (non-VCL NAL units). In order to address the presence of these two classes of NAL unit types, the hypothetical reference decoder can operated two types of bitstreams:

Type I Only the VCL NAL units and Filler Data NAL units are considered during operation of the HRD. The parameters for this type are referred to as VCL HRD parameters.

Type II VCL NAL units, Filler Data NAL units, and other types of NAL units are included for operation of the HRD, i.e. also SEI messages, parameter sets, and other non-VCL NAL units are considered as well. The parameters for this type are referred to as NAL HRD parameters.

Type II can be further distinguished to be using the byte stream format or to be directly operating on the NAL unit stream. Type I stream or the variants of Type II stream are referred to as the *conformance points* of the bitstream. The parameters for the two types can be signaled collectively in the HRD parameter syntax. The Type I bitstream can be extracted from the Type II bitstream by removing the NAL units of the respective types. For example, the two available types of filler data (Filler Data NAL units and Filler Payload SEI messages) are treated differently if the bitstream is fed to a Type I or Type II HRD. Thereby, conformance checking of the stream with and without non-VCL NAL unit data (besides potential filler data) is achieved.

5.6.6 Signaling HRD Parameters in the VPS and SPS

The HRD parameters can be signaled in the VPS and in the SPS. If included in the SPS, the HRD parameters are signaled as a part of the timing information in the Video Usability Information (detailed in the next section). For single-layer operation, the VPS signaling does not provide more information than what is provided in the SPS. When the planned or further extensions become normative, the VPS provides a general overview of the available layers and their relations. In turn, the VPS syntax structure also provides the capability to include the HRD parameters for all available layers and temporal sub-layers in the bitstream. By definition, the signaling of HRD parameters in the SPS applies only to the coded video sequence for which the SPS is activated. This coded video sequence may correspond to only a subset of the available layers, e.g. a specific spatial resolution in a scalable bitstream.

Special treatment is further foreseen for HRD operation in the context of BLA or CRA pictures. These new picture types induce a modified decoding process when occurring within a coded bitstream or at the beginning of a coded bitstream. In these cases, pictures associated with the BLA or CRA picture may be discarded. Correspondingly, alternative HRD parameters can be signaled for these cases.

5.7 Video Usability Information

The *Video Usability Information* (VUI) syntax structure collects information that is useful for preparing the decoded video for output and display. The VUI can be sent as part of the SPS, but can also be passed to the decoder by other means (e.g. hard coded or by out-of-band transmission).

The information contained in the VUI is not required by the decoding process itself. The decoding process generates the sample values of the reconstructed pictures and determines the order in which the pictures are output. The VUI provides the possibility to signal information like the sample aspect ratio (i.e. the shape of the samples), the original color space and representation of the encoded video, or, as another very important aspect, picture timing information. The different parts of the VUI are briefly described in the following subsections. The inclusion of each of the parts in the VUI syntax structure is optional an can be decided as required by the application. Default values are specified for all VUI parameters for cases where the corresponding VUI parameters have not been provided.

5.7.1 Geometrical Relations

Indicated geometrical relations include the *sample aspect ratio* (SAR) and the indication of overscan presentation.

The SAR specifies the geometry of the samples in the source material. A table including the set of predefined sample aspect ratios as specified in HEVC has been provided in Table 2.1. The applicable sample aspect ratio is signaled by its corresponding index number as indicated in the table. The indices 1 to 16 have predefined SAR values. The maximum index value 225 serves as an indication that the horizontal and vertical sample aspect ratio values are signaled explicitly.

Since the SAR type specifies the geometry of the samples, it consequently determines the geometrical relation of the absolute width and the height for the coded pictures. This relation informs the display process how the decoded picture should be treated (i.e. possibly re-sampled) to provide a geometrically undistorted representation of the video on the target display device.

If overscan presentation is indicated, the coded video material may contain areas at the left and/or right boundary of the picture which are not intended for display. With the indication of overscan, the presentation of these areas can be prevented, accordingly.

5.7.2 Video Signal Type and Color Information

The video signal type information includes an indication of the original format of the source material. This can be Component, PAL, NTSC, SECAM, MAC, or Unspecified.[10] Besides this explicit video format indication which includes a corresponding color format definition, together with the respective format specifications, explicit color primaries, transfer characteristics, and the used color conversion matrix coefficients can be transmitted.

Additionally, the location of the chroma samples in relation to the location of the luma samples can be indicated. While for the decoding process specific fixed chroma locations are assumed, the actual chroma locations for the coded original material may differ. Hence, the provided information can be used to ensure a correct color presentation during display.

5.7.3 Frame/Field Indication

As indicated before, the HEVC specification does not provide specific tools at the video coding layer which are dedicated for application with interlaced source material. The decoding process does not differ for pictures representing fields or pictures representing frames. Nonetheless, provisions are included to handle field-based or

[10] Component: used e.g. in digital color component based formats like ITU-R BT.601; PAL: "Phase Alternating Line", European analogue television broadcast format; NTSC: "National Television System Committee", US-American analogue television broadcast format; SECAM: "Séquentiel couleur à mémoire", french analogue television broadcast format; MAC: "Multiplexed Analogue Components", multiplex format for analogue satellite transmission. A review of these formats can be found in [10].

frame-based sequences. If field-based sequence is indicated, each picture of the coded video sequence is representing a field and a Picture Timing SEI message is required to be sent with each access unit. This SEI message includes the indication if the corresponding picture represents a top field or bottom field.[11]

5.7.4 Default Display Window

The SPS includes the specification of a normative cropping window[12] which must be applied to the decoded picture when being output from the decoded picture buffer. In the VUI, an additional window can be specified within the area of the normative cropping window that represents the suggested display area if no other indication is given. Such indication could e.g. be provided by the Pan-Scan Rectangle SEI message or by other means of the given application, e.g. to indicate the region of a 16:9 video to be displayed on a 4:3 screen.

5.7.5 Timing Information

Timing information is specifically important to ensure the correct play-out speed of the decoded video sequence. Since the specification of the decoding process does not include the concept of time, no timing information is provided by this process itself. As an important part of the VUI, the syntax structure for the parameters of the hypothetical reference decoder (see Sect. 5.6) is placed in the timing information section.

The provided timing information includes the necessary parameters to install a timing scheme. For this purpose a basic clock rate and the length of a clock tick are defined. The example given in the specification is a clock operating at $f_c = 27\,\text{MHz}$ and a video signal with a picture rate of 25 Hz, i.e. the temporal distance between the pictures is $t_p = 0.04\,\text{s}$. This picture rate is represented by $n_{tu} = 1080\,000$ time units, since

$$t_p = \frac{n_{tu}}{f_c} = \frac{1080\,000}{27\,000\,000\,\text{Hz}} = 0.04\,\text{s}.$$

Correspondingly, e.g. a 30 Hz sequence could be represented by $n_{tu} = 900\,000$. The timing information includes a flag that further allows to indicate that the POC is proportional to the output time of the picture (relative to the beginning of the coded video sequence, e.g. an IDR, where the POC is reset). With this indication, the picture output timing can be directly derived from the POC.

---

[11] The semantics of the picture structure syntax element in the Picture Timing SEI message include a large variety of possible variants, including frame, top/bottom field, combined fields, or e.g. indications of field or frame repetitions.

[12] This is the conformance window, see Sect. 5.6.

Note that the timing information can also be provided in the VPS. In this case, it is required that the timing information in both, VPS and VUI, is identical.

5.7.6 Bitstream Restrictions

As the last part of the VUI, restrictions on the bitstream can be indicated that may help the decoder to determine the proper configuration for decoding the associated coded video sequence, e.g. in terms of setup for parallel processing.

5.7.6.1 Fixed Tiles Structure

If a fixed tile structure is indicated, the decoder can rely on the indicated tile structure throughout the sequence, e.g. for purpose of allocation of tiles to processing threads in a multi-thread environment.

5.7.6.2 Motion Vector Picture Boundary Restriction

The motion vector picture boundary restriction indicates that the motion vectors do not induce prediction blocks to be partly outside of the picture region. Thereby, the decoder knows that no picture boundary padding is necessary for the decoding process. For example, in application scenarios where multiple smaller video sequences are merged into a large side-by-side presentation, this restriction can be used to indicate that no special processing at former picture (and now tile-) boundaries is required.

5.7.6.3 Reference Picture Lists Restriction

If reference picture list restriction is indicated, all slices within a picture are forced to not only use the same reference picture set, but are further required to use the same reference picture lists.

5.7.6.4 Minimum Spatial Segmentation

The minimum spatial segmentation indication provides a measure to the decoder to determine the size of the picture area which can be independently decoded as a spatial segment (the number of CTUs in a slice). This information can be used to make an assumption on the number of parallel processes which the decoder could employ for decoding the given coded video sequence.

5.7.6.5 Maximum Number of Bytes per Picture and Maximum Number of Bits per Coding Unit

The maximum number of bytes per picture gives an indication on the maximum expected dynamics in the number of bytes used to encode the VCL NAL units of a picture in the associated coded video sequence. The maximum number of bits per coding unit provides a corresponding measure on the CU level. Such information can be used for more accurate characterization of the operation of the coded picture buffer and also the operation of the parsing and decoding processes at the decoder.

5.7.6.6 Maximum Motion Vector Length

The indication of the maximum motion vector length in both, horizontal and vertical direction, can be used to refine the constraints on the maximum motion vector length values that are specified in the applicable profile, tier and level, see Chap. 11. Since the expected motion vector vector length can be interpreted as an indication on the irregularity of memory accesses for motion compensation, this information can be utile to the decoder to estimate the computational effort required to decoded the associated coded video sequence.

5.8 Comparison to H.264 | AVC

The basic high-level syntax concept of H.264 | AVC has been transferred to HEVC. This includes the separation between a network abstraction layer and a video coding layer as well as the basic concept of NAL units, access units, and parameter sets [11, 12]. The video usability information has been mostly transferred from H.264 | AVC. Also the concept of the hypothetical reference decoder is very similar for both specifications. The new concept of decoding units allows for operation of the HDR on a sub-picture basis for ultra-low delay processing.

In terms of high-level syntax the fundamentally new element of HEVC compared to H.264 | AVC is the video parameter set. While not much needed in the case of single-layer single-view coding, this parameter set carries the essential information on all scalable layers or all views in an HEVC stream. In H.264 | AVC, such information has to be represented in scalability information SEI messages as no comparable syntax structure had been foreseen before the respective scalable and multiview extensions were specified.

With the new two-byte NAL unit header, HEVC circumvents limitations in the NAL unit type space which became an issue with the increasing number of extensions and application spaces for H.264 | AVC. In the new NAL unit type structure, a large number of types is available for reserved usage by ITU-T and ISO/IEC. At the same time a large number of unspecified types is available which can be occupied by external applications and specifications such as the emerging RTP payload format for HEVC [1].

In H.264 | AVC, one of three different methods may be used to signal the POC of a picture to the decoder (explicit, relative to a predefined order, derived from the encoded frame number). In HEVC, the importance of POC is increased relative to H.264 | AVC and only explicit signaling is used for error resilience reasons.

With regards to error resilience, H.264 | AVC included the support for data partitioning into three partitions in the Extended profile. This profile also includes the additional SI and SP slices as indicated in Sect. 4.2.7. The support of data partitioning has not been transferred to HEVC.

The new possibility to distinguish between prefix and suffix SEI messages in HEVC enables more flexibility in the organization of NAL units within an access unit. It specifically enables the encoder to send a decoded picture hash SEI messages after the complete picture to enable verification of the decoder output. Without the specification of suffix SEI messages, this information would have to be interspersed before the encoded VCL NAL units, which would only be possible after the encoding of the picture has been completed. Sending the SEI message after the VCL NAL units synchronizes better with a regular encoder processing flow. Some of the SEI message types of H.264 | AVC have been transferred to HEVC, some have been discarded either because they are not applicable anymore or not deemed useful in HEVC. It should be noted that the SEI message type identifiers of transferred SEI messages have been aligned between the two specifications in order to avoid misinterpretation. It should be noted however that the syntax and semantics of the HEVC SEI messages are modified and adapted to the new specification compared to the H.264 | AVC SEI messages.

References

1. Wang, Y.-K., et al.: RTP payload format for high efficiency video coding. Internet draft. IETF. https://datatracker.ietf.org/doc/draft-ietf-payload-rtph265/ (2014). Accessed 14 Apr 2014
2. RTP payload format for H.264 video. IETF RFC6184. http://tools.ietf.org/html/rfc6184 (2011). Accessed 14 Apr 2014
3. Schierl, T., et al.: System layer integration of HEVC. IEEE Trans. Circ. Syst. Video Technol. **22**(12), 1871–1884 (2012). doi:10.1109/TCSVT.2012.2223054
4. RTP payload format for scalable video coding. IETF RFC6190. http://tools.ietf.org/html/rfc6190 (2011). Accessed 14 Apr 2014
5. Sjöberg, R., et al.: Overview of HEVC high-level syntax and reference picture management. IEEE Trans. Circ. Syst. Video Technol. **22**(12), 1858–1870 (2012). doi:10.1109/TCSVT.2223052
6. Sullivan, G.J., Ohm, J.-R.: Meeting report of the tenth meeting of the Joint Collaborative Team on Video Coding (JCT-VC), Stockholm, SE, 11–20 July 2012. Doc. JCTVC-J1000. 10th meeting: Joint Collaborative Team on Video Coding (JCT-VC) of ITU-T SG 16 WP 3 and ISO/IEC JTC 1/SC 29/WG 11, Stockholm, SE (2012)
7. High efficiency video coding. ITU-T Rec. H.265 (HEVC). http://www.itu.int/rec/T-REC-H.265/en (2013). Accessed 14 Apr 2014

8. Information technology—High efficiency coding and media delivery in heterogeneous environments—Part 2: High efficiency video coding. ISO/IEC 23008–2:2013 (HEVC). http://www.iso.org/iso/home/store/catalogue_tc/catalogue_detail.htm?csnumber=35424 (2013). Accessed 14 Apr 2014
9. Ribas-Corbera, J., et al.: A generalized hypothetical reference decoder for H.264/AVC. IEEE Trans. Circ. Syst. Video Technol. **13**(7), 674–687 (2003). doi:10.1109/TCSVT.814965
10. Poynton, C.: Digital Video and HD: Algorithms and Interfaces. Morgan Kaufman Publishers, Waltham, MA, USA (2012)
11. Advanced video coding for generic audiovisual services. ITU-T Rec. H.264 (AVC). Feb. 2014. http://www.itu.int/rec/T-REC-H. 264/en. Accessed 14 Apr 2014
12. Information technology—Coding of audio-visual objects—Part 10: Advanced video coding. ISO/IEC 14496–10:2012 (AVC). http://www.iso.org/iso/home/store/catalogue_tc/catalogue_detail.htm?csnumber=61490 (2012). Accessed 14 Apr 2014

Chapter 6
Intra Prediction

This chapter reviews the intra coding tools of HEVC. The large set of available prediction modes is presented including a description of the functionality and examples. Furthermore, the coded representation of the applied intra prediction modes is detailed.

The decision to apply intra or inter prediction is drawn on a CU basis and is indicated by a prediction mode flag for the CU. CUs in intra mode are predicted from reconstructed neighboring samples within the same slice. In I slices, only intra prediction is enabled for the CUs. In P and B slices, CUs may be in both, intra or inter prediction mode. Pictures which serve as random access points only comprise I slices as these pictures must not reference previous pictures. In P or B slices, the selection of the intra mode may be advisable if prediction from previously decoded pictures is not indicated, e.g. for previously not visible content in the current picture.

A set of 35 intra prediction modes is available in HEVC, including a DC, a planar, and 33 angular prediction modes. The numbering of the intra prediction modes is shown in Fig. 6.1. The prediction modes 2–18 are denoted as horizontal prediction modes, as the predominant source of prediction is in horizontal direction. The modes 19–34 are denoted as vertical prediction modes accordingly. The modes are available for prediction block sizes from 4×4 to 32×32 samples. For luma and chroma blocks, the same prediction modes are applied. Some of the smoothing operations applied for luma intra prediction are omitted for chroma blocks as further detailed below. The prediction reference is constructed from the sample row and column adjacent to the predicted block. The reference extends over two times the block size in horizontal and vertical direction using the available sample from previously reconstructed blocks.

The availability of neighboring blocks can be configured to depend on the prediction mode. For best prediction performance, reference to samples from inter or intra predicted neighboring blocks can be applied. Thereby, dependencies of intra prediction on inter predicted blocks are introduced for slices which are not intra-only coded. If the application requires complete independence of intra predicted blocks in P or B slices, constrained intra prediction can be configured by a corresponding flag in the picture parameter set. With this flag activated, only reference to previously intra predicted neighboring blocks is permitted and any dependencies of intra blocks on inter blocks are eliminated.

M. Wien, *High Efficiency Video Coding*, Signals and Communication Technology,
DOI 10.1007/978-3-662-44276-0_6

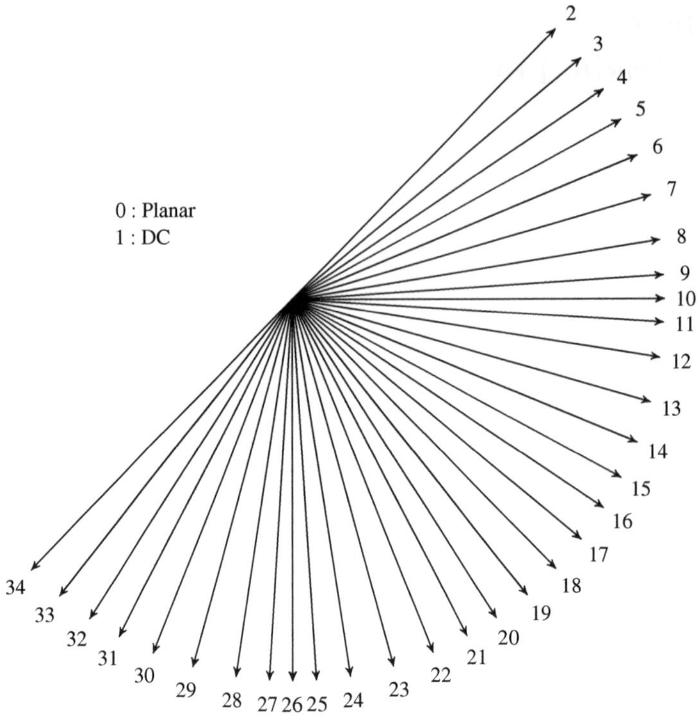

Fig. 6.1 Intra prediction modes and prediction directions for angular intra prediction. The numbering of the angular intra prediction modes is organized from *diagonal-up* to *diagonal-down*

This may be helpful e.g. to prevent error propagation from previously corrupted pictures in lossy environments.[1]

6.1 Prediction Mode and Prediction Block

In the case of CUs in intra mode, the meaning and representation of prediction units are different from the inter mode. While for inter CUs, the parameters for a prediction unit are represented in a dedicated syntax structure, such a structure does not exist for intra CUs and the intra prediction parameters are coded on the CU level.

[1] It should be noted that constrained intra prediction may induce strongly visible isolated artifacts in otherwise inter predicted regions and therefore may have impact on the overall coding performance. Constrained intra prediction may further complicate encoder implementations as the decision on intra or inter prediction for a CU strongly affects the availability of the intra prediction reference for subsequent neighboring CUs and CTUs which may already have been pre-processed for fast mode decision.

The partitioning of a CTU into CUs results in a quadtree structure of square blocks as detailed in Sect. 4.2.4. Intra prediction is applied on square prediction block sizes as well. In order to limit the signaling overhead, the prediction block size for intra prediction generally corresponds to the coding block size and the applicable intra prediction mode is coded once for the CU. Only for the smallest CB size specified in the sequence parameter set, a sub-partitioning into four squared prediction blocks may be applied. In this case, a dedicated intra prediction mode for each of these prediction blocks is provided. As the smallest configurable coding block size is 8×8 this construction allows for the indication of 4×4 intra prediction blocks with dedicated intra prediction modes.

While the applicable intra prediction mode thereby is tied to the CB size, the applicable size of the intra prediction blocks further depends on the partitioning of the transform tree. As detailed in Sect. 4.2.6, the transform tree splits the coding block into a quadtree of square transform blocks. For intra prediction, the signaled intra prediction mode of a CB is successively applied to each of the transform blocks along the leaves of the transform tree. Thereby, the effective prediction block size is equal to the transform block size. The predicted and reconstructed transform blocks are used for prediction of the subsequent transform blocks within the same CB. This concept induces the smallest possible distance between the reference samples and the predicted samples. The signaling cost are kept small for improved prediction performance and compression efficiency.

6.2 Reference Samples for Intra Prediction

In the following, the current block to be intra predicted is referred to as $\mathbf{B}_c$. Let the block be of $N_c \times N_c$ samples, with $N_c \in \{4, 8, 16, 32, 64\}$ specifying the derived prediction block size. The reference sample set is denoted by $\mathbf{p}_{ref}$, the sample coordinates in $\mathbf{B}_c$ and $\mathbf{p}_{ref}$ are given relative to the top-left sample of the block.

The reference samples may be filtered by a smoothing filter depending on N_c and the applicable intra prediction mode. For 32×32 blocks, the smoothing of the reference sample set can be switched to bi-linear interpolation of the reference sample values for a strong smoothing effect. The activation of this strong filter depends on the dynamic range of the reference sample values. The usage of this filters is controlled by a flag in the sequence parameter set.

6.2.1 Reference Construction

The reference sample set for intra prediction extends over $2 \cdot N_c$ samples in horizontal and vertical direction plus the top left neighboring corner sample next to the current block. Figure 6.2 shows a visualization for the reference sample set. In the figure, the reference samples used for the decision on strong reference smoothing are marked.

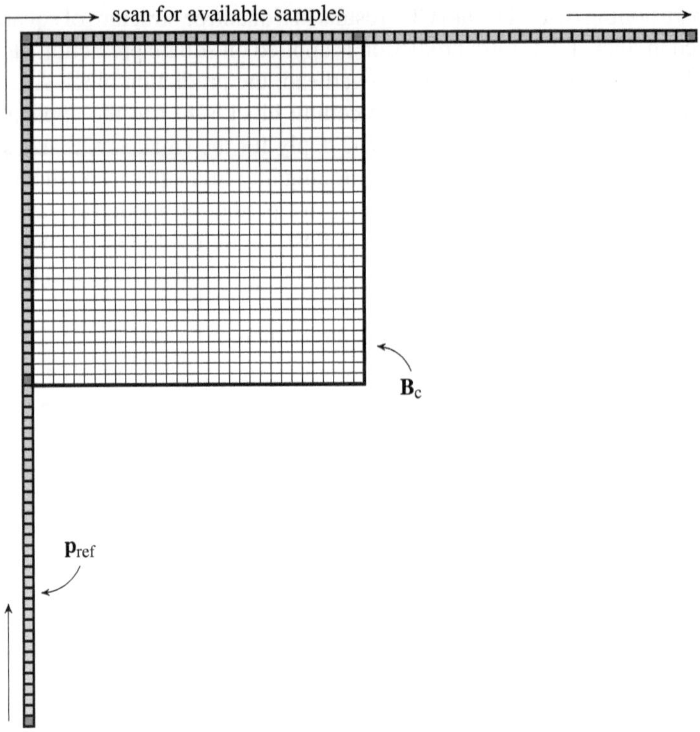

Fig. 6.2 Reference samples for intra prediction (example of an 32 × 32 block). The locations of reference samples checked for activation of 32 × 32 strong reference smoothing are marked *gray*

For construction of the set of reference samples $\mathbf{p}_{\text{ref}}$, the availability of the neighboring samples has to be tested. Samples which would lie outside of the picture and samples which do not belong to the same slice as the current block are marked as not available. If samples belong to blocks which have been inter coded and constrained intra prediction is indicated in the PPS, these samples are marked as not being available as well.

Reference sample locations which have been marked as not available are filled with the nearest available reference sample value along $\mathbf{p}_{\text{ref}}$ as indicated in Fig. 6.2. If none of the reference sample locations is available, the set of prediction samples $\mathbf{p}_{\text{ref}}$ is filled with the mid-value of the specified dynamic range of the encoded video (i.e. the value $2^{B_{\text{d}}-1}$, e.g. 128 for 8-bit video).

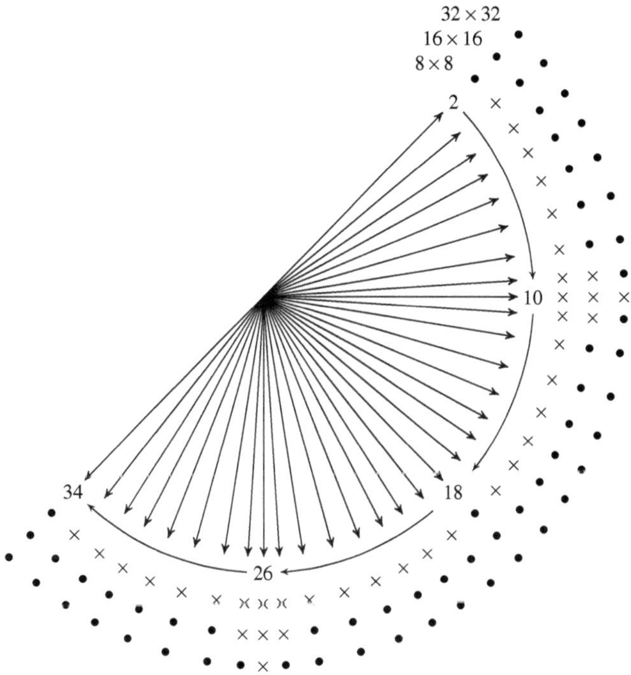

Fig. 6.3 Indication of reference sample smoothing depending on the angular intra prediction mode (filtering = *filled circle*, no filtering = *cross*)

6.2.2 Lowpass Smoothing

After the construction of $\mathbf{p}_{\text{ref}}$, lowpass filtering is conditionally applied to the reference samples of luma blocks. The activation of this smoothing filter depends on the applicable intra prediction mode and the block size N_c. For blocks using either DC, horizontal, or vertical prediction, no filtering is applied. The same holds for blocks of size $N_c = 4$.

For the block sizes $N_c > 4$, lowpass filtering is applied as visualized in Fig. 6.3. The diagonal directions (modes 2, 18, and 34) are always filtered. The number of directions where filtering is applied increases with the block size as can be seen from the figure. Reference samples for planar prediction of blocks with $N_c > 4$ are always filtered.

For the reference samples of 32 × 32 blocks, the conditions for regular or strong filtering have to be checked as detailed in the following subsection. If regular lowpass filtering is indicated a 3-tap lowpass filter $h(n) = \frac{1}{4}[1, 2, 1]$ is applied to the samples in $\mathbf{p}_{\text{ref}}$. The filter is applied along $\mathbf{p}_{\text{ref}}$ following the scan as indicated in Fig. 6.2. The first and last sample locations in the scan are not filtered.

6.2.3 Strong Smoothing for 32 × 32 Luma Reference Samples

As indicated above, strong filtering may be applied to the reference samples of selected 32 × 32 luma blocks as indicated by the corresponding flag in the sequence parameter set. If the flag is active, the local activity along the samples in the reference sample set $\mathbf{p}_{\text{ref}}$ is estimated. If the activity is below a threshold as detailed below, a bi-linear interpolation filter is applied to the reference samples for a strong smoothing effect.

The local activity is estimated by assessing the second order derivative of the sample values over the corner and center samples of both, the horizontal and vertical branches of $\mathbf{p}_{\text{ref}}$:

$$|p_{\text{ref}}(-1,-1) - 2 \cdot p_{\text{ref}}(31,-1) + p_{\text{ref}}(63,-1)| < 2^{B_{\text{d}}-5}, \tag{6.1}$$

$$|p_{\text{ref}}(-1,-1) - 2 \cdot p_{\text{ref}}(-1,31) + p_{\text{ref}}(-1,63)| < 2^{B_{\text{d}}-5}. \tag{6.2}$$

Here, B_{d} denotes the bit depth of the luma component. The respective samples have been marked in Fig. 6.2. Blocks where the conditions (6.1) and (6.2) are fulfilled are considered to have a low texture neighborhood over a constant or linearly evolving luma intensity. For these blocks, the values of the reference samples in $\mathbf{p}_{\text{ref}}$ are replaced by a linear interpolation between the corner samples as

$$p_{\text{ref}}(-1,y) = \frac{63-y}{64} p_{\text{ref}}(-1,-1) + \frac{y+1}{64} p_{\text{ref}}(-1,63), \quad \text{for } y = 0,\ldots,62, \tag{6.3}$$

$$p_{\text{ref}}(x,-1) = \frac{63-x}{64} p_{\text{ref}}(-1,-1) + \frac{x+1}{64} p_{\text{ref}}(63,-1), \quad \text{for } x = 0,\ldots,62. \tag{6.4}$$

By replacing the true sample values with the linear interpolation of the corner samples along each direction, small local intensity variations are removed. Such variations, though being small, may otherwise turn into visible structures if applied for intra prediction over the large sample area of a 32 × 32 block.

6.3 Planar Intra Prediction

With the planar intra prediction mode, gradient structures in a block can be approximated. The prediction for the block is generated by a weighted average of four reference samples, depending on the sample location as shown in Fig. 6.4a. The prediction values are derived as

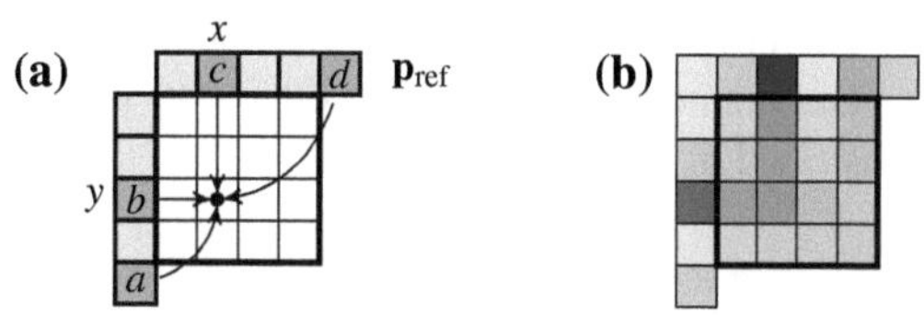

Fig. 6.4 **a** Planar intra prediction for a 4 × 4 block. **b** Example for planar prediction

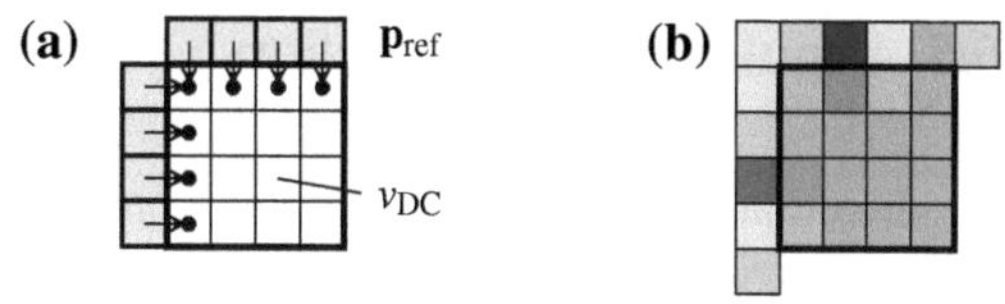

Fig. 6.5 **a** DC intra prediction for a 4 × 4 luma block. **b** Example for DC prediction

$$
\begin{aligned}
\mathbf{B}_c(x,y) =& \frac{y+1}{2N_c} p_{\text{ref}}(-1, N_c) + \frac{N_c - 1 - x}{2N_c} p_{\text{ref}}(-1, y) \\
&+ \frac{N_c - 1 - y}{2N_c} p_{\text{ref}}(x, -1) + \frac{x+1}{2N_c} p_{\text{ref}}(N_c, -1).
\end{aligned} \tag{6.5}
$$

While this prediction mode is well suited to provide smooth prediction signals with gradual changes it can be seen from the example given in Fig. 6.4b that it can also be used to generate horizontal and vertical structures from the sample boundaries.

6.4 DC Intra Prediction

For DC intra prediction, the average of the values of the adjacent horizontal and vertical neighbor samples is determined as

$$
v_{\text{DC}} = \frac{1}{2N_c} \left(\sum_{x=0}^{N_c-1} p_{\text{ref}}(x, -1) + \sum_{y=0}^{N_c-1} p_{\text{ref}}(-1, y) \right). \tag{6.6}
$$

The samples in the prediction block are assigned the average DC value v_{DC}. For luma blocks, the boundary sample (0, 0), $(x, 0)$, $(0, y)$, i.e. the samples at the left and top boundary of the block, are treated separately.

At these locations, the prediction value is a weighted average of v_{DC} and the neighboring value in $\mathbf{p}_{\text{ref}}$:

$$
\mathbf{B}_c(0,0) = \frac{1}{4} \left(p_{\text{ref}}(-1, 0) + p_{\text{ref}}(0, -1) + 2v_{\text{DC}} \right), \tag{6.7}
$$

$$
\mathbf{B}_c(x,0) = \frac{1}{4} \left(p_{\text{ref}}(x, -1) + 3v_{\text{DC}} \right), \tag{6.8}
$$

$$
\mathbf{B}_c(0,y) = \frac{1}{4} \left(p_{\text{ref}}(-1, y) + 3v_{\text{DC}} \right). \tag{6.9}
$$

Thereby, the prediction block boundary is smoothed. For chroma blocks, this special boundary filtering is omitted. DC prediction for a luma blocks is shown and illustrated in Fig. 6.5.

6.5 Angular Intra Prediction

For directional intra prediction, 33 intra prediction modes are available representing different prediction angles from diagonal-up to diagonal-down. For definition of the prediction angles, an offset value p_{ang} on a 32-sample grid is defined. The association of p_{ang} to the corresponding intra prediction mode is visualized in Fig. 6.6 for the vertical prediction modes. For the horizontal prediction modes the scheme is flipped to vertical direction and the p_{ang} values are assigned accordingly. As stated above, all angular prediction modes are available for all applicable intra prediction block sizes. They all use the same 32-sample grid for the definition of the prediction angles.

The distribution of the p_{ang} values over the 32-sample grid in Fig. 6.6 reveals an increased resolution of the prediction angles around the vertical direction and a coarser resolution of the prediction angles towards the diagonal directions. The same applies to the horizontal directions. This design stems from the observation that in lots of video content, approximately horizontal and vertical structures play an important role compared to diagonal structures [3].

6.5.1 *One-Dimensional Prediction Reference*

While for the horizontal and vertical prediction directions, the selection of samples to be used for prediction is straightforward, this task requires more effort in case of angular prediction.

For modes 11–25, when predicting the current block $\mathbf{B}_c$ from the set of prediction samples $\mathbf{p}_{ref}$ in an angular direction, samples of both, the vertical and the horizontal part of $\mathbf{p}_{ref}$ can be involved. Since the determination of the location of the respective

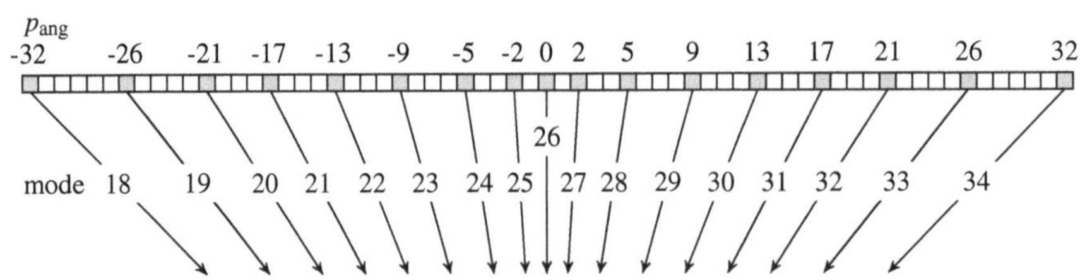

Fig. 6.6 Angular intra prediction directions and modes and the associated value of p_{ang} for vertical prediction directions

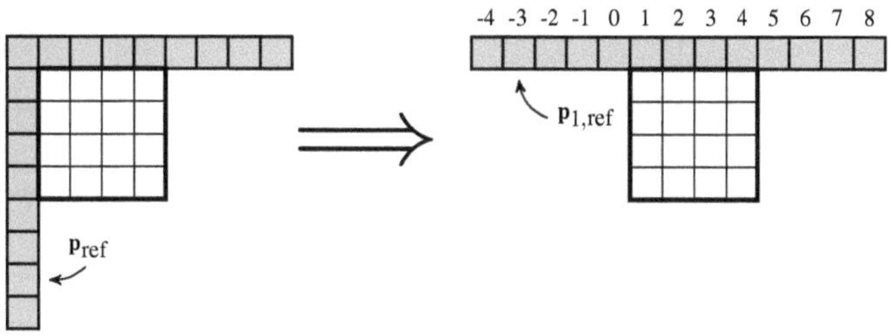

Fig. 6.7 Transformation of $\mathbf{p}_{\text{ref}}$ to $\mathbf{p}_{1,\text{ref}}$ for a 4×4 block

samples on either of the branches of $\mathbf{p}_{\text{ref}}$ requires some computational effort, a unified one-dimensional prediction reference has been designed for HEVC intra prediction [4]. The scheme is visualized in Fig. 6.7.

Before performing the actual prediction operation, the set of reference samples $\mathbf{p}_{\text{ref}}$ is mapped to a 1-dimensional vector $\mathbf{p}_{1,\text{ref}}$. The projection which is used for the mapping depends on the direction indicated by the intra prediction angle of the respective intra prediction mode. Only reference samples from the part of $\mathbf{p}_{\text{ref}}$ which is to be used for prediction are mapped to $\mathbf{p}_{1,\text{ref}}$. The actual mapping of the reference samples to $\mathbf{p}_{1,\text{ref}}$ for each angular prediction mode is depicted in Figs. 6.8 and 6.9 for horizontal and vertical angular prediction directions, respectively.

The reference samples set $\mathbf{p}_{1,\text{ref}}$ is constructed once for the predicted block. The prediction is then derived from two neighboring reference samples in the set as detailed below. As can be seen from Figs. 6.8 and 6.9 the 1-dimensional reference sample set is not completely filled for all intra prediction modes. Only the locations which are in the projection range for the corresponding intra prediction direction are included in the set.

6.5.2 Interpolated Prediction

The prediction for both, horizontal and vertical prediction modes is performed in the same manner with only swapping the x and y coordinates of the block.

The prediction from $\mathbf{p}_{1,\text{ref}}$ is performed in 1/32-pel accuracy. Depending on the value of the angle parameter p_{ang}, a sample offset i_{idx} in $\mathbf{p}_{1,\text{ref}}$ and a weighting factor i_{fact} for a sample at position (x, y) are determined. Here, the derivation for the vertical modes is provided. The derivation for the horizontal modes follows accordingly, swapping x and y.

$$i_{\text{idx}} = (y+1) \cdot \frac{p_{\text{ang}}}{32}, \qquad i_{\text{fact}} = \left[(y+1) \cdot p_{\text{ang}}\right] \bmod 32. \tag{6.10}$$

If i_{fact} is not equal to 0, i.e. the prediction does not fall exactly on a full sample location in $\mathbf{p}_{1,\text{ref}}$, a linear weighting between the two neighboring sample locations

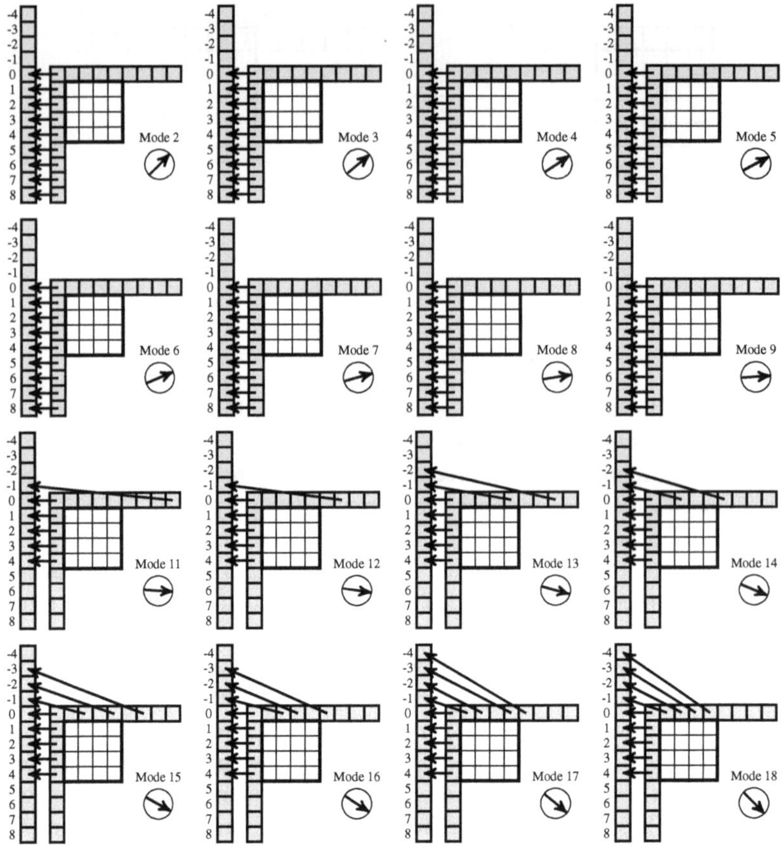

Fig. 6.8 Construction of $\mathbf{p}_{1,\text{ref}}$ for *horizontal angular* prediction

in $\mathbf{p}_{1,\text{ref}}$ is performed as

$$\mathbf{B}_\text{c}(x, y) = \frac{32 - i_\text{fact}}{32} \cdot p_{1,\text{ref}}(x + i_\text{idx} + 1) + \frac{i_\text{fact}}{32} \cdot p_{1,\text{ref}}(x + i_\text{idx} + 2), \quad (6.11)$$

with $0 \leq x, y < N_\text{c}$.

It should be noted that the values of i_idx and i_fact only depend on y and therefore only need to be calculated once per row (for vertical prediction modes).

6.5.3 *Horizontal and Vertical Intra Prediction*

As mentioned above, for blocks predicted in horizontal or vertical mode, no reference sample filtering is applied. The sample values of the same row (or column) in $\mathbf{p}_\text{ref}$

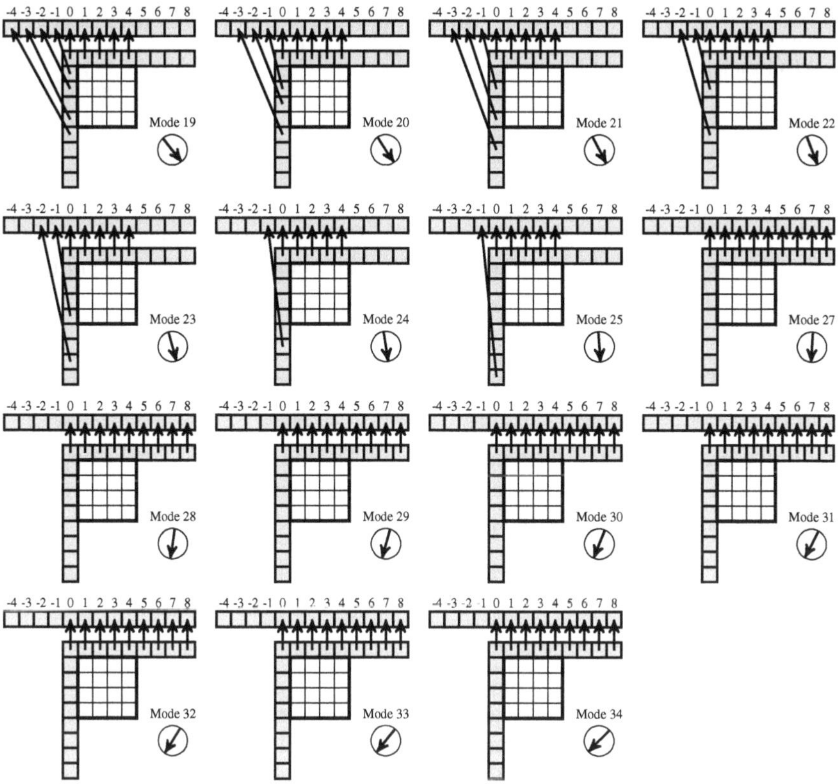

Fig. 6.9 Construction of $\mathbf{p}_{1,\text{ref}}$ for *vertical angular* prediction

(or $\mathbf{p}_{1,\text{ref}}$ correspondingly) are copied to the predicted block. As only samples from one part of the reference sample set are used for prediction, no specific construction of $\mathbf{p}_{1,\text{ref}}$ is required.

At the boundary parallel to the prediction direction ($y = 0$ for horizontal, $x = 0$ for vertical), a modified prediction is applied to reduce the visibility of the prediction block edge. For the boundary samples, a local update is added to the prediction value at each sample location. For the update, the difference between the value of the neighboring prediction block sample and the corner sample $p(-1, -1)$ is added to the prediction signal. For vertical prediction the sample values are derived as

$$\mathbf{B}_{\text{c}}(0, y) = p_{\text{ref}}(0, -1) + \frac{1}{2}\left(p_{\text{ref}}(-1, y) - p_{\text{ref}}(-1, -1)\right). \tag{6.12}$$

For horizontal prediction the method is applied accordingly.

This operation adapts the boundary samples of the predicted block to the values of the neighboring block, and thereby reduces potential blocking artifacts at such boundaries.

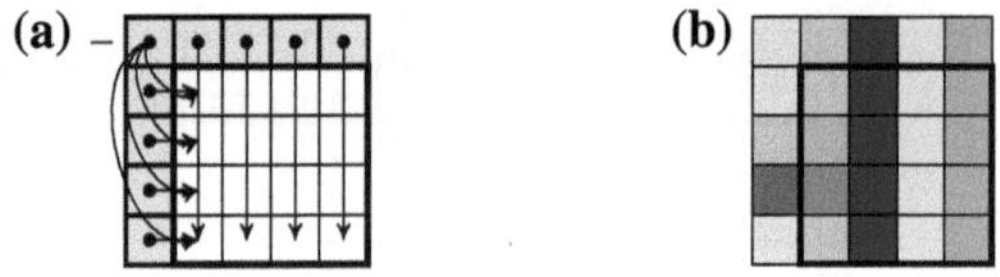

Fig. 6.10 **a** Vertical intra prediction for a 4 × 4 luma block. **b** Example for vertical prediction

The prediction scheme is visualized in Fig. 6.10 for the example of vertical prediction (mode 26). For chroma blocks, this special boundary filtering is omitted.

6.6 Signaling and Predictive Coding of Intra Prediction Modes

Due to the high number of intra prediction modes, the signaling for the applicable mode must be very efficient in order to preserve rate-distortion gain over the induced signaling overhead for coding the respective mode.

6.6.1 Luma Intra Prediction Mode

The applicable luma intra prediction mode for the current block $\mathbf{B}_c$ is coded using two different options. Either the mode is signaled by an index into a constructed list of three most probably modes (MPM) or by a fixed-length binarization of the mode index, in the case that the applicable mode was not included in the MPM list. The syntax elements to indicate the applicable intra prediction mode into the bitstream are entirely encoded using CABAC in bypass mode (see Sect. 10.2.2.6 for CABAC details). Therefore, no context adaptation is required and the bins can be grouped for efficient parallel bypass coding [5].

The three most probable modes are derived from the intra prediction modes of the top and left neighboring blocks, referred to as $\mathbf{B}_a$ and $\mathbf{B}_b$ as illustrated in Fig. 6.11. The neighboring blocks are specified by the marked sample locations (x_a, y_a) and (x_b, y_b) next to the top left corner sample at location (x_c, y_c), respectively. The blocks are marked as available if they are located in the same slice and if the locations (x_a, y_a) and (x_b, y_b) do not lie outside of the picture boundaries.

Depending on the availability and the prediction modes of the neighboring blocks, a candidate intra prediction mode list with three entries is derived. If one of the neighboring blocks is not available or if the block mode is not intra, the candidate intra prediction mode for this block is set to intra DC. For intra blocks, the candidate prediction mode m_x, with $x \in \{a, b\}$, is set to the respective intra prediction mode of the block.

If the candidate intra prediction modes of the two blocks are identical and equal to either DC or planar prediction, the MPM candidate list is set to

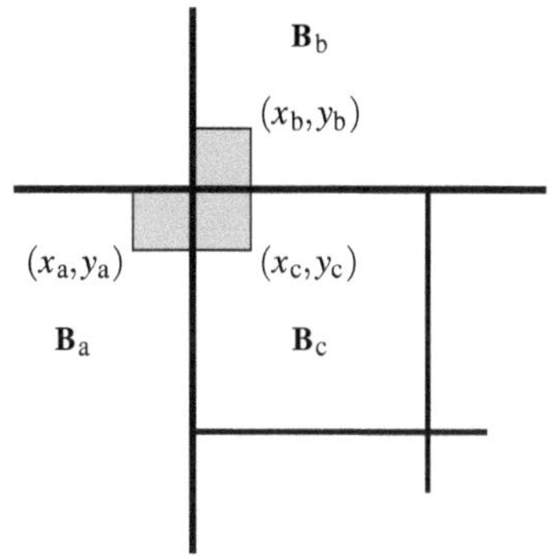

Fig. 6.11 Example for a neighborhood for intra prediction mode derivation for block $\mathbf{B}_c$

$$\begin{aligned} \text{candModeList}[0] &= 1 \text{ (planar prediction)} \\ \text{candModeList}[1] &= 0 \text{ (DC prediction)} \\ \text{candModeList}[2] &= 26 \text{ (vertical prediction)}. \end{aligned} \tag{6.13}$$

If the candidate intra prediction modes are identical and not equal to DC or planar prediction, the candidate list is derived as

$$\begin{aligned} \text{candModeList}[0] &= m_a \\ \text{candModeList}[1] &= 2 + [m_a + 29] \bmod 32 \\ \text{candModeList}[2] &= 2 + [m_a - 1] \bmod 32. \end{aligned} \tag{6.14}$$

For the angular prediction modes between 3 and 32, this implies a resulting MPM candidate list of the neighbor intra prediction mode m_a and the two adjacent prediction modes $m_a \pm 1$. For the diagonal-up prediction mode 2, modes 3, 33 are included in the list. For the diagonal-down modes 33 and 34, the modes 32, 2 and the modes 33, 3 are included, respectively. In these corner cases, the additional candidate modes are thereby selected using the corresponding direction in the opposite orientation compared to candModeList[0]. Thereby, the prediction for these candidates is performed from the other branch of the reference sample set. This change of the prediction orientation may thereby provide an additional variation of the available predictors in the candidate list for identical modes $m_a = m_b$.

If the candidate modes m_a and m_b are not equal, the two modes are assigned to the MPM candidate list as

$$\begin{aligned} \text{candModeList}[0] &= m_a \\ \text{candModeList}[1] &= m_b. \end{aligned} \tag{6.15}$$

The third entry in the candidate list is derived by the following ordered steps:

1. If planar prediction is not in the list, candModeList[2] is set to 1 (planar prediction),

Table 6.1 Derivation table for the chroma intra prediction mode m_C

m_{cp}	m'_C	if $m'_C == m_Y$
0	0 (planar)	34 (diagonal-down)
1	26 (vertical)	34
2	10 (horizontal)	34
3	1 (DC)	34
4	m_Y	–

2. Otherwise, if DC prediction is not in the list, candModeList[2] is set to 0 (DC prediction),
3. Otherwise, candModeList[2] is set to 26 (vertical prediction).

If the applicable luma intra prediction mode m_Y for the current block is in the MPM candidate list, it is coded as an index into this list. Otherwise, the mode is coded as a number m_{rem} with a 5 bitbinarization. Five bits are sufficient as the three modes of the MPM list do not need to be represented, leaving 32 possible alternative modes. In order to derive the applicable mode from m_{rem}, the modes of the MPM candidate list are sorted in ascending order and compared to m_{rem}. For each entry in the MPM list which is smaller or equal to the coded number, m_{rem} is successively increased by one leading to the target luma intra prediction mode m_Y.

6.6.2 Derivation of the Chroma Intra Prediction Mode

The chroma intra prediction mode m_C is applied to both chroma components identically. Since the selected chroma mode often corresponds to the luma intra prediction mode [3], the coding of the chroma intra prediction relies on the luma mode. The chroma mode is coded by a syntax element m_{cp} with five different values. The value of the applicable chroma intra prediction mode m_C is derived from m_{cp} and m_Y as shown in Table 6.1.

The table is constructed to avoid the occurrence of redundancies. From m_{cp}, a temporary chroma intra prediction mode m'_C is derived. If m'_C is equal to the corresponding luma intra prediction mode m_Y, the alternative mode 34 is assigned for the chroma intra prediction mode m_C. Otherwise, m_C is set to m'_C.

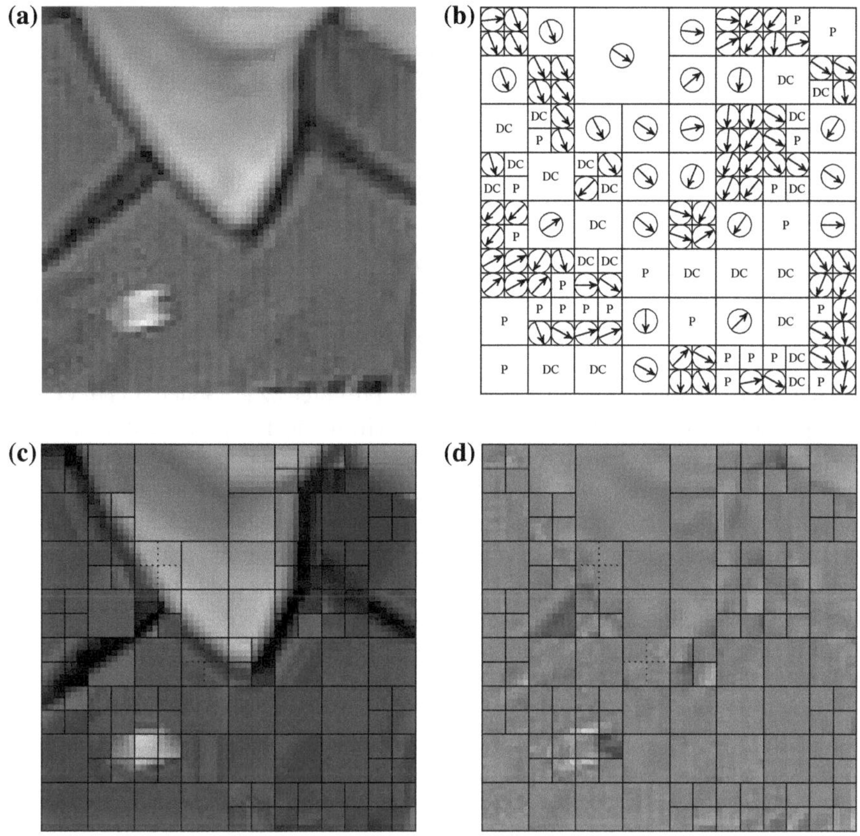

Fig. 6.12 Intra coding example of one 64 × 64 CTB with all intra coding. **a** Original. **b** Applicable intra prediction modes on the CU basis, with angular prediction (*arrows*), planar prediction (*P*), and DC prediction (*DC*). **c** Prediction signal. **d** Residual with the corresponding CB (*solid lines*) and TB partitioning (*dashed lines*)

6.7 Intra Coding Example

In Fig. 6.12, the application of intra prediction to an example region of the 1080p test sequence BasketballDrive is shown.[2] The coded intra prediction modes are coded on a CU basis as shown in Fig. 6.12b. In Fig. 6.12c the resulting prediction signal is shown with an overlay of the CB and TB partitioning of the four depicted CTBs. The partitioning reveals a TB sub-partitioning of some CBs. For these CBs, the intra prediction of the mode signaled at the CU level is applied on a TB basis.

For some of the prediction blocks, the directional structure of the prediction signal resulting from the corresponding intra prediction mode is clearly visible. In the

[2] Frame 308 of the 1080p sequence BasketballDrive, HM11.0, all intra configuration of the JCT-VCcommon testing conditions [6], QP = 27.

smooth area of the cheek, planar prediction is piece-wise applied on a TB basis. Fig. 6.12d shows the corresponding reconstructed residual signal (with an offset of 2^{B_d-1} for visualization) which is added to the intra prediction signal.

6.8 Comparison to H.264 | AVC

In H.264 | AVC, eight directional intra prediction modes and DC intra prediction are specified for 4×4 and 8×8 blocks. For 16×16 intra predicted blocks, only horizontal, vertical, DC, and plane prediction are available. The directions included horizontal, vertical, left and right diagonal, and half diagonal directions [7, 8]. For 8×8 blocks, pre-filtering similar to the HEVC pre-filtering is applied. In HEVC, the intra prediction scheme has been generalized. The number of possible directions for intra prediction has increased to 33. The directional prediction is further applicable to all possible intra prediction block sizes of 4×4 to 64×64. Also, the prediction modes are harmonized for the luma and chroma components. The harmonized construction of a uniform prediction reference helps for a lean implementation of the intra prediction process. The application of the planar intra prediction mode is also available for all prediction block sizes.

In order to reduce the blocking artifacts at the boundaries of horizontally, vertically, or DC predicted blocks, HEVC introduces an additional processing of the samples at the boundary parallel to the prediction direction and both, the top and left boundaries for DC prediction. Thereby, the blocking artifacts are already mitigated by the intra prediction process. The planar prediction process, which was only available for the 16×16 intra blocks in H.264 | AVC, has been simplified and improved to provide a better adaptation to the structures at the block boundaries [3].

For signaling of the applicable intra prediction mode, H.264 | AVC employs the lowest of the available top and left neighbor intra prediction modes as predictor. Otherwise, the number the remaining prediction mode is coded explicitly. Due to the high number of intra prediction modes in HEVC, this scheme is extended to a set of three most probable modes. In the CABAC entropy coding design, special care has been taken in HEVC to reduce the number of employed context models during the parsing process of the intra prediction syntax elements. Thereby, the complexity of the entropy coding stage can be reduced. For a detailed analysis of the HEVC entropy coding design, the reader is referred to [5]; see also Chap. 10.

References

1. High efficiency video coding. ITU-T Rec. H.265 (HEVC). http://www.itu.int/rec/T-REC-H.265/en (2013). Accessed 14 Apr 2014
2. Information technology—High efficiency coding and media delivery in heterogeneous environments—Part 2: High efficiency video coding. ISO/IEC 23008–2:2013 (HEVC). http://www.iso.org/iso/home/store/catalogue_tc/catalogue_detail.htm?csnumber=35424 (2013). Accessed 14 Apr 2014
3. Lainema, J., et al.: Intra coding of the HEVC standard. IEEE Trans. Circ. Syst. Video Technol. **22**(12), 1792–1801 (2012). doi:10.1109/TCSVT.2012.2221525
4. Bossen, F., et al.: Simplified angular intra prediction. 2nd Meeting: Joint Collaborative Team on Video Coding (JCTVC) of ITU-T SG 16 WP 3 and ISO/IEC JTC 1/SC 29/WG, Doc. JCTVC-B093, Geneva, CH (2010)
5. Sze, V., Budagavi, M.: High throughput CABAC entropy coding in HEVC. IEEE Trans. Circ. Syst. Video Technol. **22**(12), 1778–1791 (2012). doi:10.1109/TCSVT.2221526
6. Bossen, F.: Common test conditions and software reference configurations. 11th Meeting: Joint Collaborative Team on Video Coding (JCT-VC) of ITU-T SG 16 WP 3 and ISO/IEC JTC 1/SC 29/WG, Doc. JCTVC-K1100, Shanghai, CN (2012)
7. Advanced video coding for generic audiovisual services. ITU-T Rec. H.264 (AVC). http://www.itu.int/rec/T-REC-H.264/en (2014). Accessed 14 Apr 2014
8. Information technology—Coding of audio-visual objects—Part 10: Advanced video coding. ISO/IEC 14496–10:2012 (AVC). http://www.iso.org/iso/home/store/catalogue_tc/catalogue_detail.htm?csnumber=61490 (2012). Accessed 14 Apr 2014

Chapter 7
Inter Prediction

This chapter presents the inter prediction tools specified in HEVC. Inter prediction uses available previously reconstructed pictures as reference for motion compensation. Motion compensated prediction generally is the key tool for efficient representation of video content. Changes between successive pictures in the video sequence are approximated by a block-wise displacement between areas in the current picture and previously encoded pictures that are available in the decoded picture buffer at both the encoder and decoder side. Since motion compensated prediction is performed *between* pictures, this prediction mode is called *Inter Prediction* mode.

The specified precision of the motion information for motion compensation is one quarter-sample for the luma component and one eighth-sample for the chroma components. The focus in this presentation is laid on the design of the sub-sample interpolation filters and on the representation of the motion information. The implemented advanced motion vector derivation and motion vector prediction methods play a key role in the compression efficiency of HEVC. The sub-sample interpolation filters provide an improved prediction signal at sub-sample locations compared to the interpolation filters of previous video coding specifications.

7.1 Motion Compensated Prediction

In HEVC, inter prediction is performed on the prediction block (PB) level [1, 2]. The corresponding prediction unit (PU) contains the information how inter prediction is performed. Inter prediction is called motion compensated prediction since shifted areas of the reference pictures are used for prediction of the current PB. The resulting displacement between the area in the reference picture and the current PB is interpreted as the motion of the area between the reference picture and the current picture. Since the encoded displacement or motion vectors are usually determined by application of some rate-distortion criterion, these vectors do not necessarily represent the 'true' motion of the area but the most efficient representation in the sense of the applied cost criterion. However, vectors determined on this basis might be

M. Wien, *High Efficiency Video Coding*, Signals and Communication Technology,
DOI 10.1007/978-3-662-44276-0_7

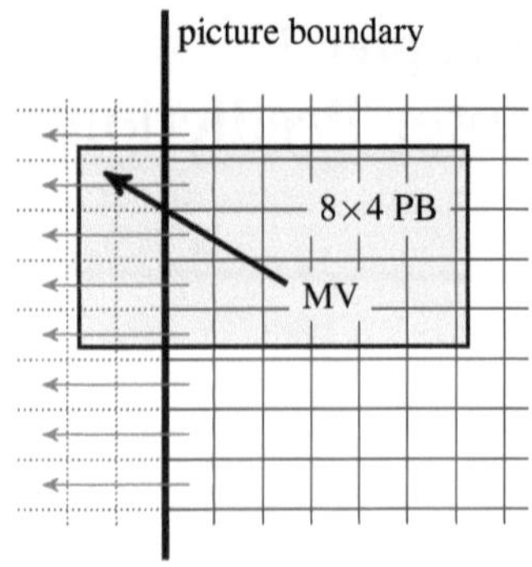

Fig. 7.1 Illustration of constant boundary extension for motion compensation over a picture boundary

regarded as a good approximation of the true motion. An example for an HEVC motion vector field is provided in Sect. 7.4 at the end of this chapter.

In HEVC, one or a combination of two predictors may be used for motion compensated prediction, similar to previous standards. The application of two or more predictors for motion compensation for multi-hypothesis prediction has been studied in the context of the development of H.264 | AVC [3]. In the final specification design, bi-prediction was selected as the best trade-off between improved compression performance and required implementation complexity. The amount of required memory access to fetch the prediction blocks from multiple reference pictures is considered to be a major complexity concern in this context, specifically for hardware implementations.

As stated above, motion vectors are applied in quarter-sample accuracy for luma inter prediction. The same motion vectors are applied for the chroma components. Since for YCbCr 4:2:0 video, the chroma components are down-sampled by a factor of two in both directions, the chroma motion vectors are applied with eighth-sample accuracy. If a motion vector (MV) points outside of the picture area, constant boundary extension is applied. For this purpose, the value of the last sample in the corresponding row or column is assigned to the respective locations outside of the picture as shown in Fig. 7.1.[1]

7.1.1 *Uni-prediction and Bi-prediction*

Motion compensated prediction can be performed using one or two reference pictures as the prediction source. The number of available prediction sources depends on the slice type of the slice the current PU belongs to. In P slices, only a single prediction reference can be used for inter prediction, enabling *uni-prediction* for the PB.

[1] The motion vectors are coded as the displacement of the reference block relative to the target location as shown in Fig. 7.1. In this book, the illustration of displacements and prediction directions is otherwise pointing from the prediction source into the target direction, as shown e.g. in Fig. 7.2.

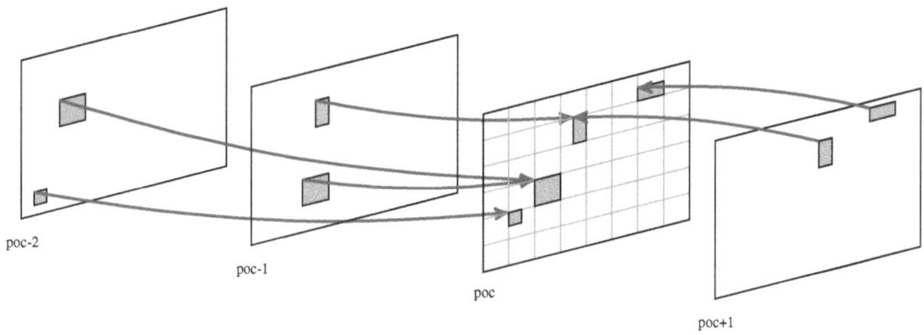

Fig. 7.2 Illustration of uni- and bi-prediction for the current picture at time instance poc and adjacent reference pictures

Here, the encoder can select one of the available reference pictures of reference picture list List 0 as described in Sect. 4.3.2. In B slices, one or two prediction sources may be applied and two reference picture lists are available. A single prediction or a combination of a prediction from each of the two lists can be used. Here, uni-prediction can be performed either from reference picture list 0 or reference picture list 1. With *bi-prediction*, one reference picture from each of the lists is employed. For both uni- and bi-prediction, the construction of the prediction block can be further controlled by applying a configurable weight for each applicable prediction source and an additional offset value. The concept of this *weighted prediction* is further detailed below.

A selection of available variations of uni and bi prediction are illustrated in Fig. 7.2. The size and structure of the reference picture set and the construction of the applicable reference picture list is up to the encoder, with restrictions resulting from the target profile and level for the encoded stream. A picture may be listed in both reference picture lists, which allows for the application of two predictors from the same reference picture for a PB. This could be used to virtually increase the motion vector accuracy if deemed beneficial. For example, the weighted prediction of two motion vectors from the same reference picture which are only a quarter-sample location apart can be interpreted as the application of a joint motion vector at eighth-sample accuracy.

7.1.2 Coding Block Partitioning into Prediction Blocks

The decision for inter or intra coding is made on a CU basis. For inter CUs, the corresponding coding blocks are further partitioned into one or more prediction blocks. The available partitioning of a CB into PBs has been discussed in Sect. 4.2.5. Here, the available inter partitioning types are shown in Fig. 7.3. For each PB in a B slice, the application of uni- or bi-prediction can be selected according to the encoder decision.

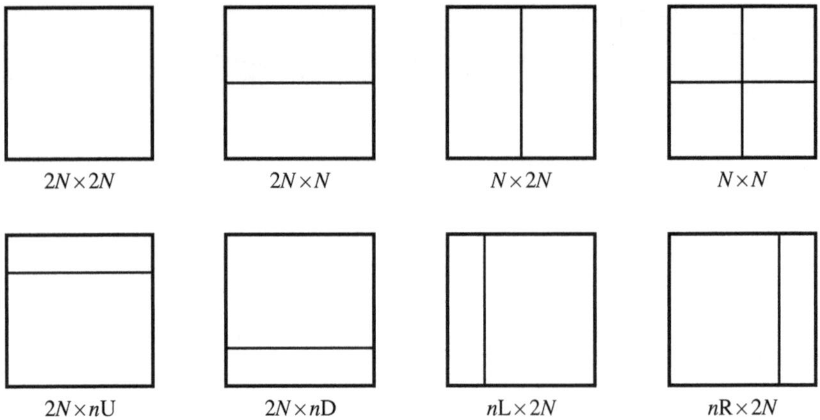

Fig. 7.3 Symmetric and asymmetric partitioning of a $2N \times 2N$ coding block

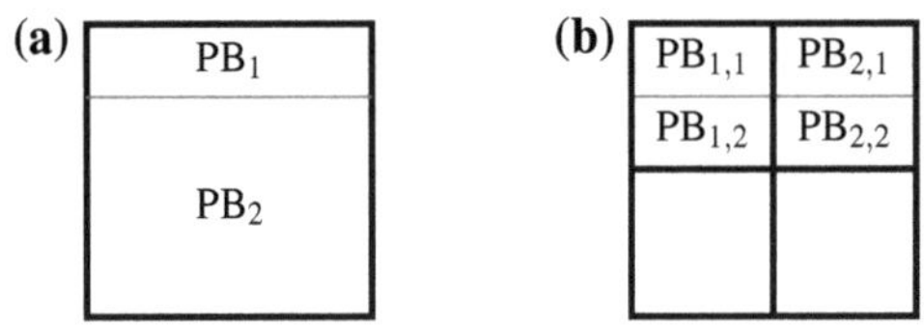

Fig. 7.4 **a** Asymmetric motion partitioning of a CB with two PBs. **b** Corresponding symmetric partitioning including four CBs and more PBs

Besides the regular (symmetric) partitioning of a CB into PBs with the partitioning types already known from H.264 | AVC, HEVC allows for an additional set of four asymmetric partitionings as shown in the lower row of Fig. 7.3. These partitioning types are referred to as *asymmetric motion partitioning* (AMP). They can e.g. be used for an efficient partitioning representation in situations with a partial overlap of a foreground object and the background within the area of one coding block. An equivalent representation using symmetric partitioning would require more signaling overhead for the respective CB and PB structure. An illustration of the issue is provided in Fig. 7.4. The usage of AMP can be configured by a corresponding enabling flag in the sequence parameter set.

The smallest block sizes for motion compensated prediction are 4×8 and 8×4. Thereby, inter prediction with 4×4 blocks is precluded. This limit has been introduced for decoder complexity reasons to reduce the maximum possible number of different motion compensation operations.

7.1.3 Weighted Prediction

Weighted prediction can be applied as a tool for prediction in P or B slices. For each reference picture, a weighting factor and an offset value can be assigned which are applied if motion compensated prediction from the reference picture is performed.

In P slices, weighted prediction can be used to fade in or fade out the intensity of the color components. In B slices, weighted prediction can be additionally used to blend over between the prediction from two reference pictures.

The application of weighted prediction is controlled by flags separately for P and B slices in the PPS. If either of the two is activated, the weighting prediction parameters are encoded in the slice segment header. A separate offset value and weight value for luma and for chroma for each picture in the reference picture list(s) is encoded. For P slices, no weighted prediction is applied if the corresponding PPS flag is off. For B slices, the flag switches between prediction with implicit and explicit weights. If the flag is off, implicit weighting with an equal weight of the two prediction blocks from the different reference pictures is applied. Otherwise, the coded explicit weights are used.

The motion compensation process with sub-sample interpolation is specified in a fixed-point implementation with a bit depth of 14 bit as previously discussed. The scaling back to the original bit depth B_{d} is part of the weighted prediction process. Let $p_{\mathrm{L}0}$ and $p_{\mathrm{L}1}$ be samples of a prediction block from the reference pictures used for the current block in reference picture lists List 0 and List 1, respectively. The sample values in the prediction blocks have been derived by the sub-sample interpolation process described in Sect. 7.3. The scaling factor for reaching the original bit depth is $s_{\mathrm{bd}} = 2^{14-B_{\mathrm{d}}}$. Accordingly in the case of implicit weighting, the resulting prediction sample p_{p} for uni- or bi-prediction is derived by

$$p_{\mathrm{p}} = \left\lfloor \frac{p_{\mathrm{L}i}}{s_{\mathrm{bd}}} + \frac{1}{2} \right\rfloor, \text{ for } i = 0, 1 \tag{7.1}$$

$$p_{\mathrm{p}} = \left\lfloor \frac{p_{\mathrm{L}0} + p_{\mathrm{L}1}}{2 \cdot s_{\mathrm{bd}}} + \frac{1}{2} \right\rfloor. \tag{7.2}$$

If explicit weighting is used, a weighting factor $w_{\mathrm{L}i}$ and an offset value $o_{\mathrm{L}i}$, with $i = 0, 1$ for each list, are derived from the parameters coded in the slice segment header according to the reference picture indices of the used reference pictures, each. The coded parameters include a weight scaling factor d_{w} which is applicable to all weighting factors in the current slice segment, and dedicated weight delta Δ_{w} and the offset value o_{L}. The weighting factor w_{L} is derived as

$$w_{\mathrm{L}} = 2^{d_{\mathrm{w}}} + \Delta_{\mathrm{w}},$$

with $-128 \leq \Delta_{\mathrm{w}} \leq 127$. The value range for the offsets is the same. The scaling factor for the weighting operation is set to $s_{\mathrm{bd\,e}} = 2^{14-B_{\mathrm{d}}+d_{\mathrm{w}}}$. This factor may turn to be equal to 1 for a bit depth of $B_{\mathrm{d}} = 14$. The resulting prediction sample p_{p} for uni- or bi-prediction is then derived by

$$p_{\mathrm{P}} = \left\lfloor \frac{w_{\mathrm{L}i} \cdot p_{\mathrm{L}i}}{s_{\mathrm{bde}}} + \frac{1}{2} \right\rfloor + o_{\mathrm{L}i}, \text{ for } i = 0, 1 \tag{7.3}$$

$$p_{\mathrm{P}} = \left\lfloor \frac{w_{\mathrm{L0}} \cdot p_{\mathrm{L0}} + w_{\mathrm{L1}} \cdot p_{\mathrm{L1}} + s_{\mathrm{bde}} \cdot (o_{\mathrm{L0}} + o_{\mathrm{L1}})}{2 \cdot s_{\mathrm{bde}}} + \frac{1}{2} \right\rfloor . \tag{7.4}$$

Note that for uni-prediction the offset value is left outside of the rounding operation while for bi-prediction, the two offset factors are included in the rounding operation.

7.2 Motion Vector Representation

The applicable motion vectors for motion compensation can be derived in two ways. The motion information (i.e., motion vector and reference index) can be encoded, where the applicable motion vector is constructed from a motion vector predictor and a motion vector difference. The applicable motion vector predictor is selected from two candidates derived from the spatial and temporal neighborhood of the current block. This predictor selection method is called *advanced motion vector prediction*. Furthermore, the motion information can be derived by selection from a configurable set of candidates, without encoding a motion vector difference. The derivation of motion vectors from a candidate set is called *merge mode*. For both, the derived representation and the predictive coding of motion vectors, a large set of potential predictors can be accounted. The two methods provide very efficient means for motion representation and thereby substantially contribute to the performance increase of HEVC.

7.2.1 Motion Data Storage Reduction

For both, merge mode and advanced motion vector prediction, motion vectors from the collocated region in a reference picture may be employed. The motion information of these reference pictures therefore must be stored and made available for prediction purposes. In order to limit the storage requirements for this type of information, the resolution of the motion information in the reference pictures is reduced to a 16×16 block grid [4]. This method of temporal motion vector memory compression is illustrated in Fig. 7.5. For prediction blocks of sizes smaller than 16×16, only the motion information from the top left block on the 16×16 grid is stored and provided for later prediction purposes. Blocks that are intra coded are marked to be unavailable for motion vector reference. As detailed further in Sect. 7.2.2, the collocated block in a reference picture may be determined by the sample to the bottom right of the current block. In the case that there is a slice boundary between the location of the

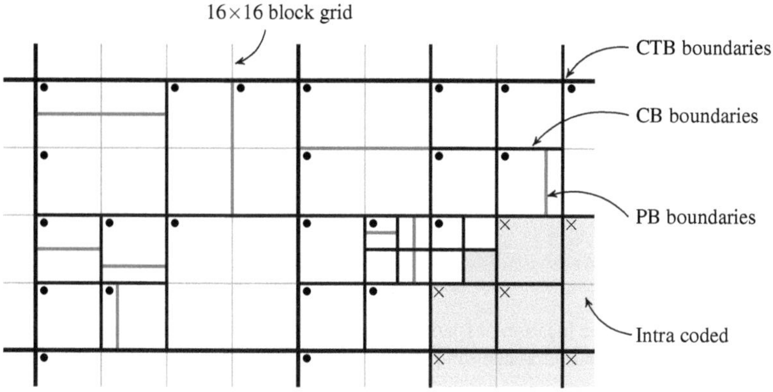

Fig. 7.5 Illustration of reference picture motion vector memory compression to a 16 × 16 block grid. The motion vectors for the 16 × 16 grid are taken from the prediction blocks marked with a "•", blocks marked with a "×" are unavailable

current block in the reference picture and the derived collocated block, such blocks are also marked unavailable for temporal motion vector prediction.

7.2.2 Merging Motion Vectors

The merging of motion information, i.e. sharing of identical motion vectors across a (potentially large) set of connected PUs, provides the ability to encode homogeneous motion for arbitrarily shaped regions of a picture in a very efficient manner.

In merge mode, the applicable motion information for the current PU is derived from a set of candidates of configurable size. The candidate motion information to be used is indicated in the PU syntax by a merge index into the candidate list. The maximum number of merge candidates to be used is specified in the slice header. It can be in the range 0, ..., 5. Based on the availability of neighbors, a list of merging candidates is constructed, including multiple spatial and potentially a temporal candidate.

Potential candidate PUs which are not available are ignored (e.g., if outside of the current slice or intra coded). The candidate derivation process ensures that the set of candidates is always filled up to the number of candidates indicated in the slice segment header. Thereby, the length of the candidate list does not depend on the result of the availability check which improves error robustness of the design (e.g., if the reference picture with the collocated block was lost and the availability check would therefore not come to the correct result) [5].

In order to enable control of encoder and decoder complexity, and also to enable parallel processing of PUs in merge mode, the granularity of the merge information can be controlled by a parameter in the PPS [6]. The parameter specifies the block size grid below which all PUs share the same merge candidate list.

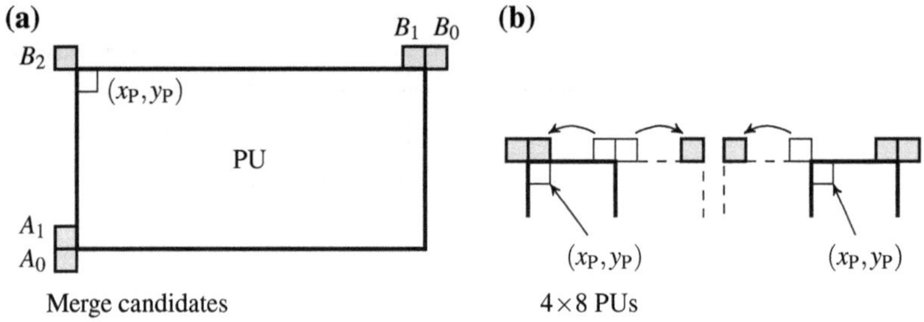

Fig. 7.6 Illustration of the location of spatial merge candidates. **a** Reference samples $A_0 - B_2$ for neighboring PUs. **b** Shifted reference sample locations for PUs of width 4 if the PU is at the top boundary of an CTU

If the number of entries in the list allows for further additions, combined bi-predictive merge candidates are added for B slices. If the size of the merge candidate list is still below the defined maximum, the merge candidate list is filled with zero motion vector merge candidates with increasing reference index i_{ref} for each empty position. If the number of available reference pictures is too small to completely fill the list, additional zero vectors with reference index $i_{\text{ref}} = 0$ are appended.

The location of the candidates from the spatial and temporal neighborhood of the current PU is indicated by a sample location within each candidate PU. These sample locations and the corresponding labeling are shown in Fig. 7.6. The derivation and location of the applicable candidate PUs is detailed in the following.

7.2.2.1 Spatial Merging Candidates

In Fig. 7.6, the reference sample locations for spatial merge candidate PUs are shown. They are denoted as A_0, A_1, B_0, B_1, and B_2, respectively. The locations A_i indicate the neighboring candidates to the left, the neighboring locations B_j indicate the candidates at the top of the current block. Each candidate PUs is specified by a reference sample belonging to the candidate PU. If this PU is available (i.e., it is not outside the current slice and is coded in inter mode), the motion information of the PU is placed in the candidate list. Merging candidates are derived for both, List 0 and List 1. Additional candidates for bi-prediction are generated from the set of candidates for the two lists as detailed below.

For PUs of size 4×8 at the top CTU boundary, the locations of the reference samples are modified as shown in Fig. 7.6b. Thereby, the number of motion vectors that have to be stored in each CTU line for prediction is reduced [6].

The processing order of the spatial candidate locations is A_1, B_1, B_0, A_0, B_2. If multiple spatial merging candidates share the same motion information (motion vector and reference index), the duplicates are eliminated from the candidate list. The locations A_1 and B_1 are only included in the set of possible candidates if the current PU is not the right or bottom PU in a CU with two partitions, respectively. These cases could lead to identical motion vectors for the two PUs of the current CU. Since identical motion would be represented more effectively by indicating a single PU for the CU instead of a split, these cases are excluded. A maximum of four spatial neighbors can be added to the candidate list.

7.2.2.2 Temporal Merging Candidates

The motion vector of the collocated location in a reference picture is used for derivation of the temporal merging candidate. The applicable reference picture is selected on a slice basis and indicated in the slice header. For B slices, a flag further indicates if the picture is to be taken from reference picture list 0 or reference picture list 1. For error resilience and complexity reduction purposes, the reference index for the temporal merge candidate is always set to $i_{\text{ref}} = 0$.

If the current PU is located at the lower boundary of the CTU or if the block at location C_0 is unavailable, the collocated PU is determined by the sample location C_1 in Fig. 7.7a. Otherwise, the collocated PU is determined by the sample location C_0. Thereby, it is asserted that the processing of motion information in the reference picture is aligned to CTU rows, which reduces the burden on memory access.

If the POC distance between the picture of the collocated PU and the reference picture, which the collocated PU is predicted from, is the same as the distance between the current picture and reference picture containing the collocated PU, the collocated motion vector mv_{PU} can be directly used. Otherwise, the motion vector is scaled according to the relation of the distances between the pictures. Let t_{b} denote the difference between the POC values of the current picture and the collocated picture,

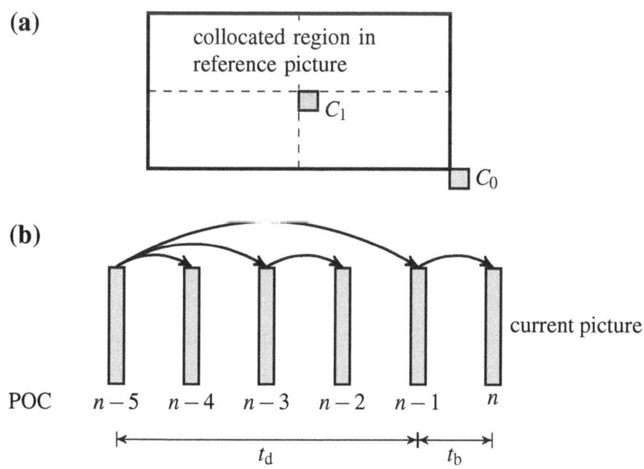

Fig. 7.7 **a** Illustration of the location of temporal merge candidates in the collocated picture. **b** Example for the temporal distance of pictures based on POC

and t_d denote the difference between the POC values of the collocated picture and the reference picture of the respective motion vector, see Fig. 7.7b. The applicable motion vector mv_col of the collocated PU is then scaled to

$$\mathrm{mv_{PU}} = \frac{t_\mathrm{b}}{t_\mathrm{d}} \cdot \mathrm{mv_{col}}. \tag{7.5}$$

7.2.2.3 Combined Bi-predictive Merging Candidates

If the current slice is a B slice, a derivation process for combined bi-predictive merging candidates is used to generate further merging candidates which combine candidates from List 0 and List 1.

Consider lists of four candidates that have been derived for List 0 and List 1. A potential bi-prediction candidate is derived by combination of one List 0 and one List 1 candidate. To this end, the single-list candidates are tested for availability according to a predefined permutation table. If both candidates of the tested pair are available for prediction, the combination of the tested candidate pair is added to the merging candidate list as a combined bi-predictive merging candidate.

7.2.2.4 Skip Mode

At the very beginning of the CU, the syntax includes a flag indicating *skip mode*. If skip mode is indicated, the applicable index for the merge candidate is indicated only if the list of merge candidates is larger than 1. No further information is coded for the CU. In skip mode, the motion vector derived by the described merge operation is applied without a residual update.

7.2.3 Predictive Motion Vector Coding

For predictive motion vector coding, a similar candidate derivation process as for the merge mode is applied. Since the derivation process is much more elaborated than in previous specifications, this method of motion vector prediction is denoted as *advanced motion vector prediction* (AMVP). In contrast to the merge mode where the usage of up to five candidates can be configured, only two candidates are derived here. Depending on their availability, two motion vector candidates from the spatial neighborhood are derived. Like in merge mode, a potential temporal candidate may be included. If the candidate list is still not filled, zero motion vectors are appended such that the presence of two motion vector predictor candidates is asserted for error resilience purposes. The applicable predictor is indicated by an index (in this case with only two possible candidates, the index is a flag). In contrast to the merge mode where the applicable reference index was inferred, the index into the reference picture list is explicitly coded in the PU syntax in this case.

7.2.3.1 Motion Vector Predictor Candidates

The derivation steps for generation of the motion vector candidates are performed using reference picture list 0 and reference picture list 1 (if available) with the spatial neighborhood as shown for the merge mode in Fig. 7.6a.

Each candidate is derived using the following steps. First the candidate from the current reference picture list is tested if the reference picture list index of the neighboring block matches the coded reference picture index of the current block. The same test is performed for the other reference picture list of the neighboring block, if available. If the previous test was not successful and both, the current and the neighboring block reference a long-term reference picture, the corresponding motion vector is used. Finally, if the reference index of the candidate does not correspond to the reference index encoded for the current PU, the respective motion vector is scaled according to the relation of the POC differences according to (7.5). Like for merge mode, the distance between the current picture and the specified reference picture is denoted by t_b and the POC difference between the current picture and the picture indicated by the collocated PU is denoted by t_d for this purpose.

The first available candidate on locations A_0, A_1 is derived according to these steps as motion vector predictor A. The same is applied for locations B_0, B_1, and B_2 for motion vector predictor B.

If the set of spatial neighbors is not completed by the described steps, an additional temporal candidate can be derived from the collocated PU in the temporal neighbor picture as described for the merge mode in Sect. 7.2.2.2. If the temporal candidate is not available, the candidate position is filled with the zero motion vector.

7.2.3.2 Motion Vector Construction

Based on the candidate derivation process described above, the motion vector predictor $\mathrm{mv}_{\mathrm{pL}i}$ is indicated by a flag in the PU syntax. The applicable motion vector displacement $\mathrm{mv}_{\mathrm{L}i}$ for the reference picture list i is then constructed from this predictor and the coded motion vector difference $\mathrm{mv}_{\mathrm{dL}i}$ as

$$\mathrm{mv}_{\mathrm{L}i} = \mathrm{mv}_{\mathrm{pL}i} + \mathrm{mv}_{\mathrm{dL}i}. \tag{7.6}$$

Together with the coded reference picture index i_{ref}, the applicable motion vector is thereby specified. The motion vectors are represented in luma quarter sample precision. They are clipped to a maximum range of

$$-2^{15} \leq \mathrm{mv}_{\mathrm{L}i} \leq 2^{15} - 1. \tag{7.7}$$

7.2.4 *Signaling*

The general usage of the temporal motion vector predictor candidate is enabled in the SPS. The granularity of the merge operation is configured in the PPS as indicated in Sect. 7.2.2. The actual application of the temporal motion vector predictor candidate as well as the length of the merge candidate list is indicated in the slice segment header. For all syntax elements on the PU level, including the motion information, arithmetic coding is applied. For a description of entropy coding and the binarization of syntax elements see Chap. 10.

7.2.4.1 Merge Mode

The syntax element for the candidate index for merge mode is coded with truncated unary binarization where the first bin is coded with context update and the remaining bins are coded in bypass mode (see Sect. 10.2.3). The maximum value of the binarization corresponds to the number of merge candidates indicated in the slice segment header.

7.2.4.2 Predictive Motion Vector Coding

If motion vector prediction is used, the selection of the applicable predictor out of the two predictor candidates has to be indicated. Therefore, a binary flag syntax element is used to indicate the applicable candidate. The flag is coded with context update. The flag is always sent, even if the derivation of candidates yields only one candidate. This feature helps to reduce parsing dependencies and thereby reduces the implementation complexity of the scheme [5].

The next syntax element to be coded is the reference picture set index. The value range of the reference picture index depends on the size of the reference picture set and can vary with different coder configurations. For arithmetic coding, the index is binarized using a truncated Rice binarization. The first two bins are coded with a context update, potential remaining bins are coded in bypass mode.

A similar approach is used for the signaling of the motion vector difference, with the usage of dedicated flags instead of an integration with the remaining binarization. Here, a flag indicating the absolute motion vector difference to be $|\mathrm{mv_d}| > 0$ and (conditionally) a flag indicating $|\mathrm{mv_d}| > 1$ are coded using context updates. If the motion vector difference is greater than 1, the remaining $|\mathrm{mv_d}|$ value is coded with a Exp-Golomb binarization in bypass mode. The sign of the motion vector difference is coded as a flag in bypass mode as well. It should be noted that the context for the two bins is selected based on the bin index only. Therefore, no storage of motion vector differences is required. The x and y components of the motion vector difference flags are coded subsequently to enable grouping of bypass bins.

7.3 Sub-Sample Interpolation

In HEVC, the same motion vectors are applied for the luma and chroma components. Fractional offsets of 1/4, 1/2, and 3/4 full sample locations in both horizontal and vertical directions can be addressed for luma. For YCbCr 4:2:0 video, the motion vectors scale for chroma accordingly and are thereby presented in eighth-chroma-sample.

In H.264 | AVC, a 16-bit sub-sample interpolation scheme was specified. It has been shown that under some circumstances intermediate bit shifts and rounding operations could induce visible artifacts in the reconstructed video [7]. In order to reduce such effects and to enable sub-sample interpolation at sufficient precision, the dynamic range used for inter prediction has been increased in HEVC.

The interpolation filters all share the same normalization factor of 64. After the filtering operation, normalization of the filtered value is achieved by a right-shift of 6 bits. For motion compensation, the following steps are performed to generate the prediction signal for motion compensated prediction: First, the sample values in the reference picture are scaled to a dynamic range of 14 bit by performing a left-shift using the appropriate number of bits depending on the bit depth B_d of the encoded video (i.e. 6 bits for the Main Profiles with a bit depth of 8 bit). If the motion vector indicates a sub-sample location in horizontal direction, filtering is performed in horizontal direction and a normalizing right-shift by 6 bits is applied, bringing the dynamic range back to 14 bit. The same type of filtering and right-shifting is performed if a vertical sub-sample location is needed. If bi-prediction is applied, the two prediction signals are added and the result is right-shifted by 1 bit. Finally, the resulting prediction scaled back from 14 bit to the original dynamic range. In contrast to the previous down-shifting operations, rounding is applied in this final scaling step.[2]

7.3.1 Luma Sub-Sample Interpolation Filtering

The sample values for sub-sample locations are interpolated from the full-sample luma values by applying fixed interpolation filters depending on the current sub-sample location. The applicable sub-sample location is derived by a modulo-operation on the derived motion vector which is presented in quarter-sample accuracy,

$$\mathrm{mv}_{\mathrm{frac}} = \mathrm{mv} \bmod 4 \tag{7.8}$$

The coefficients for the luma component filter are as follows, representing the $\frac{1}{4}$, $\frac{1}{2}$, and $\frac{3}{4}$ sub-sample locations, respectively (and omitting the previously described normalization by a factor of 64):

[2] In the specification, the final scaling step with rounding is incorporated in the weighted sample prediction processes.

$$h_{\mathrm{a}}^{\mathrm{y}}(n) = [-1, 4, -10, 58, 17, -5, 1], \quad n = -3, \ldots, 3 \tag{7.9}$$

$$h_{\mathrm{b}}^{\mathrm{y}}(n) = [-1, 4, -11, 40, 40, -11, 4, -1], \quad n = -3, \ldots, 4 \tag{7.10}$$

$$h_{\mathrm{c}}^{\mathrm{y}}(n) = h_{\mathrm{a}}^{\mathrm{y}}(-n) \tag{7.11}$$

These filters are applied in horizontal or vertical direction, depending on the sub-sample location to be filtered. The full-sample and sub-sample locations are shown in Fig. 7.8. The filtering process is described for the sub-sample locations relative to a luma sample $\mathrm{A}_{0,0}$, as shown in Fig. 7.8. Each sub-sample location is referenced by a unique name sharing the location index (0, 0). In the motion compensation process, this luma sample location corresponds to the integer resolution part $\mathrm{mv}_{\mathrm{int}}$ of the applicable motion vector,

$$\mathrm{mv}_{\mathrm{int}} = \left\lfloor \frac{\mathrm{mv}}{4} \right\rfloor. \tag{7.12}$$

The sub-sample interpolation filtering applies for all samples $(i, j) \neq (0, 0)$ accordingly.

For the sub-sample locations $\mathrm{a}_{0,0}$, $\mathrm{b}_{0,0}$, and $\mathrm{c}_{0,0}$, the filters $h_{\mathrm{a}}^{\mathrm{y}}(n)$, $h_{\mathrm{b}}^{\mathrm{y}}(n)$, and $h_{\mathrm{c}}^{\mathrm{y}}(n)$ are applied, respectively, with n denoting the luma locations $\mathrm{A}_{n,0}$, $n = -3, \ldots, 4$. For the sample locations $\mathrm{d}_{0,0}$, $\mathrm{h}_{0,0}$, and $\mathrm{n}_{0,0}$, the same filtering is applied in vertical direction, i.e. n denoting the luma locations $\mathrm{A}_{0,n}$, $n = -3, \ldots, 4$. For sub-sample locations with fractional offset in both, horizontal and vertical direction, the same filtering as for sample locations $\mathrm{d}_{0,0}$, $\mathrm{h}_{0,0}$, and $\mathrm{n}_{0,0}$ is applied with the full sample locations $\mathrm{A}_{0,n}$ being replaced by $\mathrm{a}_{0,n}$ for the locations $\mathrm{e}_{0,0}$, $\mathrm{i}_{0,0}$, $\mathrm{p}_{0,0}$, by $\mathrm{b}_{0,n}$ for the locations $\mathrm{f}_{0,0}$, $\mathrm{j}_{0,0}$, $\mathrm{q}_{0,0}$, and by $\mathrm{c}_{0,n}$ for the locations $\mathrm{g}_{0,0}$, $\mathrm{k}_{0,0}$, $\mathrm{r}_{0,0}$, respectively. i.e., the filtering is applied using the sub-sample values of the same column as the target sample location and the sub-sample locations of the same rows as the luma samples.

The order of operation is specified as horizontal first and then vertical. By specifying this normative processing order, deviations in interpolated results due to intermediate down-scaling operations if horizontal and vertical filtering were exchanged are omitted.

7.3.2 Chroma Sub-Sample Interpolation Filtering

The chroma sub-sample interpolation is performed similarly to the luma sub-sample interpolation. Since the motion vectors are applied in eighth-sample resolution, filters for the corresponding additional sub-sample locations have to be specified. The applicable filters are given as follows (again omitting normalization):

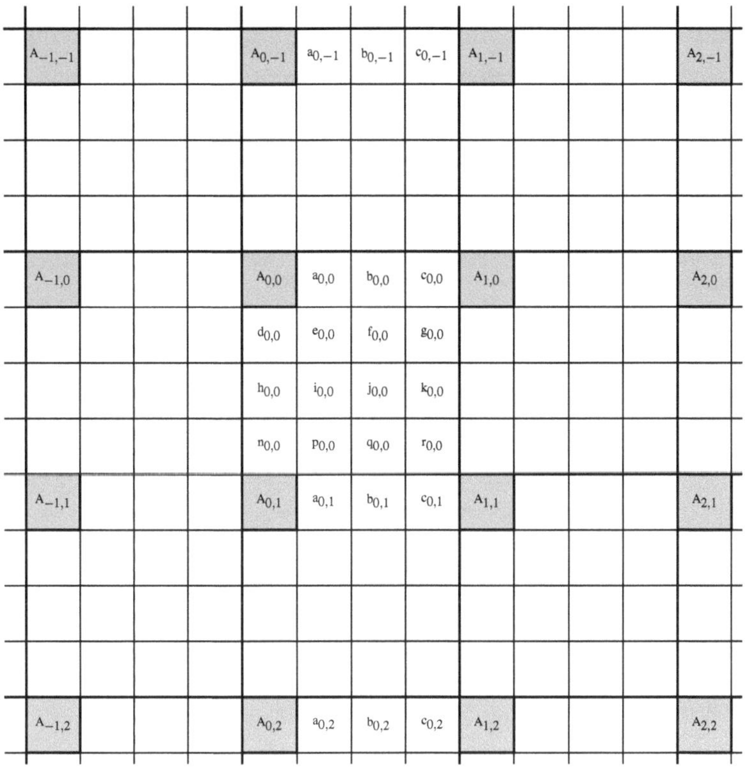

Fig. 7.8 Arrangement of luma sample locations $\mathrm{A}_{i,j}$ and quarter-sample locations $\mathrm{a}_{i,j}$ to $\mathrm{r}_{i,j}$ sharing the same location index (i, j) [1]

$$h^{\mathrm{c}}_{\mathrm{b}}(n) = [-2, 58, 10, -2], \quad n = -1, \dots, 2 \tag{7.13}$$
$$h^{\mathrm{c}}_{\mathrm{c}}(n) = [-4, 54, 16, -2], \quad n = -1, \dots, 2 \tag{7.14}$$
$$h^{\mathrm{c}}_{\mathrm{d}}(n) = [-6, 46, 28, -4], \quad n = -1, \dots, 2 \tag{7.15}$$
$$h^{\mathrm{c}}_{\mathrm{e}}(n) = [-4, 36, 36, -4], \quad n = -1, \dots, 2 \tag{7.16}$$
$$h^{\mathrm{c}}_{\mathrm{f}}(n) = h^{\mathrm{c}}_{\mathrm{d}}(1-n) \tag{7.17}$$
$$h^{\mathrm{c}}_{\mathrm{g}}(n) = h^{\mathrm{c}}_{\mathrm{c}}(1-n) \tag{7.18}$$
$$h^{\mathrm{c}}_{\mathrm{h}}(n) = h^{\mathrm{c}}_{\mathrm{b}}(1-n) \tag{7.19}$$

As for luma, these filters are applied in horizontal or vertical direction, depending on the sub-sample location to be filtered. Since all filter coefficients for chroma interpolation are even, the interpolation process could have been specified with a dynamic range reduced by 1 bit for the chroma component. In order to achieve a harmonized description, a single joint scaling operation has been specified, see [8].

The eighth-sample chroma locations are depicted in Fig. 7.9. The naming of the sub-sample locations uses the labels a,b,…,h in horizontal and vertical direction for reference.

$hh_{-1,-1}$	$ha_{0,-1}$	$hb_{0,-1}$	$hc_{0,-1}$	$hd_{0,-1}$	$he_{0,-1}$	$hf_{0,-1}$	$hg_{0,-1}$	$hh_{0,-1}$	$ha_{1,-1}$
$ah_{-1,0}$	$B_{0,0}$	$ab_{0,0}$	$ac_{0,0}$	$ad_{0,0}$	$ae_{0,0}$	$af_{0,0}$	$ag_{0,0}$	$ah_{0,0}$	$B_{1,0}$
$bh_{-1,0}$	$ba_{0,0}$	$bb_{0,0}$	$bc_{0,0}$	$bd_{0,0}$	$be_{0,0}$	$bf_{0,0}$	$bg_{0,0}$	$bh_{0,0}$	$ba_{1,0}$
$ch_{-1,0}$	$ca_{0,0}$	$cb_{0,0}$	$cc_{0,0}$	$cd_{0,0}$	$ce_{0,0}$	$cf_{0,0}$	$cg_{0,0}$	$ch_{0,0}$	$ca_{1,0}$
$dh_{-1,0}$	$da_{0,0}$	$db_{0,0}$	$dc_{0,0}$	$dd_{0,0}$	$de_{0,0}$	$df_{0,0}$	$dg_{0,0}$	$dh_{0,0}$	$da_{1,0}$
$eh_{-1,0}$	$ea_{0,0}$	$eb_{0,0}$	$ec_{0,0}$	$ed_{0,0}$	$ee_{0,0}$	$ef_{0,0}$	$eg_{0,0}$	$eh_{0,0}$	$ea_{1,0}$
$fh_{-1,0}$	$fa_{0,0}$	$fb_{0,0}$	$fc_{0,0}$	$fd_{0,0}$	$fe_{0,0}$	$ff_{0,0}$	$fg_{0,0}$	$fh_{0,0}$	$fa_{1,0}$
$gh_{-1,0}$	$ga_{0,0}$	$gb_{0,0}$	$gc_{0,0}$	$gd_{0,0}$	$ge_{0,0}$	$gf_{0,0}$	$gg_{0,0}$	$gh_{0,0}$	$ga_{1,0}$
$hh_{-1,0}$	$ha_{0,0}$	$hb_{0,0}$	$hc_{0,0}$	$hd_{0,0}$	$he_{0,0}$	$hf_{0,0}$	$hg_{0,0}$	$hh_{0,0}$	$ha_{1,0}$
$ah_{-1,1}$	$B_{0,1}$	$ab_{0,1}$	$ac_{0,1}$	$ad_{0,1}$	$ae_{0,1}$	$af_{0,1}$	$ag_{0,1}$	$ah_{0,1}$	$B_{1,1}$

Fig. 7.9 Arrangement of chroma sample locations $B_{i,j}$ and eighth-sample locations. 64 eighth-sample locations are allocated with each chroma sample, sharing the same location index (i, j). Notation from [1]

The filtering is described for the chroma location $B_{0,0}$. The filtering for all other locations follows accordingly. For the locations $ab_{0,0}$, $ac_{0,0}$, ..., $ah_{0,0}$, the filters $h_{\mathrm{b}}^{\mathrm{c}}(n), h_{\mathrm{c}}^{\mathrm{c}}(n), \ldots, h_{\mathrm{h}}^{\mathrm{c}}(n)$ are applied, respectively, with n denoting the chroma locations $B_{n,0}, n = -1, \ldots, 2$. Accordingly, the filters are applied at the locations $ba_{0,0}$, $ca_{0,0}$, ..., $ha_{0,0}$, with the chroma locations $B_{0,n}$, $n = -1, \ldots, 2$. For locations with sub-sample location in both, horizontal and vertical direction, the filtering is applied using the sub-sample values of the same column as the target sample location and the sub-sample locations of the same rows as the chroma samples.

7.3.3 Derivation of the Interpolation Filter Coefficients

The HEVC interpolation filters have been derived using the DCT [9, 10]. The DCT has been used for interpolation tasks before, e.g. in [11]. Here, the basic concept of the filter derivation process is summarized. A detailed analysis of the filters is provided in [12].

With DCT-based interpolation, the basic steps for a one-dimensional interpolation operation are as follows:

- Perform the forward DCT over N neighboring samples where N defines the applicable filter length.
- Reconstruct the shifted center location from the DCT coefficients with a shift δ.

By merging the two steps, filter coefficients for a direct filter can be derived as briefly outlined in the following.

Let $\mathbf{p} = [p_0, \ldots, p_{N-1}]^{\mathsf{T}}$ denote a one-dimensional set of sample values. The transform coefficients $\mathbf{c} = [c_0, \ldots, c_{N-1}]^{\mathsf{T}}$ are derived by

$$\mathbf{c} = \mathbf{T}_{\mathrm{DCT}} \cdot \mathbf{p}, \tag{7.20}$$

with the DCT matrix as defined in Sect. 2.4.5.2. Accordingly, the sample values can be reconstructed by

$$\mathbf{p} = \mathbf{T}^{\mathrm{i}}_{\mathrm{DCT}} \cdot \mathbf{c} = \underbrace{\mathbf{T}^{\mathrm{i}}_{\mathrm{DCT}} \cdot \mathbf{T}_{\mathrm{DCT}}}_{=\mathbf{I}} \cdot \mathbf{p}. \tag{7.21}$$

Here, the notation $\mathbf{T}^{\mathrm{i}}_{\mathrm{DCT}}$ denotes the inverse transform $\mathbf{T}^{\mathrm{i}}_{\mathrm{DCT}} = \mathbf{T}^{\mathsf{T}}_{\mathrm{DCT}}$ with base vectors $\mathbf{t}^{\mathrm{i}}_n$.

Now, the DCT shall be applied for interpolation of intermediate values as stated above. Let $\mathbf{p}$ be a set of $N = N_{\max} - N_{\min} + 1$ samples p_k, $k = N_{\min}, \ldots, N_{\max}$. It is assumed that $N_{\min} < 0$ and $N_{\max} > 0$. The reconstruction of the center location $k = 0$ is regarded, included a shift δ to the target intermediate location. The setting of the samples is shown in Fig. 7.10.

For reconstruction of the intermediate location δ in Fig. 7.10, the inverse transform base function at $n = N - N_{\max} - 1$ is shifted to $n' = N - N_{\max} - 1 + \delta$. The corresponding base function becomes

$$t^{\mathrm{i}}_{\delta}(m) = \sqrt{\frac{2}{N}} \cdot a(m) \cdot \cos\left(\pi m \cdot \frac{2N + 2\delta - 2N_{\max} - 1}{2N}\right), \quad m = 1, \ldots, N - 1, \tag{7.22}$$

with $a(m)$ defined in (2.11). According to (7.21), a sample value as location n is reconstructed by $p_n = \mathbf{t}^{\mathrm{i}}_n \cdot \mathbf{T}_{\mathrm{DCT}} \cdot \mathbf{p}$. In the same manner, the sample value p_δ at the intermediate location δ can be calculated by

$$p_\delta = \mathbf{t}^{\mathrm{i}}_{\delta} \cdot \mathbf{T}_{\mathrm{DCT}} \cdot \mathbf{p}, \tag{7.23}$$

where $\mathbf{t}^{\mathrm{i}}_{\delta}$ denotes the row vector of the corresponding modified base function. This provides the formulation for a DCT-based interpolation filter,

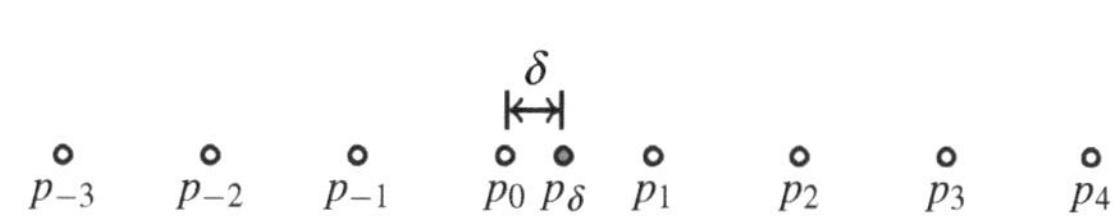

Fig. 7.10 Sample locations p_k used for derivation of the sample value at location p_δ

$$\mathbf{h} = \mathbf{t}_\delta^{\mathrm{i}} \cdot \mathbf{T}_{\mathrm{DCT}} \tag{7.24}$$

with filter coefficients $h(n)$, $n = N_{\mathrm{min}}, \ldots, N_{\mathrm{max}}$.

The transfer function of the filters derived by this method shows some ripple in the pass-band, which may be considered undesirable. In order to reduce this effect, weighting functions have been applied to derive the final form of the filters. A cos-weighted window function with a configurable window size N_{w} has been selected in the design of the luma interpolation filters [10],

$$h_{\mathrm{w,l}}(n) = \cos\left(\pi \frac{n-\delta}{N_{\mathrm{w}}}\right), \quad n = N_{\mathrm{min}}, \ldots, N_{\mathrm{max}}. \tag{7.25}$$

For the chroma interpolation filters, a different parametrized weighting function was selected as [12]

$$h_{\mathrm{w,c}}(n) = \frac{a}{1 - \sigma \cdot (n-b)^2}, \quad n = N_{\mathrm{min}}, \ldots, N_{\mathrm{max}}. \tag{7.26}$$

For an integer implementation, the filter coefficients finally need to be scaled and rounded to integer values. The parameter values for the design of the interpolation filters including the scaling factor ν_{scal} for integer rounding are provided in Tables 7.1 and 7.2. Due to rounding to the next integer, the norm of the integer filter might not be equal to the scaling factor ν_{scal}. In these cases, an amplification or an attenuation would occur with the integer implementation. To resolve this issue, some coefficients are modified by ± 1 to meet the normalization constraint [13].

Table 7.1 Parameter values for the derivation of the luma interpolation filter coefficients

Position	δ	N_{min}	N_{max}	N_{w}	ν_{scal}
1/4[a]	0.25	−3	3	12	64
1/2	0.5	−3	4	12	64

[a] Re-adjustment for normalization necessary

Table 7.2 Parameter values for the derivation of the chroma interpolation filter coefficients

Position	δ	N_{min}	N_{max}	σ	a	b	ν_{scal}
1/8	0.125	−1	2	0.26	1.2	1.0	64
1/4	0.25	−1	2	0.28	1.0	0.25	64
3/8	0.375	−1	2	0.12	1.0	0.125	64
1/2	0.5	−1	2	0.28	1.0	0.5	64

7.4 Inter Coding Examples

Figures 7.11 and 7.12 show an example for inter coding of four CTBs in a test sequence of the JCT-VC common testing conditions.[3] The partitioning of the CTBs into CBs and PBs, the distribution of the reference indices over the blocks, and the assignment of skip, merge mode, and predictive motion vector coding are shown in Fig. 7.11. The example shows that a significant part of the depicted area is coded using either skip or merge mode, where no further motion information is transmitted besides the applicable merge index. In Fig. 7.12, the motion vectors with the corresponding motion vector predictors and motion vector differences are shown. It can be seen from the figure that only small motion vector differences have to be encoded in the given example. It can further be seen that the motion vectors of List 0 and List 1 tend to be oriented into approximately opposite directions. This is due to the fact that the reference pictures of List 0 refer to previous pictures while List 1 includes pictures which follow the current picture in display order.

The shown scene further includes camera motion which overlays the motion of the persons in the scene. For illustration, the motion vectors for a complete picture are depicted in Fig. 7.13.[4]

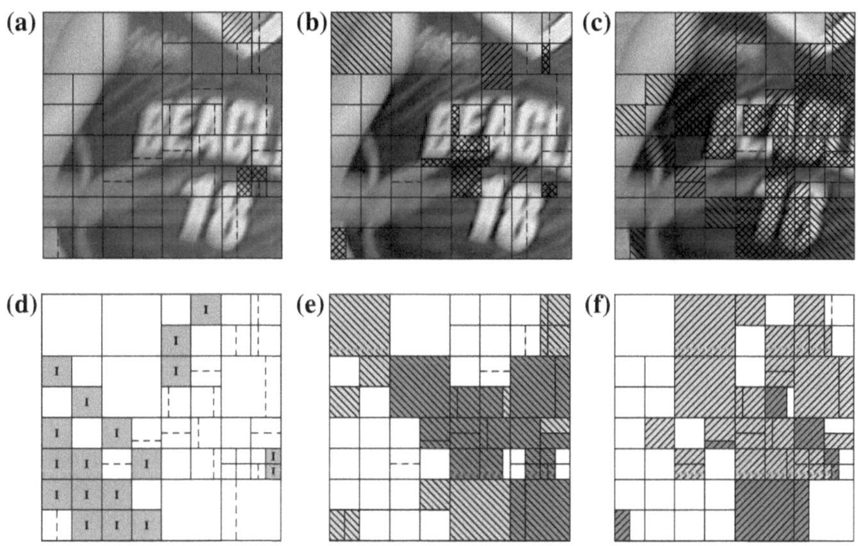

Fig. 7.11 Example for inter coding for four CTBs with uni- and bi-prediction (List 0: *diagonal down hatch*, List 1: *diagonal up hatch*, reference picture list index $i_{\text{ref LX}} = 0$: *light gray*, $i_{\text{ref LX}} = 1$: *dark gray*). **a** Skip. **b** Merge mode. **c** AMVP. **d** CB and PB grid. **e** i_{refL0}. **f** i_{refL1}

[3] Frame 308 of the 1080p sequence BasketballDrive, HM11.0, random access configuration of the JCT-VC common testing conditions [14], QP = 37.

[4] Frame 250 of the 1080p sequence BasketballDrive, HM11.0, random access configuration of the JCT-VC common testing conditions [14], QP = 37.

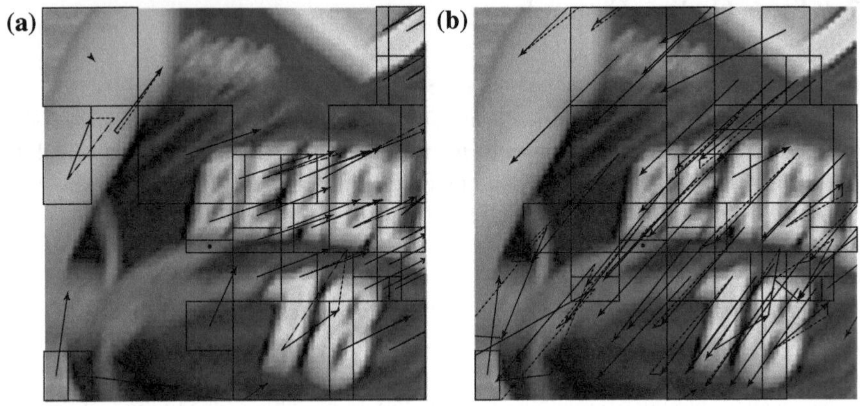

Fig. 7.12 Motion vectors for the prediction blocks in Fig. 7.11. **a** List 0. **b** List 1

7.5 Comparison to H.264 | AVC

For comparison to H.264 | AVC [15, 16], two main aspects are highlighted in the following: The way motion information is represented and coded, and the design and characteristics of the respective sub-sample interpolation processes.

7.5.1 Motion Vector Representation

Both, H.264 | AVC and HEVC apply quarter-sample precision for motion compensation in the luma component and eighth-sample precision for motion compensation in the chroma component. The representation of the motion vectors in HEVC is more sophisticated than the representation in H.264 | AVC, which leads to a more compact representation of the motion information in HEVC.

In H.264 | AVC, the *direct mode* is applied to derive motion vectors for the current block instead of sending it. This mode is only available in B slices and is applied for bi-prediction. Two variants of direct mode motion vector derivation are specified: In the spatial direct mode, the applicable motion vector is derived from the median of the motion vectors of three neighboring blocks identified by the neighboring samples A, B, C as illustrated in Fig. 7.14. In the temporal direct mode, the motion vector is derived from the collocated block in the reference picture. The HEVC merge mode can be interpreted as a generalization of the direct mode. By inheritance of merged motion information from the neighboring blocks, constant motion of an almost arbitrarily shaped region in a picture can be coded very efficiently.

For motion vector prediction, the median of the motion vectors of the spatially neighboring blocks is applied, using the same predictor blocks as in the spatial direct

(a)
(b)
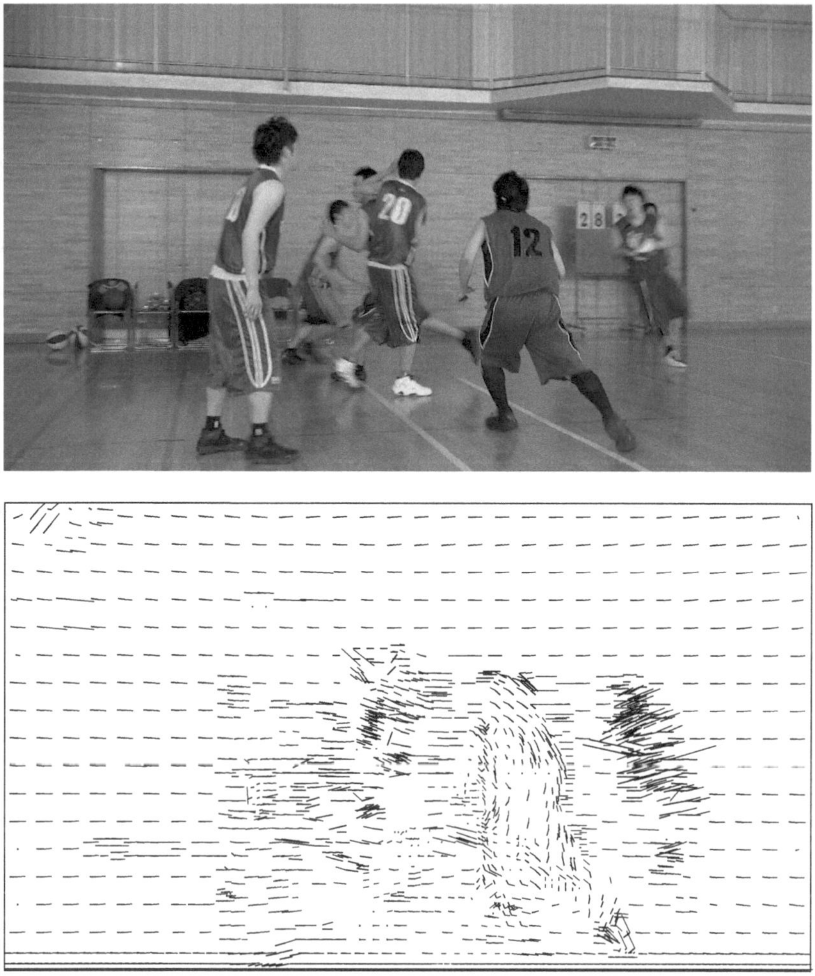

Fig. 7.13 Motion vector field for a complete picture. **a** Original. **b** Motion vectors (List 0: *black*, List 1:*gray*)

mode shown in Fig. 7.14. The derived predictor is added to the coded motion vector difference to form the applicable motion vector. Similar to the merge mode and the direct mode, the advanced motion vector prediction of HEVC can be seen as a generalization of the motion vector prediction in H.264 | AVC. The option of selecting from two available predictors and the larger set of predictor candidates in HEVC improves the coding efficiency of the scheme.

It should be noted that in H.264 | AVC, the motion vectors of the reference pictures, which are used temporal direct mode are accessed using a 4×4 sample grid. Thereby, the number of motion vectors which has to be stored for access with the temporal direct mode is much larger than in HEVC where the motion vector storage

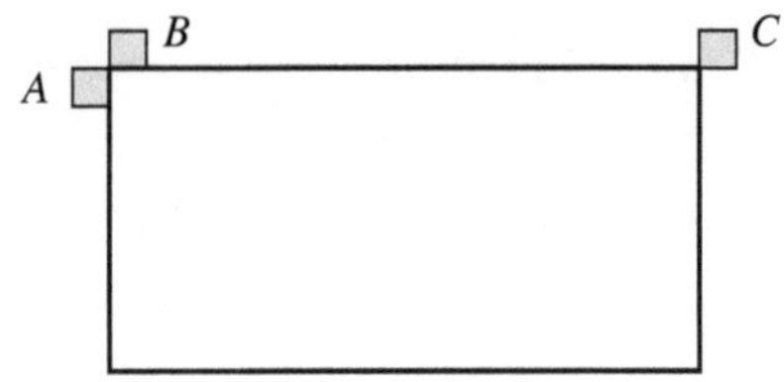

Fig. 7.14 Illustration of neighboring blocks for motion vector predictor derivation in H.264 | AVC

is provided only on a 16 × 16 sample grid. As a further difference, H.264 | AVC requires a line buffer for storage of the coded motion vector differences as the sum of neighboring differences is needed for CABAC context selection [5].

7.5.2 Sub-Sample Interpolation

In H.264 | AVC, a six-tap filter is specified for interpolation of the values for the half-sample locations in Fig. 7.15 (omitting normalization),

$$h_{\mathrm{b}}(n) = [1, -5, 20, 20, -5, 1], \quad n = -2, \ldots, 3. \tag{7.27}$$

The filter is applied to the full-sample locations around the target sub-sample location. For derivation of the value for location $\mathrm{j}_{0,0}$ as shown in Fig. 7.15a, the half-sample interpolation filter is applied in horizontal and vertical direction. The values for the quarter-sample locations $p_{\mathrm{y}}(x, y)$ are derived by averaging from the two nearest full- and half-sample locations with rounding, e.g.

$$p_{\mathrm{y}}(\mathrm{a}_{0,0}) = \left\lfloor \frac{1}{2}\big(p_{\mathrm{y}}(\mathrm{A}_{0,0}) + p_{\mathrm{y}}(\mathrm{b}_{0,0}) + 1\big) \right\rfloor. \tag{7.28}$$

The applicable full- and half-sample locations for filtering of each quarter-sample luma location are shown in Fig. 7.15a. The averaging operation keeps the number of samples which are used for interpolation small. However, the interpolated signal reveals strong lowpass characteristics.

In Fig. 7.16, the transfer function of the half sample and quarter sample luma interpolation filters in H.264 | AVC and HEVC are shown for comparison. The depicted transfer functions of the H.264 | AVC quarter-sample filter has been derived from (7.28). The curves reveal a higher cut-off frequency for the HEVC interpolation filters and a reduced overshoot of the transfer function in the higher frequency range for the HEVC half-sample filter. For the quarter-sample locations, the attenuation for high-frequency components is strongly reduced compared to the H.264 | AVC filter.

For chroma motion compensation, the luma motion vectors are applied at eighth-sample resolution for YCbCr 4:2:0 video in both specifications. In H.264 | AVC, the sample values for the sub-sample locations $p_{\mathrm{c}}(x, y)$ are derived by bi-linear

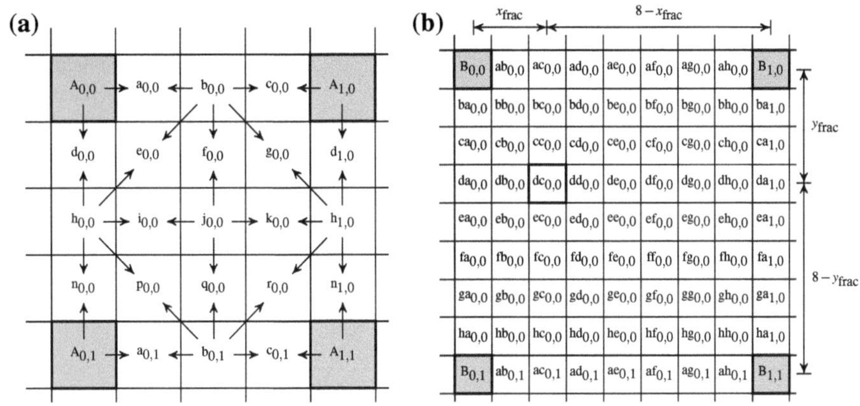

Fig. 7.15 Interpolation for sub-sample luma locations in H.264 | AVC. **a** Luma quarter-sample locations. **b** Chroma sub-sample locations

interpolation from the four nearest full-sample locations using the fractional sample locations $0 \leq x_{\text{frac}}, y_{\text{frac}} \leq 7$. An example is provided in Fig. 7.15b, with the corresponding interpolation equation

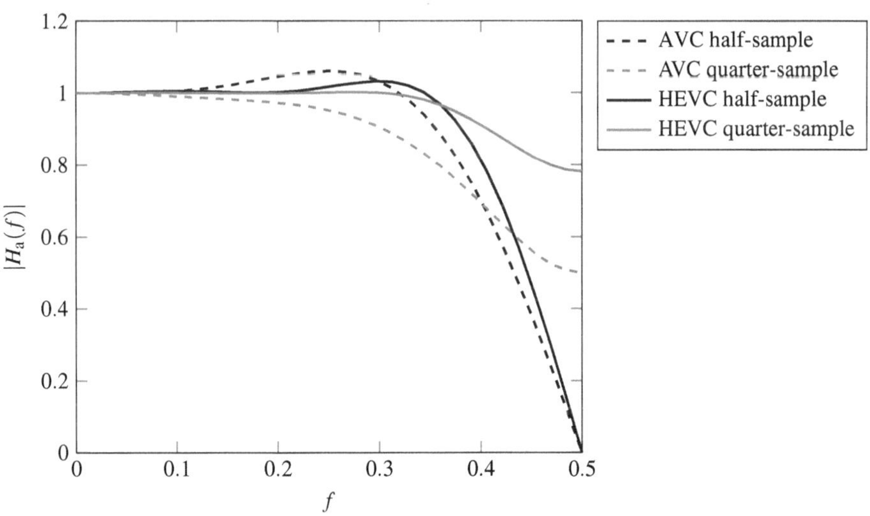

Fig. 7.16 Transfer functions of the luma interpolation filters for half-sample and quarter-sample locations in H.264 | AVC and HEVC

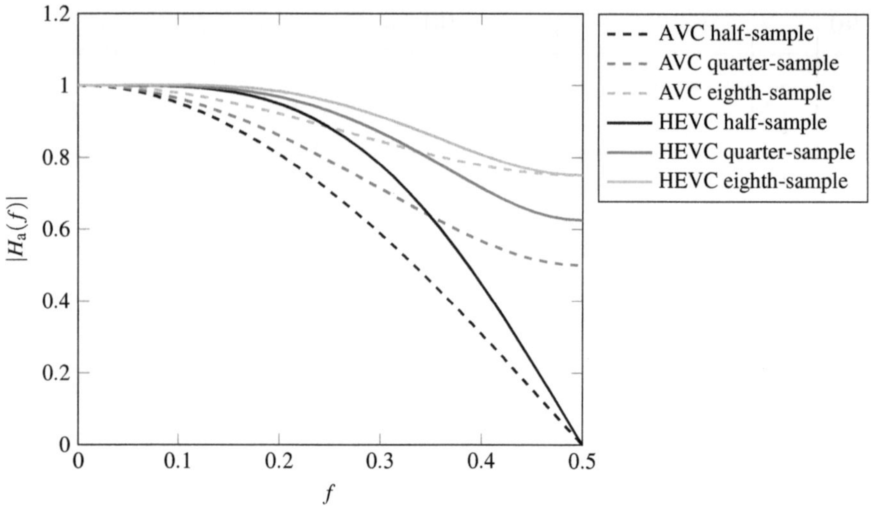

Fig. 7.17 Transfer functions of the chroma interpolation filters for half-, quarter-, and eighth-sample locations in H.264 | AVC and HEVC

$$p_c(\mathrm{dc}_{0,0}) = \left\lfloor \frac{1}{64}\Big((8 - x_{\mathrm{frac}})(8 - y_{\mathrm{frac}}) \cdot p_c(\mathrm{B}_{0,0}) + x_{\mathrm{frac}}(8 - y_{\mathrm{frac}}) \cdot p_c(\mathrm{B}_{1,0}) + (8 - x_{\mathrm{frac}})y_{\mathrm{frac}} \cdot p_c(\mathrm{B}_{1,0}) + x_{\mathrm{frac}}y_{\mathrm{frac}} \cdot p_c(\mathrm{B}_{1,0}) + 32\Big)\right\rfloor. \quad (7.29)$$

In HEVC, four-tap filters are applied in both, horizontal and vertical direction for derivation of the interpolated sample value as detailed in (7.13)–(7.19).

Figure 7.17 shows the transfer functions of the H.264 | AVC and HEVC chroma interpolation filters for the half-, quarter-, and eighth-sample locations. Like for the luma interpolation filters, a shifted cut-off frequency can be observed such that more detail is preserved with the HEVC interpolation filters.

References

1. High efficiency video coding. ITU-T Rec. H.265 (HEVC). http://www.itu.int/rec/T-REC-H.265/en (2013). Accessed 14 Apr 2014
2. Information technology–High efficiency coding and media delivery in heterogeneous environments—Part 2: High efficiency video coding. ISO/IEC 23008-2:2013 (HEVC). http://www.iso.org/iso/home/store/catalogue_tc/catalogue_detail.htm?csnumber=35424 (2013). Accessed 14 Apr 2014
3. Flierl, M., Girod, B.: Generalized B pictures and the draft H.264/AVC video-compression standard. IEEE Trans. Circ. Syst. Video Technol. **13**(7), 587–597 (2003). doi:10.1109/TCSVT.2003.814963

4. Su, Y., Segall, A.: CE9: Reduced resolution storage of motion vector data. Doc. JCTVC-D072. 4th Meeting: Joint Collaborative Team on Video Coding (JCT-VC) of ITU-T SG 16 WP 3 and ISO/IEC JTC 1/SC 29/WG 11, Daegu, KR (2011)
5. Sze, V., Budagavi, M.: High throughput CABAC entropy coding in HEVC. IEEE Trans. Circ. Syst. Video Technol. **22**(12), 1778–1791 (2012). doi:10.1109/TCSVT.2012.2221526
6. Helle, P., et al.: Block merging for quadtree-based partitioning in HEVC. IEEE Trans. Circ. Syst. Video Technol. **22**(12), 1720–1731 (2012). doi:10.1109/TCSVT.2012.2223051
7. Kamp, S., Wien, M.: Error accumulation in motion compensation in P and B slices. Doc. JVT-AA039. 27th Meeting: Joint Video Team (JVT) of ISO/IEC MPEG & ITU-T VCEG, Geneva, CH (2008)
8. Sullivan, G.J., Ohm, J.-R.: Meeting report of the seventh meeting of the Joint Collaborative Team on Video Coding (JCT-VC), Geneva, CH. Doc. JCTVC-G1100. 7th Meeting: Joint Collaborative Team on Video Coding (JCT-VC) of ITU-T SG 16 WP 3 and ISO/IEC JTC 1/SC 29/WG 11, Geneva, CH (2011)
9. McCann, K., et al.: Samsung's Response to the Call for Proposals on Video Compression Technology. Doc. JCTVC-A124. 1st Meeting: Joint Collaborative Team on Video Coding (JCT-VC) of ITU-T SG 16 WP 3 and ISO/IEC JTC 1/SC 29/WG 11, Dresden, Germany (2010)
10. Alshina, E., Alshin, A.: CE3: DCT derived interpolation filter test by Samsung. Doc. JCTVC-F247. 6th Meeting: Joint Collaborative Team on Video Coding (JCT-VC) of ITU-T SG 16 WP 3 and ISO/IEC JTC 1/SC 29/WG 11, Torino, IT (2011)
11. Agbinya, J.I.: Interpolation using the Discrete Cosine Transform. Electron. Lett. **28**(20), 1927–1928 (1992). doi:10.1049/el:19921233
12. Ugur, K., et al.: Motion compensated prediction and interpolation filter design in H.265/HEVC. IEEE J. Sel. Top. Sig. Process. **7**(6), 946–956 (2013). doi:10.1109/JSTSP.2013.2272771
13. Alshina, E., et al.: CE3: experimental results of DCTIF by Samsung. Doc. JCTVC-D344. 4th Meeting: Joint Collaborative Team on Video Coding (JCT-VC) of ITU-T SG 16 WP 3 and ISO/IEC JTC 1/SC 29/WG 11, Daegu, KR (2011)
14. Bossen, F.: Common test conditions and software reference configurations. Doc. JCTVC-K1100. 11th Meeting: Joint Collaborative Team on Video Coding (JCT-VC) of ITU-T SG 16 WP 3 and ISO/IEC JTC 1/SC 29/WG 11, Shanghai, CN (2012)
15. Advanced video coding for generic audiovisual services. ITU-T Rec. H.264 (AVC). http://www.itu.int/rec/T-REC-H.264/en (2014). Accessed 14 Apr 2014
16. Information technology—Coding of audio-visual objects—Part 10: Advanced video coding. ISO/IEC 14496-10:2012 (AVC). http://www.iso.org/iso/home/store/catalogue_tc/catalogue_detail.htm?csnumber=61490 (2012). Accessed 14 Apr 2014

Chapter 8
Residual Coding

In this chapter, the residual coding in HEVC is detailed. The reconstructed residual signal in the transform blocks is added to the prediction signal which is represented in prediction blocks for reconstruction of the coding block. Conceptually, the specification provides four different methods to code the residual.

As the first and most commonly used option, the residual which remains between the original block and its prediction is transformed and the transform coefficients are quantized and encoded into the bitstream at the encoder side. At the decoder side, the quantization is inverted and by inverse transformation, the quantized coefficients are reconstructed to the decoded residual block which is added to the given prediction. This method generally provides the most compact representation of the coded residual signal. The specification only includes the definition of the inverse transforms at the decoder. Here both, the encoder and decoder side, are discussed.

As a second option, the specification includes the option to skip the transform and apply quantization directly to the non-transformed residual signal on a 4×4 TU basis. This method allows for a residual representation without ringing artifacts which can be introduced by inverse transformation of the quantized transform coefficients. As the residual signal is not decorrelated for coding by the transform, this method may be suitable only at higher bitrates and for blocks with very small prediction error.

In the third case, transform and quantization are bypassed. I.e., the residual is coded directly without the application of the transform or quantization processes. This option allows for perfect reconstruction of the residual signal of inter or intra prediction, and thereby for a lossless representation of the coded block.

The fourth and last option is direct PCM coding of the sample levels. In this case, no prediction is applied and the coding unit samples are directly coded into the bitstream. As this representation allows for perfect reconstruction of the original input sample values, the corresponding bitrate cost can be interpreted as the upper bound for coding the respective coding unit.

In the following, the transform, quantization and coding process is detailed in Sects. 8.1–8.2. The transform skip, transform and quantization bypass, and PCM coding modes are described in Sects. 8.3–8.5.

M. Wien, *High Efficiency Video Coding*, Signals and Communication Technology,
DOI 10.1007/978-3-662-44276-0_8

8.1 Transforms and Quantization

The HEVC transform and quantization process is specified using fixed-point integer operations with output and intermediate values not exceeding 16-bit word length. This constraint is achieved by the application of multiple bit-shift operations which are inserted between the major process steps.

The construction of the transform matrices is presented in Sects. 8.1.1 and 8.1.2. The fixed-point implementation to achieve the 16-bit constraint is described in Sect. 8.1.3.

8.1.1 Integer DCTs

The HEVC specification includes four integer inverse transform matrices of sizes 4×4, 8×8, 16×16, and 32×32 [1, 2]. These transform matrices are integer approximations of the DCT-II matrix of the same size, aiming at the preservation of the DCT coefficient structure. An additional 4×4 DST matrix is specified which is applied to the residual of intra predicted 4×4 blocks. For distinction from the DST, the four DCTs are referred to as the HEVC *core transforms* [3].

Starting from the goal to preserve the structure and symmetry properties of the DCT matrices as much as possible, the HEVC transform matrices are defined following the construction rules as described in Chap. cha:basics:hc:trafo:dct. The resulting matrices are shown in (8.1)–(8.5). Here, the coefficient values of the 4×4 transform $\mathbf{T}_4$ are denoted by a_0 for the even rows, and b_0, b_1 for the odd rows. For the larger transform matrices $\mathbf{T}_8$ to $\mathbf{T}_{32}$, coefficients c_i, d_j, and e_k are added, with $i = 0, \ldots, 3$, $j = 0, \ldots, 7$, and $k = 0, \ldots, 15$, respectively. The specified values of the transform matrix coefficients are given in Table 8.1.

As in the specification, the transform matrices are given in the DCT-II form, i.e. the matrices are transposed for application in the inverse transform operation. The matrices are constructed as follows:

$$\mathbf{T}_4 = \begin{bmatrix} \mathbf{t}_0 \\ \mathbf{t}_1 \\ \mathbf{t}_2 \\ \mathbf{t}_3 \end{bmatrix} = \begin{bmatrix} a_0 & a_0 & a_0 & a_0 \\ b_0 & b_1 & -b_1 & -b_0 \\ a_0 & -a_0 & -a_0 & a_0 \\ b_1 & -b_0 & b_0 & -b_1 \end{bmatrix}, \tag{8.1}$$

$$\mathbf{T}_8 = \begin{bmatrix} \mathbf{t}_0 \\ \mathbf{t}_1 \\ \vdots \\ \mathbf{t}_7 \end{bmatrix} = \begin{bmatrix} a_0 & a_0 & a_0 & a_0 & a_0 & a_0 & a_0 & a_0 \\ c_0 & c_1 & c_2 & c_3 & -c_3 & -c_2 & -c_1 & -c_0 \\ b_0 & b_1 & -b_1 & -b_0 & -b_0 & -b_1 & b_1 & b_0 \\ c_1 & -c_3 & -c_0 & -c_2 & c_2 & c_0 & c_3 & -c_1 \\ a_0 & -a_0 & -a_0 & a_0 & a_0 & -a_0 & -a_0 & a_0 \\ c_2 & -c_0 & c_3 & c_1 & -c_1 & -c_3 & c_0 & -c_2 \\ b_1 & -b_0 & b_0 & -b_1 & -b_1 & b_0 & -b_0 & b_1 \\ c_3 & -c_2 & c_1 & -c_0 & c_0 & -c_1 & c_2 & -c_3 \end{bmatrix}, \tag{8.2}$$

$$\mathbf{T}_{16} = \begin{bmatrix} \mathbf{t}_0 \\ \mathbf{t}_1 \\ \vdots \\ \mathbf{t}_{15} \end{bmatrix} = \left[\begin{array}{cccccccc|c} a_0 & a_0 & a_0 & a_0 & a_0 & a_0 & a_0 & a_0 & \\ d_0 & d_1 & d_2 & d_3 & d_4 & d_5 & d_6 & d_7 & \\ c_0 & c_1 & c_2 & c_3 & -c_3 & -c_2 & -c_1 & -c_0 & \\ d_1 & d_4 & d_7 & -d_5 & -d_2 & -d_0 & -d_3 & -d_6 & \\ b_0 & b_1 & -b_1 & -b_0 & -b_0 & -b_1 & b_1 & b_0 & \\ d_2 & d_7 & d_3 & d_1 & d_6 & d_4 & d_0 & d_5 & \\ c_1 & -c_3 & -c_0 & -c_2 & c_2 & c_0 & c_3 & -c_1 & \\ d_3 & -d_5 & -d_1 & d_7 & d_0 & d_6 & -d_2 & -d_4 & \cdots \\ a_0 & -a_0 & -a_0 & a_0 & a_0 & -a_0 & -a_0 & a_0 & \\ d_4 & -d_2 & -d_6 & d_0 & -d_7 & -d_1 & d_5 & d_3 & \\ c_2 & -c_0 & c_3 & c_1 & -c_1 & -c_3 & c_0 & -c_2 & \\ d_5 & -d_0 & d_4 & d_6 & -d_1 & d_3 & d_7 & -d_2 & \\ b_1 & -b_0 & b_0 & -b_1 & -b_1 & b_0 & -b_0 & b_1 & \\ d_6 & -d_3 & d_0 & -d_2 & d_5 & d_7 & -d_4 & d_1 & \\ c_3 & -c_2 & c_1 & -c_0 & c_0 & -c_1 & c_2 & -c_3 & \\ d_7 & -d_6 & d_5 & -d_4 & d_3 & -d_2 & d_1 & -d_0 & \end{array}\right], \tag{8.3}$$

and finally, the largest transform $\mathbf{T}_{32}$ shown in Fig. 8.1.

In (8.3) and (8.5), only the left half of the columns of the transform matrices is given due to the large size. The missing right half of the colums is generated by mirroring the left side at the indicated boundary with even symmetry for the even lines and odd symmetry for the odd lines, as indicated in (2.12).

As can be seen, the complete set of core transform matrices can be specified by 31 coefficient values, which are present in the 32×32 transform matrix. For orthogonal transforms of this structure, the norms of the transforms relate as

$$\|\mathbf{T}_{2n}\| = \sqrt{2}\,\|\mathbf{T}_n\|, \quad \text{with} \quad n = 4, 8, 16. \tag{8.5}$$

Hence, in the block-wise application, the norms of $\mathbf{T}_n$ and $\mathbf{T}_{2n}$ relate by a factor of 2. The norms have to be taken into account in the design of the quantization and normalization.

For realization in integer arithmetic, integer coefficient values have to be found which approximate the suitably scaled DCT coefficients. At the same time, the base vectors of the integer matrices should have the same—or at least very similar—norm to allow for joint normalization and for application of the same quantizer step sizes to the resulting transform coefficients, regardless of their position in the transformed

$$\mathbf{T}_{32} =$$

$$\left[\begin{array}{cccccccccccccccc|c}
a_0 & a_0 & a_0 & a_0 & a_0 & a_0 & a_0 & a_0 & a_0 & a_0 & a_0 & a_0 & a_0 & a_0 & a_0 & a_0 & \\
e_0 & e_1 & e_2 & e_3 & e_4 & e_5 & e_6 & e_7 & e_8 & e_9 & e_{10} & e_{11} & e_{12} & e_{13} & e_{14} & e_{15} & \\
d_0 & d_1 & d_2 & d_3 & d_4 & d_5 & d_6 & d_7 & -d_7 & -d_6 & -d_5 & -d_4 & -d_3 & -d_2 & -d_1 & -d_0 & \\
e_1 & e_4 & e_7 & e_{10} & e_{13} & -e_{15} & -e_{12} & -e_9 & -e_6 & -e_3 & -e_0 & -e_2 & -e_5 & -e_8 & -e_{11} & -e_{14} & \\
c_0 & c_1 & c_2 & c_3 & -c_3 & -c_2 & -c_1 & -c_0 & -c_0 & -c_1 & -c_2 & -c_3 & c_3 & c_2 & c_1 & c_0 & \\
e_2 & e_7 & e_{12} & -e_{14} & -e_9 & -e_4 & -e_0 & -e_5 & -e_{10} & -e_{15} & e_{11} & e_6 & e_1 & e_3 & e_8 & e_{13} & \\
d_1 & d_4 & d_7 & -d_5 & -d_2 & -d_0 & -d_3 & -d_6 & d_6 & d_3 & d_0 & d_2 & d_5 & -d_7 & -d_4 & -d_1 & \\
e_3 & e_{10} & -e_{14} & -e_7 & -e_0 & -e_6 & -e_{13} & e_{11} & e_4 & e_2 & e_9 & -e_{15} & -e_8 & -e_1 & -e_5 & -e_{12} & \\
b_0 & b_1 & -b_1 & -b_0 & -b_0 & -b_1 & b_1 & b_0 & b_0 & b_1 & -b_1 & -b_0 & -b_0 & -b_1 & b_1 & b_0 & \\
e_4 & e_{13} & -e_9 & -e_0 & -e_8 & e_{14} & e_5 & e_3 & e_{12} & -e_{10} & -e_1 & -e_7 & e_{15} & e_6 & e_2 & e_{11} & \\
d_2 & d_7 & -d_3 & -d_1 & -d_6 & d_4 & d_0 & d_5 & -d_5 & -d_0 & -d_4 & d_6 & d_1 & d_3 & -d_7 & -d_2 & \\
e_5 & -e_{15} & -e_4 & -e_6 & e_{14} & e_3 & e_7 & -e_{13} & -e_2 & -e_8 & e_{12} & e_1 & e_9 & -e_{11} & -e_0 & -e_{10} & \\
c_1 & -c_3 & -c_0 & -c_2 & c_2 & c_0 & c_3 & -c_1 & -c_1 & c_3 & c_0 & c_2 & -c_2 & -c_0 & -c_3 & c_1 & \\
e_6 & -e_{12} & -e_0 & -e_{13} & e_5 & e_7 & -e_{11} & -e_1 & -e_{14} & e_4 & e_8 & -e_{10} & -e_2 & -e_{15} & e_3 & e_9 & \\
d_3 & -d_5 & -d_1 & d_7 & d_0 & d_6 & -d_2 & -d_4 & d_4 & d_2 & -d_6 & -d_0 & -d_7 & d_1 & d_5 & -d_3 & \\
e_7 & -e_9 & -e_5 & e_{11} & e_3 & -e_{13} & -e_1 & e_{15} & e_0 & e_{14} & -e_2 & -e_{12} & e_4 & e_{10} & -e_6 & -e_8 & \\
a_0 & -a_0 & -a_0 & a_0 & a_0 & -a_0 & -a_0 & a_0 & a_0 & -a_0 & -a_0 & a_0 & a_0 & -a_0 & -a_0 & a_0 & \cdots \\
e_8 & -e_6 & -e_{10} & e_4 & e_{12} & -e_2 & -e_{14} & e_0 & -e_{15} & -e_1 & e_{13} & e_3 & -e_{11} & -e_5 & e_9 & e_7 & \\
d_4 & -d_2 & -d_6 & d_0 & -d_7 & -d_1 & d_5 & d_3 & -d_3 & -d_5 & d_1 & d_7 & -d_0 & d_6 & d_2 & -d_4 & \\
e_9 & -e_3 & -e_{15} & e_2 & -e_{10} & -e_8 & e_4 & e_{14} & -e_1 & e_{11} & e_7 & -e_5 & -e_{13} & e_0 & -e_{12} & -e_6 & \\
c_2 & -c_0 & c_3 & c_1 & -c_1 & -c_3 & c_0 & -c_2 & -c_2 & c_0 & -c_3 & -c_1 & c_1 & c_3 & -c_0 & c_2 & \\
e_{10} & -e_0 & e_{11} & e_9 & -e_1 & e_{12} & e_8 & -e_2 & e_{13} & e_7 & -e_3 & e_{14} & e_6 & -e_4 & e_{15} & e_5 & \\
d_5 & -d_0 & d_4 & d_6 & -d_1 & d_3 & d_7 & -d_2 & d_2 & -d_7 & -d_3 & d_1 & -d_6 & -d_4 & d_0 & -d_5 & \\
e_{11} & -e_2 & e_6 & -e_{15} & -e_7 & e_1 & -e_{10} & -e_{12} & e_3 & -e_5 & e_{14} & e_8 & -e_0 & e_9 & e_{13} & -e_4 & \\
b_1 & -b_0 & b_0 & -b_1 & -b_1 & b_0 & -b_0 & b_1 & b_1 & -b_0 & b_0 & -b_1 & -b_1 & b_0 & -b_0 & b_1 & \\
e_{12} & -e_5 & e_1 & -e_8 & e_{15} & e_9 & -e_2 & e_4 & -e_{11} & -e_{13} & e_6 & -e_0 & e_7 & -e_{14} & -e_{10} & e_3 & \\
d_6 & -d_3 & d_0 & -d_2 & d_5 & d_7 & -d_4 & d_1 & -d_1 & d_4 & -d_7 & -d_5 & d_2 & -d_0 & d_3 & -d_6 & \\
e_{13} & -e_8 & e_3 & -e_1 & e_6 & -e_{11} & -e_{15} & e_{10} & -e_5 & e_0 & -e_4 & e_9 & -e_{14} & -e_{12} & e_7 & -e_2 & \\
c_3 & -c_2 & c_1 & -c_0 & c_0 & -c_1 & c_2 & -c_3 & -c_3 & c_2 & -c_1 & c_0 & -c_0 & c_1 & -c_2 & c_3 & \\
e_{14} & -e_{11} & e_8 & -e_5 & e_2 & -e_0 & e_3 & -e_6 & e_9 & -e_{12} & e_{15} & e_{13} & -e_{10} & e_7 & -e_4 & e_1 & \\
d_7 & -d_6 & d_5 & -d_4 & d_3 & -d_2 & d_1 & -d_0 & d_0 & -d_1 & d_2 & -d_3 & d_4 & -d_5 & d_6 & -d_7 & \\
e_{15} & -e_{14} & e_{13} & -e_{12} & e_{11} & -e_{10} & e_9 & -e_8 & e_7 & -e_6 & e_5 & -e_4 & e_3 & -e_2 & e_1 & -e_0 &
\end{array}\right] \quad (8.4)$$

Fig. 8.1 Transform matrix $\mathbf{T}_{32}$

residual block. This is beneficial for a simple and lean design of the quantization process.

For transform sizes of 4×4 and 8×8, integer coefficient values a_0 to c_3 can be found such that the resulting matrices approximate the DCT and are orthogonal with identical or different norms for all base vectors, e.g. the transform matrices given in (2.30) and (2.31).[1] For transform matrices of size 16×16 (and thereby, also for 32×32), such sets of suitable integer coefficients have not been found [4]. In order to establish an integer transform set including all specified transform sizes, the orthogonality constraint has been relaxed in the HEVC design. The coefficients were selected to make the matrices 'almost orthogonal', with a deviation so small that the effect is negligible in the overall transform and quantization process.

[1] Due to symmetry, the 4×4 matrix in (8.1) reveals orthogonality with equal or different norms of the base vectors for arbitrary real values $a_0, b_0, b_1 \neq 0$.

In consideration of the aforementioned 16-bit constraint for output and intermediate values, the applicable coefficient values have been defined with 8 bit precision (7 bit amplitude plus sign). The a_0 coefficient for the DC base vector $\mathbf{t}_0$ is set to $a_0 = 64$. Starting from a scaled and rounded $N \times N$ DCT matrix, with $\alpha = a_0\sqrt{N}$,

$$\mathbf{T}_{N,\text{pre}} = \text{round}\left\{\alpha \cdot \mathbf{T}_{\text{DCT},N}\right\}, \quad \text{with} \quad N = 4, 8, 16, 32, \tag{8.6}$$

the coefficient values have been optimized according to the following criteria [5]:

- The approximative orthogonality or leakage of the base vectors, by minimizing

$$o_{ij} = \frac{\mathbf{t}_i^{\mathsf{T}}\mathbf{t}_j}{\mathbf{t}_0^{\mathsf{T}}\mathbf{t}_0}, \tag{8.7}$$

- the closeness to the DCT coefficients, by measuring and minimizing for each coefficient

$$m_{ij} = \frac{1}{a_0}\left|\alpha \cdot t_{\text{DCT},i}(j) - t_i(j)\right|, \text{ and} \tag{8.8}$$

- the deviation of the norms, by measuring and minimizing for the base vectors

$$n_i = \left|1 - \frac{\mathbf{t}_i^{\mathsf{T}}\mathbf{t}_i}{\mathbf{t}_0^{\mathsf{T}}\mathbf{t}_0}\right|. \tag{8.9}$$

Using and minimizing these criteria, the HEVC transform coefficient values as listed in Table 8.1 have been derived.

Inspecting the resulting 8 × 8 transform matrix,

Table 8.1 Coefficients of the HEVC core transforms

$\mathbf{T}_4$		$\mathbf{T}_8$	$\mathbf{T}_{16}$	$\mathbf{T}_{32}$	
$a_0 = 64$	$b_0 = 83$	$c_0 = 89$	$d_0 = 90$	$e_0 = 90$	$e_8 = 61$
	$b_1 = 36$	$c_1 = 75$	$d_1 = 87$	$e_1 = 90$	$e_9 = 54$
		$c_2 = 50$	$d_2 = 80$	$e_2 = 88$	$e_{10} = 46$
		$c_3 = 18$	$d_3 = 70$	$e_3 = 85$	$e_{11} = 38$
			$d_4 = 57$	$e_4 = 82$	$e_{12} = 31$
			$d_5 = 43$	$e_5 = 78$	$e_{13} = 22$
			$d_6 = 25$	$e_6 = 73$	$e_{14} = 13$
			$d_7 = 9$	$e_7 = 67$	$e_{15} = 4$

Table 8.2 DCT approximation of the core transforms

Measure	4×4	8×8	16×16	32×32
o_{ij}	0	<0.0016	<0.0029	<0.0029
m_{ij}	<0.0213	<0.0213	<0.0213	<0.0213
n_i	<0.0009	<0.0009	<0.0009	<0.0013

$$\mathbf{T}_8 = \begin{bmatrix} 64 & 64 & 64 & 64 & 64 & 64 & 64 & 64 \\ 89 & 75 & 50 & 18 & -18 & -50 & -75 & -89 \\ 83 & 36 & -36 & -83 & -83 & -36 & 36 & 83 \\ 75 & -18 & -89 & -50 & 50 & 89 & 18 & -75 \\ 64 & -64 & -64 & 64 & 64 & -64 & -64 & 64 \\ 50 & -89 & 18 & 75 & -75 & -18 & 89 & -50 \\ 36 & -83 & 83 & -36 & -36 & 83 & -83 & 36 \\ 18 & -50 & 75 & -89 & 89 & -75 & 50 & -18 \end{bmatrix}, \tag{8.10}$$

the orthogonality test reveals

$$\mathbf{T}_8 \cdot \mathbf{T}_8^{\mathsf{T}} = \begin{bmatrix} 32768 & 0 & 0 & 0 & 0 & 0 & 0 & 0 \\ 0 & 32740 & 0 & -50 & 0 & 50 & 0 & 0 \\ 0 & 0 & 32740 & 0 & 0 & 0 & 0 & 0 \\ 0 & -50 & 0 & 32740 & 0 & 0 & 0 & 50 \\ 0 & 0 & 0 & 0 & 32768 & 0 & 0 & 0 \\ 0 & 50 & 0 & 0 & 0 & 32740 & 0 & 50 \\ 0 & 0 & 0 & 0 & 0 & 0 & 32740 & 0 \\ 0 & 0 & 0 & 50 & 0 & 50 & 0 & 32740 \end{bmatrix}. \tag{8.11}$$

The variation of the elements on the main diagonal of this matrix results in a very small norm deviation value. Thereby, the base vectors are considered to have approximately the same norm and can therefore be jointly treated for normalization and quantization. The deviation from true orthogonality, which is evident from the ± 50 values aside the main diagonal, is small as well. The approximation quality of the HEVC DCT matrices is summarized in Table 8.2 reporting the values of the optimization measures for all transform sizes.

For efficient implementation of the DCT based transform operations, the symmetry of the matrices can be utilized. An optimized implementation design has been proposed in [5].

8.1.2 *Integer* 4×4 *DST*

Besides the core transforms, which resemble an integer implementation of the conventionally used DCT, an additional integer DST transform is specified for 4×4 intra blocks. In literature, the application of transforms other than the DCT to the prediction error signal in directional intra prediction has been shown to be beneficial. Specifically, the suitability of the DST-VI for one-dimensional prediction by a neighbor value has been derived by Han et al. [6]. A very similar derivation was provided by Yeo et al. in a proposal to the JCT-VC at about the same time [7]. While originally, the intention of proposals was to select the applicable transform (DCT or DST) based on the direction of the intra prediction mode in 4×4 blocks, the design was simplified to only use the DST [8].

The 4×4 integer DST matrix specified in HEVC can be derived from the definition of the DST-VI in (2.17) by scaling the transform coefficients by 128 and rounding to the next integer,

$$\mathbf{T}_{\text{DST}} = \begin{bmatrix} 29 & 55 & 74 & 84 \\ 74 & 74 & 0 & -74 \\ 84 & -29 & -74 & 55 \\ 55 & -84 & 74 & -29 \end{bmatrix}. \tag{8.12}$$

This DST matrix has been proposed in [7] alongside a proposal for a fast implementation. Since the scaling of the transform coefficients corresponds to the scaling of the 4×4 DCT coefficients according to (8.6), it can be seamlessly integrated into the normalization and quantization process further described below.

The matrix is approximately orthogonal, as can be seen from

$$\mathbf{T}_{\text{DST}} \cdot \mathbf{T}_{\text{DST}}^{\text{T}} = \begin{bmatrix} 16398 & 0 & -15 & 15 \\ 0 & 16428 & 0 & 0 \\ -15 & 0 & 16398 & -15 \\ 15 & 0 & -15 & 16398 \end{bmatrix}. \tag{8.13}$$

Applying the measures introduced for the DCT matrices, the approximation quality is described by

$$\begin{aligned} o_{ij} &< 0.0010, \\ m_{ij} &< 0.0030, \\ n_i &< 0.0027. \end{aligned} \tag{8.14}$$

Theses values have been calculated relative to the base vector $\mathbf{t}_0$ of the DCT, since the scaling factor α is the same for the 4×4 DST and DCT matrices. Comparing (8.14) to the values reported in Table 8.2 it can be seen that the approximation quality of the DST is comparable to the DCT design.

8.1.3 Dynamic Range and Transform Normalization

In order to implement the 16-bit requirement on stored input and output values as well as the limitation to 32-bit operations at maximum, scaling operations at all intermediate steps are specified. In the specification text, only the decoder-side scaling operations are included. Here, the complete chain of forward and inverse transform operations is considered as realized in the HEVC reference software [9].

Consider the processing of a $N \times N$ prediction residual block $\mathbf{X}$ and the application of the corresponding $N \times N$ transform $\mathbf{T}$ with $N = 2^M = 4, 8, 16, 32$ and $\mathbf{T}$ being one of the previously defined transform matrices. Following the deduction by Budagavi et al. [5], let

$$\mathbf{X} = -2^{B_\mathrm{d}} \cdot \begin{bmatrix} 1 & \dots & 1 \\ \vdots & \ddots & \vdots \\ 1 & \dots & 1 \end{bmatrix}, \tag{8.15}$$

with B_d denoting the bit depth and -2^{B_d} denoting the assumed maximum possible negative signal value in $\mathbf{X}$. First, the transform in vertical direction is calculated as

$$\mathbf{C}_1 = \mathbf{T} \cdot \mathbf{X} = -2^{B_\mathrm{d}} \cdot 64 \cdot N \cdot \begin{bmatrix} 1 & 1 & \dots & 1 \\ 0 & 0 & \dots & 0 \\ \vdots & & \ddots & \vdots \\ 0 & & \dots & 0 \end{bmatrix}. \tag{8.16}$$

In order to keep the output result of this operation within 16 bit, a scaling factor $s_{\mathrm{f}1}$ is introduced such that

$$\left(-2^{B_\mathrm{d}} \cdot 64 \cdot N\right) \cdot 2^{-s_{\mathrm{f}1}} = -2^{B_\mathrm{d}+M+6-s_{\mathrm{f}1}} = -2^{15}. \tag{8.17}$$

This leads to

$$s_{\mathrm{f}1} = B_\mathrm{d} + M - 9. \tag{8.18}$$

Accordingly, the horizontal transform operation is performed as

$$\mathbf{C}_2 = \mathrm{round}\left\{2^{-s_{\mathrm{f}1}} \cdot \mathbf{C}_1\right\} \cdot \mathbf{T}^\mathrm{T} = -2^{15+M+6} \cdot \begin{bmatrix} 1 & 0 & \dots & 0 \\ 0 & 0 & \dots & 0 \\ \vdots & & \ddots & \vdots \\ 0 & & & 0 \end{bmatrix} = -2^{21+M} \cdot \mathbf{O}_1. \tag{8.19}$$

Following the considerations which led to (8.17), the second scaling factor after the horizontal transform is found as

$$s_{\mathrm{f}2} = 21 + M - 15 = 6 + M, \tag{8.20}$$

and the output becomes $\mathbf{C} = \mathrm{round}\left\{2^{-s_{\mathrm{f}2}} \cdot \mathbf{C}_2\right\}$, i.e. $\mathbf{C} = -2^{15} \cdot \mathbf{O}_1$. This concludes the scaling operations with respect to dynamic range considerations at the encoder side. The subsequent quantization operation is detailed in Sect. 8.1.4.

Now, taking the vertical inverse transform of **C** for reconstruction of the original input at the decoder side,

$$\mathbf{R}_1 = \mathbf{T}^\mathsf{T} \cdot \mathbf{C} = -2^{15} \cdot 64 \cdot \begin{bmatrix} 1 & 0 & \dots & 0 \\ 1 & 0 & \dots & 0 \\ \vdots & & \ddots & \vdots \\ 1 & & \dots & 0 \end{bmatrix}, \tag{8.21}$$

hence, a scaling by 64 is needed after the first inverse transform step, i.e.

$$s_{\mathrm{i1}} = 6. \tag{8.22}$$

The second inverse transform, which is in horizontal direction, leads to

$$\mathbf{R}_2 = \mathrm{round}\left\{2^{-s_{\mathrm{i1}}} \cdot \mathbf{R}_1\right\} \cdot \mathbf{T} = -2^{15} \cdot 64 \cdot \begin{bmatrix} 1 & \dots & 1 \\ \vdots & \ddots & \vdots \\ 1 & \dots & 1 \end{bmatrix}. \tag{8.23}$$

A final scaling with a factor s_{i2} is applied to re-establish the original bit depth B_d, as

$$s_{\mathrm{i2}} = 15 + 6 - B_\mathrm{d} = 21 - B_\mathrm{d}, \tag{8.24}$$

leading to the reconstructed block **R**

$$\mathbf{R} = \mathrm{round}\left\{2^{-s_{\mathrm{i2}}} \cdot \mathbf{R}_2\right\}. \tag{8.25}$$

Note that in this trivial example, perfect reconstruction $\mathbf{R} = \mathbf{X}$ is achieved, which is not generally the case for arbitrary input signals.

The above discussion reveals scaling factors as required for the $\mathbf{t}_0$ base vector of the DCT transforms as only the DC coefficient has been involved in the example. However, an inspection of the norms of the base vectors of the DCT and DST transform matrices shows that for some transforms the maximum norm of the base vectors exceeds the norm of the $\mathbf{t}_0$ base vector. Therefore, the deduction is approximative for the other bases. For illustration, the squared norms and the relative deviation from the squared $\mathbf{t}_0$ norm are summarized in Table 8.3. It can be seen that observed overshoot is small and will therefore hardly affect real coding applications.

8.1.4 Quantizer Design

The applicable quantizer is indicated by a quantization parameter QP which serves as an integer index to derive the applicable quantizer step size Δ_q. As in H.264 | AVC, the quantization design in HEVC follows a logarithmic structure with a duplication

Table 8.3 Maximum squared transform norms

Transform	$\|\mathbf{t}_0\|^2$	$\max_{n=0,\ldots N-1} \|\mathbf{t}_n\|^2$	$\Delta_{\max}$ [a]
4 × 4 DST	16398	16428	0.27
4 × 4 DCT	$2^{14} = 16384$	16384	0
8 × 8 DCT	$2^{15} = 32768$	32768	0
16 × 16 DCT	$2^{16} = 65536$	65546	0.02
32 × 32 DCT	$2^{17} = 131072$	131244	0.13

[a] $\Delta_{\max} = \frac{\|\mathbf{t}_n\|^2_{\max}}{\|\mathbf{t}_0\|^2} - 1$, for DST relative to 2^{14}

of the quantizer step size between QP and QP + 6. For an even distribution of the quantizer step sizes according to this design, the relation between the step sizes is

$$\Delta_q(\text{QP} + 1) = \sqrt[6]{2} \cdot \Delta_q(\text{QP}). \tag{8.26}$$

Defining $\Delta_q = 1$ for QP = 4 [5], the first six quantizer step sizes are

$$\Delta_{q,0} \in \left\{2^{-\frac{4}{6}}, 2^{-\frac{3}{6}}, 2^{-\frac{2}{6}}, 2^{-\frac{1}{6}}, 1, 2^{\frac{1}{6}}\right\}. \tag{8.27}$$

The quantizer step sizes for QP > 5 are derived by scaling,

$$\Delta_q(\text{QP}) = \Delta_{q,0}(\text{QP} \bmod 6) \cdot 2^{\left\lfloor \frac{\text{QP}}{6} \right\rfloor}. \tag{8.28}$$

With the quantizer range defined to

$$\text{QP} = 0, \ldots, 51, \tag{8.29}$$

the quantizer step size is in the range of $0.630 \leq \Delta_q \leq 228.1$. Thereby, the maximum quantizer step size is in the range of the maximum amplitude of the 8 bit input signal. For higher bit-depths, a QP offset is specified, increasing the QP parameter range by 6 per additional bit of signal bit depth towards finer quantization (i.e. towards lower QP values).

8.1.4.1 Integer Quantization Implementation

In the required integer design, the quantizer step sizes are approximated by corresponding integer quotients, see also (2.36). In the HEVC design, the correspondingly scaled integer quantizer step sizes are generated by

$$g_q = \text{round}\left\{2^6 \cdot \Delta_q\right\}. \tag{8.30}$$

This precision provides a deviation from the relation of successive quantizer steps in (8.26) of less than 2 %. Thereby, the set of integer quantizer steps according to (8.27) becomes

$$g_{q,0} \in \{40, 45, 51, 57, 64, 72\}. \tag{8.31}$$

On the encoder side, the division by the quantizer step size can be approximated by a scaling factor and a corresponding bit shift, see (2.36). In the HEVC reference software, the factors are set to

$$f_q = \text{round}\left\{\frac{2^{14}}{\Delta_q}\right\}, \tag{8.32}$$

resulting in

$$f_{q,0} \in \{26{,}214,\ 23{,}302,\ 20{,}560,\ 18{,}396,\ 16{,}384,\ 14{,}564\}. \tag{8.33}$$

This choice results in $f_{q,0} \cdot g_{q,0} \approx 2^{20}$ with a deviation of 0.003 % at maximum.

The full specification of the forward and inverse quantization operations can be written as follows, including additional scaling factors 2^{-s_q} and $2^{-s_{iq}}$ on the encoder and decoder side, respectively. At the encoder side, the quantizer level for the coefficient c is derived as

$$n_q = \text{round}\left\{c \cdot \left[f_{q,0}(\text{QP} \bmod 6) \cdot 2^{-\left\lfloor \frac{\text{QP}}{6} \right\rfloor}\right] \cdot 2^{-s_q}\right\}. \tag{8.34}$$

On the decoder side, the quantizer level is scaled to the quantized coefficient value as

$$c_q = \text{round}\left\{n_q \cdot \left[g_{q,0}(\text{QP} \bmod 6) \cdot 2^{\left\lfloor \frac{\text{QP}}{6} \right\rfloor}\right] \cdot 2^{-s_{iq}}\right\}. \tag{8.35}$$

The scaling factors 2^{-s_q} and $2^{-s_{iq}}$ are introduced to keep track of the transform dynamic range constraints and the scaling introduced by quantization. Multiplication of a block with the $N \times N$ transform matrix for horizontal and vertical transformation scales the result by $a_0^2 \cdot N = 2^{12+M}$, with $N = 2^M$. Taking into account the scaling factors s_{f1}, s_{f2}, and s_q on the encoder side, it can be seen that the output of the transform and quantization stage is scaled by

$$2^{(12+M)/2} \cdot 2^{-s_{f1}} \cdot 2^{(12+M)/2} \cdot 2^{-s_{f2}} \cdot 2^{14} \cdot 2^{-s_q} = 2^{29-B_d-M} \cdot 2^{-s_q}, \tag{8.36}$$

compared to the usage of a normalized transform and non-integer quantization. Accordingly, the scaling factor s_q must be set to

$$s_q = 29 - B_d - M, \tag{8.37}$$

to level out the effect of the integer implementation.

On the decoder side, the same consideration reveals a scaling by

$$2^{(12+M)/2} \cdot 2^{-s_{i1}} \cdot 2^{(12+M)/2} \cdot 2^{-s_{i2}} \cdot 2^{6} \cdot 2^{-s_{iq}} = 2^{B_d+M-9} \cdot 2^{-s_{iq}}, \quad (8.38)$$

compared to the operation with normalized transforms. Accordingly, the scaling factor s_{iq} is set to

$$s_{iq} = B_d + M - 9. \quad (8.39)$$

8.1.4.2 Dynamic Range Considerations

It may happen that the intended 16-bit dynamic range is violated due to upward rounding to a high level for large transform coefficient values at high QPs. In order to reduce the danger of exceeding the dynamic range after the first inverse transform at the decoder side, the scaling factor s_{i1} is increased by 1 to [5]

$$s_{i1} = 7. \quad (8.40)$$

At the same time, the scaling factor s_{i2} is reduced by 1

$$s_{i2} = 20 - B_d, \quad (8.41)$$

to keep the normalization of the overall transform operation. The increased scaling after the first transform enlarges the headroom for the intermediate transform step of the dequantized coefficient values. In order to prevent any possibility of a dynamic range overflow at this intermediate stage, an additional clipping operation is specified which sets a hard limit to the intermediate transform coefficients c_{r1} of the scaled $\mathbf{R}_1$ matrix to[2]

$$c_{r1c} = \text{clip}\left\{-2^{15}, 2^{15}-1, c_{r1}\right\}. \quad (8.42)$$

Thereby, it is asserted that the output of the second inverse transform stage will not exceed the range of 16-bit dynamic range.

8.1.5 *Quantizer Weighting Matrix*

The quantization scheme as discussed in the previous section applies a constant quantizer step size to all transform coefficients regardless of the frequency of the corresponding base function. As discussed in Sect. 2.4.5.6, a quantization weighting

---

[2] Using $\text{clip}\{a, b, v\} = \begin{cases} a, & \text{if } v < a \\ v, & \text{if } a \le v \le b \\ b, & \text{if } b < v \end{cases}$

(a)

16	16	16	16	17	18	21	24
16	16	16	16	17	19	22	25
16	16	17	18	20	22	25	29
16	16	18	21	24	27	31	36
17	17	20	24	30	35	41	47
18	19	22	27	35	44	54	65
21	22	25	31	41	54	70	88
24	25	29	36	47	65	88	115

(b)

16	16	16	16	17	18	20	24
16	16	16	17	18	20	24	25
16	16	17	18	20	24	25	28
16	17	18	20	24	25	28	33
17	18	20	24	25	28	33	41
18	20	24	25	28	33	41	54
20	24	25	28	33	41	54	71
24	25	28	33	41	54	71	91

Fig. 8.2 Illustration of the default quantizer weights $w_q(x, y)$ for an 8×8 coefficient block. **a** Default intra weights. **b** Default inter weights

matrix can be used to allow for modification of the quantizer step size in dependency of the coefficient position, i.e. the base function frequency, according to the application needs. In HEVC, three options can be configured for the operation of the quantizer:

- Flat quantization: no application of frequency dependent coefficient weighting
- Default weighting matrix: the weighting of the coefficients is performed according to a selectable predefined weighting table
- Custom weighting matrix: the applicable weighting is signaled relative to one of the predefined weighting tables

The default quantization weighting matrix is initialized from reference scaling lists on a picture level. The weighting values are scaled by 16, which is taken into account in the overall decoder-side scaling process, see the next section. Six applicable quantization weighting matrices are established, discriminating between inter and intra predicted CUs for each of the three color components Y, Cb, and Cr. The default matrices for 4×4 blocks have all entries set to 16, hence, no specific weighting is specified here. For the other block sizes, two different scaling lists are used, which are differentiating between intra and inter predicted blocks. Thereby, identical weighting matrices for luma and chroma blocks are applied when using the default scaling lists. The initialization is visualized in Fig. 8.2, where the values of the scaling lists are assigned to the elements $m(x, y)$ of the quantizer weighting matrix along the diagonal scan of an 8×8 block.

For 16×16 blocks, the same scaling lists are applied, with the values of the 8×8 quantization scaling matrix being assigned to 2×2 sets of coefficients. The same approach is applied for the 32×32 transform blocks, where the values are applied to 4×4 sets of coefficients, accordingly. Note that for 32×32 blocks, quantization weighting is only available for the luma component. The default scaling sets for both, intra and inter predicted 32×32 blocks, are initialized according to Fig. 8.2a. The

32×32 quantization weighting matrix shares the matrix identifier with the first two (Y, Cb) intra matrices for the other block sizes.

Dedicated custom quantization matrices can be signaled in the SPS or the PPS for each of the six quantization matrix indices. A dedicated scaling list entry for the DC coefficient can be signaled for the 16×16 and 32×32 blocks with a scaling factor range of $1, \ldots, 255$ (signaled with an offset of 8). For each coefficient position along the scan the delta to the previous coefficient is signaled (in the range of $-128, \ldots, 127$), where all resulting scaling list entries must be positive.

8.1.6 Decoder-Side Weighting and Level Scaling Operation

The scaling process to derive the quantized coefficient value as specified in (8.35) needs to be extended to include the weighting by the quantizer weighting matrix as applicable. Furthermore, clipping needs to be implemented to assert that the stored values do not exceed the required 16-bit word length. For a coefficient at the position (x, y) in the transform coefficient matrix, $m(x, y)$ denotes the value of the quantizer weighting matrix at the coefficient position (x, y). If the quantizer weighting matrix is not applied, $m(x, y) = 16$ is used. The scaling factor s_{iq} is updated with regard to this additional factor to

$$s_{\mathrm{iq}} = B + M - 5. \tag{8.43}$$

The scaling of the quantizer levels can then be written with bit shift operations,

$$\tilde{c}_{\mathrm{q}}(x, y) = \left[n_{\mathrm{q}}(x, y) \cdot m(x, y) \cdot \left(g_{\mathrm{q},0}(\mathrm{QP} \bmod 6) \ll \left\lfloor \frac{\mathrm{QP}}{6} \right\rfloor \right) + \left(1 \ll (s_{\mathrm{iq}} - 1)\right) \right] \gg s_{\mathrm{iq}}, \tag{8.44}$$

$$c_{\mathrm{q}}(x, y) = \mathrm{clip}\left\{ -2^{15}, 2^{15} - 1, \tilde{c}_{\mathrm{q}}(x, y) \right\}.$$

8.1.7 Signaling of the Quantization Parameter

Within a slice or tile (or a row in case of wavefront parallel processing), the quantization parameter for the first CU is derived from the initial QP indicated in the PPS. This is updated by a QP delta value in the slice segment header, and further refined by a potential CU level QP delta. The applicable QP for a CU is derived from the QP applied in the previous CU in decoding order and, if available, the current CU level QP update. The value of the applicable QP delta is transmitted in the first TU of the given CU. The delta QP values at the PPS and slice segment level are coded with a signed Exp-Golomb code (see Sect. 2.4.8.4). The delta QP value at the CU level is coded with CABAC (for the binarization see Sect. 10.2.3.9). Methods for derivation of rate-distortion optimized quantization parameters and rate-distortion optimized selection of quantization levels on a block basis have been proposed by Karczewicz et al., see e.g. [10]. Such methods are available in the HEVC test model software as well [11].

The applicable QP value for the chroma components is derived from the luma QP value. Finer quantization is applied for chroma quantization from QP $>$ 30 on to prevent complete cancellation of the chroma transform coefficients. The relation between the luma and chroma QP values has been shown in Fig. 2.16.

The granularity in which the quantization parameter can be modified at the CU level is controlled by coding unit quantization groups which are configured in the PPS. The granularity can be configured between the luma coding tree block size and the smallest luma coding block size configured in the SPS.

8.2 Coded Representation of Transform Blocks

As introduced in Sect. 4.2.6, a CB is partitioned into a residual quadtree of transform blocks for adaptation of the applicable transforms for the residual. Here, the coded representation of a transform block is detailed.

The residual quadtree is jointly signaled for the three color components down to a luma block size of 8×8. While the residual quadtree for the luma component may further split into a transform block size of 4×4, the depth of the chroma transform tree remains at this decomposition level as the chroma components are already using the smallest transform size of 4×4 in 4:2:0 video.

Due to the high efficiency of the prediction tools and depending on the selected quantizer step size, the array of quantized residual can be sparse with a very low number of coefficients, which in turn may only have small values each. The coded representation of the transform blocks is designed to efficiently encode such sparse blocks. As a second design criterion, the scheme has been optimized to allow a high throughput rate for the CABAC entropy coding stage, enabling as much parallelism as possible.[3]

8.2.1 *Transform Sub-Blocks*

Since the number of existing quantized transform coefficients often is small or even zero, and since it is further likely that local areas of the transform coefficient array do not contain any significant coefficients (i.e., coefficients unequal to zero), a hierarchical approach is chosen for the coded representation of the transform coefficients.

For each color component, a separate flag indicates if the transform block contains any significant coefficients. Transform blocks larger than 4×4 are partitioned into

[3] Detailed treatments of the design and a throughput analysis can be found in [12, 13].

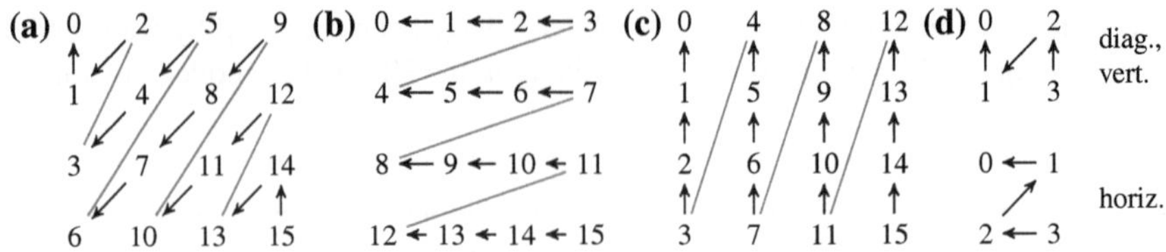

Fig. 8.3 Example coefficient scan patterns in transform sub-blocks. **a** 4×4 diagonal. **b** 4×4 horizontal. **c** 4×4 vertical. **d** Transform sub-block scans in 8×8 transform blocks

non-overlapping transform sub-blocks (TSBs) of 4×4 coefficients each. A dedicated coded sub-block flag indicates the existence of at least one significant transform coefficient in the TSB. For the top-left TSB containing the DC coefficient, the flag is not coded but implicitly set to 1 as the probability of all coefficients in this transform sub-block being quantized to zero is very low.

A set of three different scan patterns is available for scanning the transform coefficients in a transform sub-block. The scans are shown in Fig. 8.3. For inter coded blocks, a diagonal scan is used for all block sizes. For intra coded blocks of size 4×4 or 8×8, the three scanning patterns diagonal, horizontal, and vertical are available. The applicable scanning pattern depends on the direction of the intra prediction mode.[4] For blocks of size 16×16 and 32×32, only the diagonal scan is applied. For all block sizes, the ordering of the transform sub-blocks in the transform block follows the same pattern as the coefficients scan within the TSB.

In Fig. 8.3, the position numbers for the scans of a 4×4 block are indicated. The arrows indicate the direction which is followed along the scan pass. It can be seen that for coding, the ordering of the coefficients along the pass is inverted. I.e. the scan starts at the last coefficient position and follows the scan towards the first scan position. This orientation of the scans has already been used for transform coefficient coding in H.264 | AVC. According to the distribution of transform coefficient levels in a transform block, it is likely that significant coefficients of high frequency base functions only have levels of ± 1 while it is more likely to also have larger values towards lowest frequencies. Along the forward scan direction, such sets of just ± 1 levels at the end of the forward scan are denoted as 'trailing ones'. In H.264 | AVC, the number of the 'trailing ones' is used in the VLC-based entropy coding process for transform coefficients. In HEVC, this property is used in the CABAC context selection, see Sect. 10.2.5.

8.2.2 *Last Significant Coefficient Position*

As mentioned before, it is likely that significant coefficients concentrate towards the low frequency bases in the transform block. In turn, transform sub-blocks at higher frequency positions may only contain non-significant zero coefficients. In order to

---

[4] Intra prediction directions are shown in Fig. 6.1. Horizontal modes 22–30 use the vertical scan; vertical modes 6–14 use the horizontal scan, other modes use the diagonal scan.

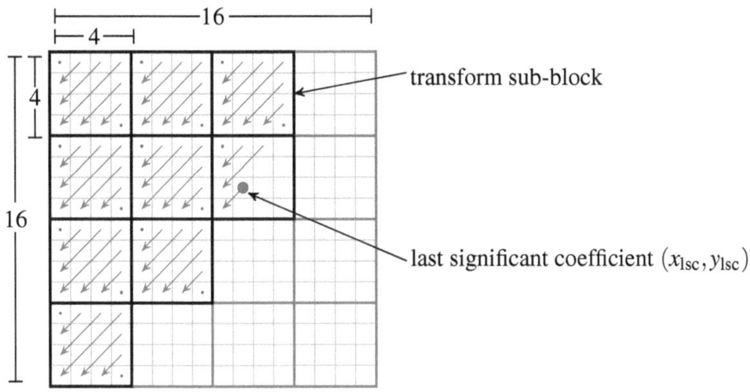

Fig. 8.4 Example sub-block scanning of a 16×16 transform block. The 4×4 transform sub-blocks with inverse coefficient scan and the last significant coefficient are indicated

efficiently indicate the set of significant transform sub-blocks, the position of the last significant coefficient relative to the top-left corner is signaled for the transform block.

Starting from the transform sub-block containing the last significant coefficient position TSB_{lsc}, the transform sub-blocks along the inverse scan are coded. For all transform sub-blocks subsequent to TSB_{lsc} but the top left one including the DC coefficient, the coded sub-block flag indicates the need of TSB coefficient coding. Within the transform sub-block containing the last significant coefficient position, the coefficient scan starts at this already coded position. An example is given for a 16×16 transform block in Fig. 8.4.

8.2.3 Transform Block Coefficient Coding

For transform sub-blocks that are indicated to be significant as a block, the quantizer levels n_q of the coefficients are encoded. The information on the significance of a coefficient, the level, as well as the sign of the coefficient are coded sequentially in separate scans. Up to five scan passes are employed for coding as follows:

- Significance Map—Indicates the position of significant coefficients. A significance flag is coded for each coefficient position in the transform sub-block.
- Level Greater than 1—Indicates if the level is greater than 1 for each significant coefficient. The flag is only sent for the first eight significant coefficients in order to limit the number of non-bypass coded symbols in transform coding [12].
- Level Greater than 2—Indicates if the coefficient level is greater than 2 up to and including the first coefficient with this property. The flag is only sent for coefficients larger than 1 as indicated by the previous scan. After the first coefficient which is

Fig. 8.5 Example 4 × 4 transform sub-block with coefficient levels

13	2	8	1
10	-5	1	0
4	-3	0	0
-1	0	1	0

Table 8.4 Transform sub-block scans for the TSB in Fig. 8.5

Coefficient	1	0	0	0	1	1	−3	−1	8	−5	4	2	10	13
Significance flag	–	0	0	0	1	1	1	1	1	1	1	1	1	1
Coeff. > 1	0	–	–	–	0	0	1	0	1	1	1	1	–	–
Coeff. > 2	–	–	–	–	–	–	1	–	–	–	–	–	–	–
Sign flag	0	–	–	–	0	0	1	1	0	1	0	0	0	0
Remaining level	–	–	–	–	–	–	1	–	7	4	3	1	10	13

greater than two, the flag is not further sent and the levels of subsequent coefficients greater than 2 are coded using the scan for remaining absolute levels, see below.

- Coefficient Sign—The sign information for the significant coefficients.
- Remaining Absolute Level—Level information which has not been covered by the previous scan passes.

As can be seen from this list, the significance scan is the only scan which includes all coefficient positions (up to the indicated last significant position). Each of the succeeding scan passes provides further information for the significant coefficients in the transform block. Information that can be derived from the previously coded syntax elements is not explicitly coded.

For illustration, an example 4 × 4 TSB is given in Fig. 8.5, where $n_q(2, 3) = 1$ is assumed to be the last significant coefficient in the transform block. The corresponding scan as described above is shown in Table 8.4.

For all coefficients but the last significant coefficient, a significance flag is coded. In the next scan, the coefficient greater that 1 flag is coded for the following eight significant coefficients. The coefficient $n_q(1, 2) = -3$ is the first coefficient with an absolute level greater than 1. Therefore, the corresponding flag is coded for this coefficient. After the set of signs for all significant coefficients, the remaining level values are coded. For the coefficient $n_q(1, 2)$, the value of 1 remains corresponding to the previously encoded greater-than-1 and greater-than-2 flags. For the coefficients for which the greater-than-1 flag is coded, the absolute value $n_q - 1$ remains. For the last two coefficients, the remaining level value corresponds to the absolute value of the coefficients.

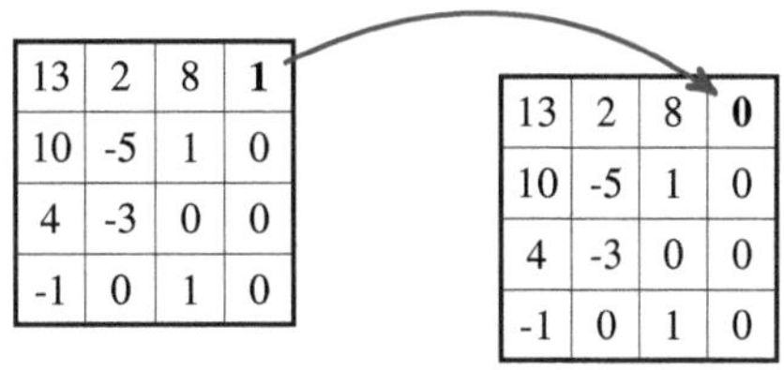

Fig. 8.6 Modified coefficient value in the example 4 × 4 transform sub-block for sign data hiding

Table 8.5 Modified transform sub-block scans for the modified TSB in Fig. 8.6 with × denoting the syntax elements which are not coded anymore

Coefficient	1	0	0	0	0	1	−3	−1	8	−5	4	2	10	13
Significance flag	–	0	0	0	0	1	1	1	1	1	1	1	1	1
Coeff. > 1	0	–	–	–	×	0	1	0	1	1	1	1	–	–
Coeff. > 2	–	–	–	–	–	–	1	–	–	–	–	–	–	–
Sign flag	0	–	–	–	–	0	1	1	0	1	0	0	0	×
Remaining level	–	–	–	–	–	–	1	–	7	4	3	1	10	13

8.2.4 Sign Data Hiding

The transform coefficient coding in HEVC includes an option to omit the coding of one sign flag per TSB if a sufficient number of significant coefficients is present in the block. The scheme is called *sign data hiding*. Since the sign flags can be considered to be uncorrelated, these flags are coded in CABAC bypass mode at the cost of 1 bit (see Sect. 10.2). Hence, sign data hiding saves one bit per TSB when applied. The use of sign data hiding is configured by a flag in the PPS.

Sign data hiding can be used if a transmission sub-block contains more than three significant coefficients. Instead of explicit coding, the sign of the first significant coefficient in the transform sub-block (i.e. the coefficient with the lowest scan index according to Fig. 8.3) is derived from the parity of the sum of the absolute level values of the significant coefficients. Let $\{|c_q(n)|\}$, $n = 0, \ldots, n_s - 1$ be the set of scaled absolute significant transform coefficient values. With

$$s_{\text{coeff}} = \sum_{n=0}^{n_s-1} |c_q(n)|, \tag{8.45}$$

the signed value of coefficient $c_q(0)$ is

$$c_q(0) = \Big(1 - 2(s_{\text{coeff}} \bmod 2)\Big) \cdot |c_q(0)|. \tag{8.46}$$

In the example of Fig. 8.5, the sum of the absolute coefficient values is $s_{\text{coeff}} = 49$. Hence, with sign data hiding the sign of the coefficient $c_q(0) = 13$ would be derived to be negative. In order to derive the correct sign of this coefficient, one of the coefficient levels would need to be changed by ± 1 here. The decision which coefficient to modify is left to the choice of the encoder which can use rate-distortion criteria to derive the best suited coefficient. Note that such a selection is required in the encoder if sign data hiding is activated and the parity of the sum of the absolute coefficient values does not match the sign of the coefficient $c_q(0)$. In Fig. 8.6, an example is given where the coefficient at position (3, 0) has been turned to 0 to achieve an even s_{coeff}. The modified scans compared to Table. 8.4 are shown in Table. 8.5.

8.3 Transform Skip

The option to signal transform skip is configured in the PPS. If activated, a transform skip flag is signaled for each transform block of size 4×4 separately for each color component.

The quantizer scaling operation for the coded transform coefficient levels is performed independently of the application of transform skip. If transform skip is indicated for a TB, the inverse transform operations are omitted. Instead, the coefficients c_{rlc} of (8.44) are further scaled to adapt to the dynamic range of the reconstructed residual $\mathbf{R}$ as described in Sect. 8.1.3,

$$r(x, y) = c_q(x, y) \cdot 2^7. \tag{8.47}$$

With transform skip, ringing and blurring artifacts which may be introduced in the reconstruction of the transformed residual signal are omitted. Instead, the quantization error instantaneously affects the residual sample values.

8.4 Transform and Quantization Bypass

Transform and quantization bypass is indicated at the CU level by a flag at the beginning of the CU syntax structure. The presence of this bypass flag is indicated in the PPS.

If bypass is indicated, the inverse transform step as well as the quantizer scaling operation are skipped and the residual signal is directly coded without any degradation. Thereby this mode enables perfect reconstruction for a lossless representation of the coded block. The sample differences are encoded as if they were quantized transform coefficient levels, i.e. reusing the transform block coding with transform sub-blocks, coefficient scans, and last significant coefficient signaling.

This mode can e.g. be useful for local coding of graphic content where quantization artifacts may be hardly or not at all tolerable. The encoder may also switch to this mode if the bit cost for transform coding at a very low QP would exceed the cost for bypass coding.

8.5 PCM Coding

If activated in the SPS, PCM coding can be indicated for coding units. In this case, no prediction and no quantization nor transformation is applied. Instead, the sample values of the samples in the corresponding coding blocks are directly coded into the bitstream at a PCM sample bit depth which is configured in the SPS.

The granularity for the application of PCM coding can be configured between the minimum of the luma coding tree block size and 32×32 on the high end and the smallest luma coding block size on the low end. If a coding unit is coded in PCM mode, the size of other coding units in the same coding tree unit must not be smaller than the size of the PCM unit.

Since by definition PCM coding enables a lossless representation of the corresponding block, the bits spent for PCM coding of a coding unit can be considered as an upper limit of the amount of bits required for encoding a CU. Hence, the encoder may switch to PCM coding in cases where the application of transform-based residual coding would exceed that limit, e.g. for exceptionally noisy content.

8.6 Comparison to H.264 | AVC

The H.264 | AVC specification includes the specification of two main inverse transform operations for 4×4 and 8×8 transform coefficient blocks, respectively [14, 15]. In both, H.264 | AVC and HEVC, the inverse transform is specified in integer arithmetic. In H.264 | AVC, the inverse transform operation includes right-shift operations in order to enable an implementation with a 16-bit dynamic range [16]. While this operation limits the dynamic range of the transform operation, it also limits the flexibility in the implementation design. As an additional transform for intra 16×16 predicted blocks, a two-stage transform operation is specified in H.264 | AVC. This transform is based on the 4×4 transform operation. For the 16×16 blocks, the DC coefficients of the 4×4 transform coefficient blocks are further transformed using a 4×4 Hadamard transform. This second stage transform aims at additional decorrelation of the transformed signal at the lowest possible complexity.

The constraint of low dynamic range induces base vectors with different norms for the 4×4 and 8×8 transforms in H.264 | AVC. These have to be taken into account in the inverse scaling operation for dequantization. Accordingly, a coefficient dependent scaling scheme is specified. The transform design in HEVC approximates single-norm transform matrices requiring 32-bit dynamic range. Rounding operations are specified outside of the inverse transform operation such that more flexibility is provided in the transform implementation. Furthermore, the inverse scaling for dequantization does not require coefficient dependency and thereby can be realized in a simpler implementation [5].

The scan of the transform coefficients in H.264 | AVC depends on the source format. For progressively coded blocks, the zigzag-scan is employed. For blocks coded in interlaced mode, a different dedicated field coefficient scan is specified. This rearrangement of the transform coefficients adapts the coefficient statistics along the scan for encoding and thereby improves the compression efficiency of the entropy coding stage. Since HEVC does not include dedicated coding tools for interlaced video such a special scan is not required. In HEVC, the traditionally used zigzag-scan is replaced by a diagonal scan. The scan is performed on a transform sub-block basis for reduced computational complexity [12]. For 4×4 and 8×8 intra predicted blocks, HEVC includes the specification of a vertical and a horizontal coefficient scan according to the intra prediction direction. In both specifications, the scan is applied in inverse direction from the highest frequency coefficients down to the DC coefficient.

In contrast to HEVC, H.264 | AVC includes the specification of two entropy coding schemes, context adaptive variable length coding (CAVLC) and context adaptive binary arithmetic coding (CABAC). The CABAC design in HEVC has been optimized to allow for streamlined CABAC operations with sets of bypass-coded bins for complexity efficient arithmetic coding. A detailed analysis of the HEVC CABAC implementation is provided in [17].

HEVC includes the specification of transform skip and the optional bypass of the transform and quantization stages as part of the Main profiles. In H.264 | AVC, a transform bypass mode is available which allows for lossless coding of the residual in 4:4:4 profiles.

References

1. High efficiency video coding. ITU-T Rec. H.265 (HEVC). http://www.itu.int/rec/T-REC-H.265/en (2013). Accessed 14 Apr 2014
2. Information technology—High efficiency coding and media delivery in heterogeneous environments—Part 2: High efficiency video coding. ISO/IEC 23008-2:2013 (HEVC). http://www.iso.org/iso/home/store/catalogue-tc/catalogue-detail.htm?csnumber=35424 (2013). Accessed 14 Apr 2014
3. Sullivan, G.J., et al.: Overview of the high efficiency video coding (HEVC) standard. IEEE Trans. Circ. Syst. Video Technol. **22**, 1649–1668 (2012). doi:10.1109/TCSVT.2012.22191
4. Wien, M.: Variable Block-Size Transforms for Hybrid Video Coding. Aachen Series on Multimedia and Communications Engineering. Shaker Verlag, Aachen (2004)
5. Budagavi, M., et al.: Core transform design in the high efficiency video coding (HEVC) standard. IEEE J. Sel. Top. Sign. Process. **7**(6), 1029–1041 (2013). doi:10.1109/JSTSP.22
6. Han, J., et al.: Towards jointly optimal spatial prediction and adaptive transform in video/image coding. In: Proceedings of IEEE International Conference on Acoustics, Speech and Signal Processing (ICASSP'10), pp. 726–729. doi:10.1109/ICASSP.2010.5495043 (2010)
7. Yeo, C., et al.: Mode-dependent fast separable KLT for block-based intra coding. Doc. JCTVC-B024. 2nd meeting: Joint Collaborative Team on Video Coding (JCT-VC) of ITU-T SG 16 WP 3 and ISO/IEC JTC 1/SC 29/WG 11, Geneva, CH (2010)

8. Ugur, K., Saxena, A.: CE1: Summary report of Core Experiment on intra transform mode dependency simplications. Doc. JCTVC-J0021. 10th meeting: Joint Collaborative Team on Video Coding (JCT-VC) of ITU-T SG 16 WP 3 and ISO/IEC JTC 1/SC 29/WG 11, Stockholm, SE (2012)
9. JCT-VC. HEVC Reference Software. https://hevc.hhi.fraunhofer.de/svn/svn-HEVCSoftware/ (2014). Accessed 14 Apr 2014
10. Karczewicz, M., et al.: Rate distortion optimized quantization. Doc. VCEGAH21. ITU-T SG16/Q6 VCEG, Antalya, TR (2008)
11. Bossen, F., Tan T.K.: TE12: results for experiments on max CU size, RDOQ and AIS. Doc. JCTVC-C044. 3rd meeting: Joint Collaborative Team on Video Coding (JCT-VC) of ITU-T SG 16 WP 3 and ISO/IEC JTC 1/SC 29/WG 11, Guangzhou, CN (2010)
12. Sole, J., et al.: Transform coefficient coding in HEVC. IEEE Trans. Circ. Syst. Video Technol. **22**(12), 1765–1777 (2012). doi:10.1109/TCSVT.2012.2223055
13. Nguyen, T., et al.: Transform coding techniques in HEVC. IEEE J. Sel. Top. Sig. Process. **7**(6), 978–989 (2013). doi:10.1109/JSTSP.2013.2278071
14. Advanced video coding for generic audiovisual services. ITU-T Rec. H.264 (AVC). http://www.itu.int/rec/T-REC-H.264/en (2014). Accessed 14 Apr 2014
15. Information technology—Coding of audio-visual objects—Part 10: Advanced Video Coding. ISO/IEC 14496–10:2012 (AVC). http://www.iso.org/iso/home/store/catalogue-tc/catalogue-detail.htm?csnumber=61490 (2012). Accessed 14 Apr 2014
16. Malvar, H.S., et al.: Low-complexity transform and quantization in H.264/AVC. IEEE Trans. Circ. Syst. Video Technol. **13**(7), 598–603 (2003). doi:10.1109/TCSVT.2003.814964
17. Sze, V., Budagavi, M.: High throughput CABAC entropy coding in HEVC. IEEE Trans. Circ. Syst. Video Technol. **22**, 1778–1791 (2012). doi:10.1109/TCSVT.2012.2221526

Chapter 9
In-Loop Filtering

In the processing flow of the hybrid video coding scheme, in-loop filtering is applied after reconstruction of the coding blocks. The filtered reconstructed picture is stored in the decoded picture buffer for output and for prediction if indicated.

In HEVC, two consecutive in-loop filters are specified. First, a deblocking filter is applied to prediction block and transform block edges in order to reduce the amount of visible block structures, which result from the block-based nature of the coding scheme. For this purpose, the filter operates on block edges with adaptive filter strength and adaptive filter length. In a second step, a sample-oriented filtering step is applied. The sample adaptive filter can be configured to be either driven by sample value differences of a local neighborhood, or by the value range the intensity value of the current sample falls into. This filter operates on the samples in the slice and does not only consider block edges.

Both filter types can be activated separately and can further be configured to operate across slice or tile boundaries according to the application needs. Blocks which have been encoded in PCM mode can be excluded from deblocking and sample adaptive offset processing by indication of a corresponding flag in the SPS. Blocks which have been encoded in transform and quantization bypass mode are skipped in the in-loop filtering process. Blocks coded in this mode provide perfect reconstruction of the original input signal by design and, therefore, must not be filtered in order to preserve the original signal content.

In the following, the deblocking filter process is described in Sect. 9.1. Sample adaptive offset filtering is detailed in Sect. 9.2.

9.1 Deblocking Filter

The deblocking filter applies an adaptive smoothing filter operation across the boundaries of prediction and transform blocks inside the reconstructed picture. At the left picture boundary and at the upper picture boundary, filtering is omitted for the vertical and the horizontal edge, respectively. The filtering strength depends on properties of

M. Wien, *High Efficiency Video Coding*, Signals and Communication Technology,
DOI 10.1007/978-3-662-44276-0_9

the neighboring blocks, and the local decision whether to apply the deblocking filter depends on the difference between the edge sample values. If filtering across slice or tile boundaries is disabled, these boundaries are treated as if they were picture boundaries. In the deblocking filtering process, the left and top neighboring blocks are used for determination of the deblocking filter configuration.

The specification of the deblocking filter allows for an implementation on a picture basis as well as for an implementation on a coding block basis. The processing order is to first process vertical edges and then process the horizontal edges of the vertically processed signal. Under appropriate consideration of this processing order, the picture-based or coding-block-based operation can be selected. In the CB-based implementation, the deblocking operation of the bottom and right samples of the current CTB can only be performed upon availability of the respective neighboring blocks which follow in processing order. For the deblocking of the bottom samples, a line buffer is therefore required, which holds the data until the filtering operation can be completed.

While the smallest specified prediction block and transform block size is 4×4, the operation of the deblocking filter is limited to the 8×8 block grid. Thereby, 4×4 block edges within an 8×8 block are not regarded for both, luma and chroma blocks. This restriction significantly reduces the complexity of the deblocking filter. Since neighboring horizontal or neighboring vertical 8×8 block edges do not interfere, parallel processing of all edges in one orientation is supported. With a proper organization of the processing order, also an 8×8 block parallel operation is enabled [1].

The default configuration of the deblocking filter can be modified by syntax structures in the PPS and in the slice header to allow for customized operation. In the PPS, the presence of slice-level deblocking filter parameters can be indicated [2, 3]. A specific global flag can be sent to disable the deblocking filter for all slices which refer to the given active PPS. Deblocking filter control parameters can be signaled if the deblocking filter operation is enabled. On the slice level, the syntax allows for further control of the deblocking filter. Here, the filter can be disabled or the deblocking filter control parameters can be modified as needed.

In the following, the successive specified deblocking filter processing steps are detailed.

9.1.1 Determination of Edges

In the first processing step, all potential edge samples are determined and marked by an edge flag. The sample locations at the left and top edges of prediction blocks or transform blocks are marked as potential deblocking filter edges. In the specification, this operation is applied to all prediction block sizes and for the full residual quadtree, even though the subsequent filter operation is only performed on an 8×8 block grid. Samples on the left and top CTB boundary are not marked if they are at the picture

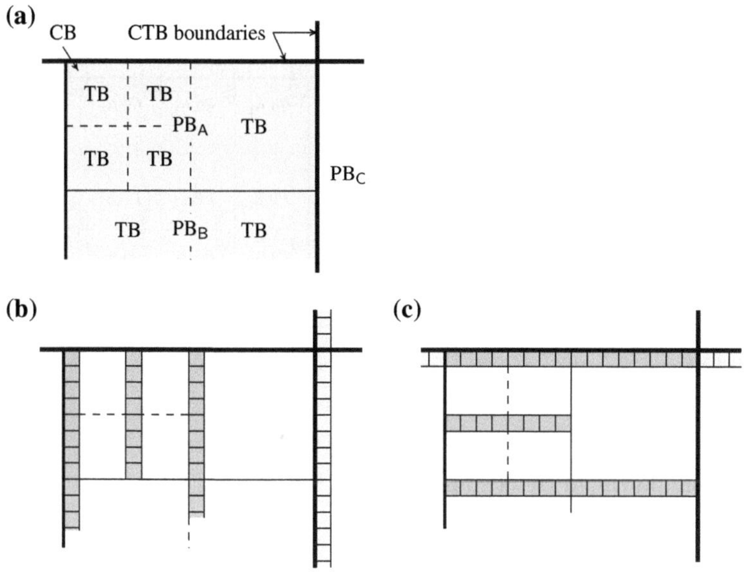

Fig. 9.1 **a** Example CTB with PBs indicated with *solid lines* and TBs indicated with *dashed lines*. PB_A and PB_B have a size of 16×8 samples. The TBs have sizes of 8×8 and 4×4. Edge flags are indicated in *gray* for **b** vertical and **c** horizontal edges

boundary. Samples on the CTB boundary may be marked if they are at slice or tile boundaries, depending on loop filter flags in the PPS or in the slice header.

An example for the marking of vertical and horizontal edges is given in Fig. 9.1 for a 16×16 coding block at the top right of a coding tree block. In the example, the coding block is split into two 16×8 prediction blocks. The residual quadtree results in transform blocks of size 8×8 and 4×4 as indicated in the figure.

9.1.2 Determination of the Deblocking Filter Strength Parameter

For samples on the 8×8 grid which have been marked with the edge flag, a boundary strength parameter b_S is determined. This parameter controls the filtering strength of the deblocking filter. The boundary strength b_S is determined for sets of four samples along the edge under consideration. It depends on the prediction modes, motion vectors, and the presence of transform coefficients in the neighboring blocks as further detailed below.

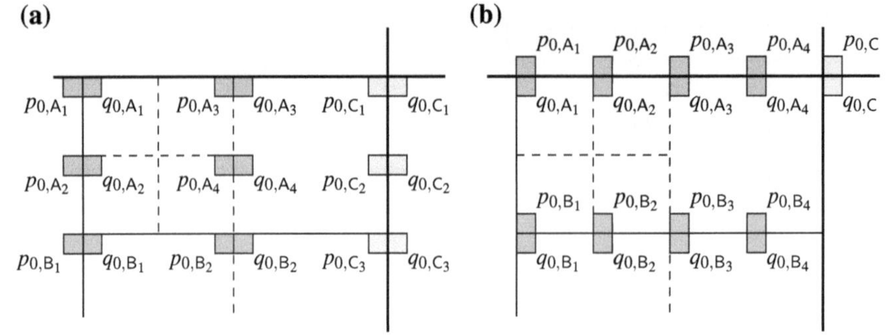

Fig. 9.2 Location of samples p_0 and q_0 for **a** vertical and **b** horizontal edges for the example CB given in Fig. 9.1

9.1.2.1 Neighborhood Determination

The neighboring blocks are determined by the location of two edge samples p_0 and q_0, where the sample q_0 lies in the current block and the location p_0 determines the neighboring block. As the deblocking filter operates on segments of four samples along the edge to be filtered, the locations p_0 and q_0 for b_S determination are taken in a four sample distance.

An example is given in Fig. 9.2 for the coding block presented in Fig. 9.1. It can be seen that only the block boundaries on the 8×8 grid are regarded. The locations p_0 and q_0 for the boundary strength derivation are assessed every four samples along the respective edge. In the figure, the edge samples which are considered during the processing of the current CB are marked in gray. The corresponding samples which are considered when processing the neighboring CB to the right are marked in light gray.

9.1.2.2 Boundary Strength

For control of the filtering strength, the boundary strength parameter b_S is used to determine the value of the threshold t_C, which in turn is used for selection of the filter type and for determination of the value range within which the deblocking filter may modify the filtered sample values. This process is further detailed in Sect. 9.1.3 below.

For determination of the boundary strength b_S, the block containing q_0 is referred to as $\mathbf{B}_q$ and the block containing p_0 is referred to as $\mathbf{B}_p$. The conditions for selection of the boundary strength parameter are as follows:

- If $\mathbf{B}_q$ or $\mathbf{B}_p$ is an intra blocks, the boundary strength parameter is set to $b_S = 2$.
- Otherwise, if the boundary between $\mathbf{B}_q$ and $\mathbf{B}_p$ is also a transform block edge and $\mathbf{B}_q$ or $\mathbf{B}_p$ contains non-zero transform coefficients, the boundary strength parameter is set to $b_S = 1$.

- Otherwise, if one the following conditions on motion vectors is fulfilled, the boundary strength parameter is set to $b_S = 1$ (the motion vectors are referred to in full sample precision here):
  - the reference pictures or the number of motion vectors for $\mathbf{B}_q$ and $\mathbf{B}_p$ differ,
  - a single motion vector is used for $\mathbf{B}_q$ and $\mathbf{B}_p$ each, and

$$|mv_{x,p} - mv_{x,q}| \geq 1 \quad \text{or} \quad |mv_{y,p} - mv_{y,q}| \geq 1 \tag{9.1}$$

  - two motion vectors are used for $\mathbf{B}_q$ and $\mathbf{B}_p$ each, from the *same two* reference pictures,[1] and for at least one of the reference pictures

$$|mv_{x,p} - mv_{x,q}| \geq 1 \quad \text{or} \quad |mv_{y,p} - mv_{y,q}| \geq 1 \tag{9.2}$$

  - two motion vectors are used for $\mathbf{B}_q$ and $\mathbf{B}_p$ each, from the *same* reference picture, and

$$|mv_{x,p0} - mv_{x,q0}| \geq 1 \quad \text{or} \quad |mv_{y,p0} - mv_{y,q0}| \geq 1 \tag{9.3}$$

$$|mv_{x,p1} - mv_{x,q1}| \geq 1 \quad \text{or} \quad |mv_{y,p1} - mv_{y,q1}| \geq 1 \tag{9.4}$$

or

$$|mv_{x,p0} - mv_{x,q1}| \geq 1 \quad \text{or} \quad |mv_{y,p0} - mv_{y,q1}| \geq 1 \tag{9.5}$$

$$|mv_{x,p1} - mv_{x,q0}| \geq 1 \quad \text{or} \quad |mv_{y,p1} - mv_{y,q0}| \geq 1 \tag{9.6}$$

- Otherwise, the boundary strength parameter is set to $b_S = 0$.

Note again that in Eqs. (9.1)–(9.6), the motion vectors are denoted in full sample resolution. Since the motion vectors are represented in quarter-sample resolution for coding, the motion vector difference in at least one of the components must be larger than 4 in quarter-sample units.

9.1.3 Deblocking Filtering

As mentioned above, the edges on a 8×8 grid are partitioned into 4-sample segments for application of the deblocking filter. Depending on the determined boundary strength value and the local features of the reconstructed picture, the deblocking filtering operation reaches up to three samples from the edge into the corresponding block.

---

[1] Note that a difference in the reference picture index is not a sufficient condition to check if reference pictures differ. The same picture may be available multiple times in the reference picture list.

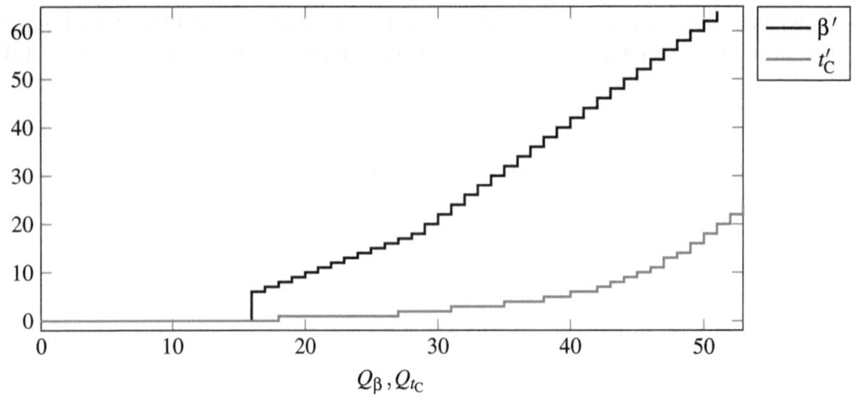

Fig. 9.3 Threshold parameters β' and t'_{C} as a function of Q_β and $Q_{t\mathrm{C}}$

In the following, the filtering process is described for vertical edges. The filtering process is applied to samples in the same line. For horizontal edges, the same process is applied in vertical direction. The sample locations are transposed accordingly in this case.

9.1.3.1 Threshold Parameters

Two threshold parameters β and t_{C} control the operation of the deblocking filter. The values of the two parameters depend on the luma quantization parameters of the coding units the neighboring blocks $\mathbf{B}_q$ and $\mathbf{B}_p$ belong to,

$$\mathrm{QP}_{\mathrm{L}} = \mathrm{round}\left\{\frac{1}{2}\left(\mathrm{QP}_q + \mathrm{QP}_p\right)\right\}. \tag{9.7}$$

The quantization parameter serves as an index to determine the applicable parameter values for β and t_{C}. The determination of the parameters can be further modified by the offsets β_{os} and $t_{\mathrm{C,os}}$ if PPS or slice based deblocking filter control is enabled and the value of t_{C} further depends on the deblocking filter strength parameter b_{S}:

$$Q_\beta = \mathrm{QP}_{\mathrm{L}} + \beta_{\mathrm{os}} \tag{9.8}$$

$$Q_{t\mathrm{C}} = \mathrm{QP}_{\mathrm{L}} + t_{\mathrm{C,os}} + 2 \cdot (b_{\mathrm{S}} - 1) \tag{9.9}$$

For the derivation of β and t_{C}, these adapted quantization parameters Q_β and $Q_{t\mathrm{C}}$ are clipped to $(0, 51)$ and $(0, 53)$, respectively. For application in the deblocking filter process, the parameters are scaled according to the video bit depth B_{d}. The evolution of the intermediate thresholds β' and t'_{C} is shown in Fig. 9.3. The applicable thresholds are derived as

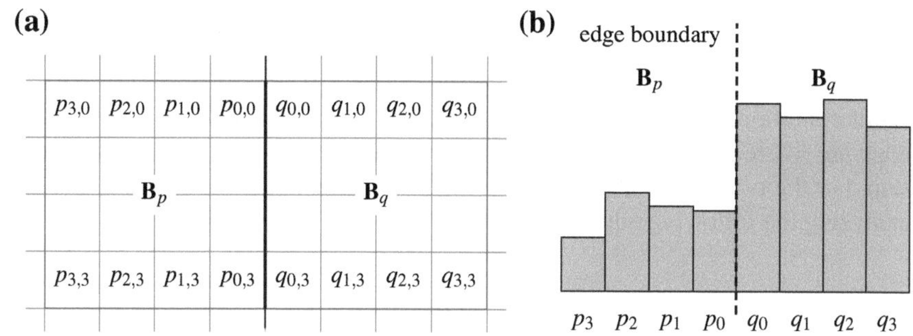

Fig. 9.4 **a** Locations of the samples $p_{i,k}$ in the 4×4 block $\mathbf{B}_p$ and $q_{i,k}$ in $\mathbf{B}_q$ for derivation of the vertical edge filtering decisions for the corresponding edge segment. **b** 1-dimensional illustration for an edge with blocking artifact

$$\beta = \beta' \cdot 2^{B_\mathrm{d}-8}, \tag{9.10}$$

$$t_\mathrm{C} = t'_\mathrm{C} \cdot 2^{B_\mathrm{d}-8}. \tag{9.11}$$

As can be seen from Fig. 9.3, the parameters are $\beta = 0$ for $Q_\beta \leq 15$ and $t_\mathrm{C} = 0$ for $Q_{t_\mathrm{C}} \leq 17$. Thereby, the deblocking filter is effectively turned off and no deblocking operation is performed for smaller QP values.

9.1.3.2 Block Filtering Decisions

The decision whether to filter an edge is derived based on the values of the samples in the first and last line of the neighboring 4×4 blocks under consideration, see Fig. 9.4a. The one-dimensional illustration of a blocking artifact in one of these lines is shown in Fig. 9.4b. In order to derive the filter configuration decisions for the current edge, the absolute value of the second order derivative is determined for the two lines across the block boundary:

$$d_{p0} = \left| p_{0,0} - 2p_{1,0} + p_{2,0} \right| \tag{9.12}$$

$$d_{p3} = \left| p_{0,3} - 2p_{1,3} + p_{2,3} \right| \tag{9.13}$$

$$d_{q0} = \left| q_{0,0} - 2q_{1,0} + q_{2,0} \right| \tag{9.14}$$

$$d_{q3} = \left| q_{0,3} - 2q_{1,3} + q_{2,3} \right| \tag{9.15}$$

These derivative values are further analyzed to derive an indication of the divergence in the sample values across the edge segment. For blocks with significant intensity variations across the edge, a block boundary—and hence a blocking artifact—can become easily visible. The decision if deblocking filtering is applied to the current edge segment is drawn using the sum of the derivatives in relation to the threshold parameter β:

$$d_{p0} + d_{p3} + d_{q0} + d_{q3} < \beta. \tag{9.16}$$

If this condition is fulfilled, a blocking artifact is indicated and the samples along the edge are filtered. The decision on the applicable filtering type is drawn in the next step. For the two lines 0 and 3, the amount of the blocking artifact is determined by analyzing the following edge conditions for $i \in \{0, 3\}$:

$$2 \cdot \left(d_{pi} + d_{qi}\right) < \left\lfloor \frac{\beta}{2} \right\rfloor, \tag{9.17}$$

$$\left|p_{3,i} - p_{0,i}\right| + \left|q_{3,i} - q_{0,i}\right| < \left\lfloor \frac{\beta}{8} \right\rfloor, \tag{9.18}$$

$$\left|p_{0,i} - q_{0,i}\right| < \text{round}\left\{\frac{5}{2} \cdot t_C\right\}. \tag{9.19}$$

As can be seen from the equations, the local activity in the blocks on both sides of the edge is checked relative to the threshold parameter β in (9.17) and (9.18). The 'intensity step' between the two blocks at the edge boundary is determined relative to the threshold parameter t_C in (9.19). If all conditions (9.17)–(9.19) hold for both, the first and the last line, a strong blocking artifact is indicated and strong filtering is applied to the edge. In this case, the three samples on each side of the edge are filtered as detailed in Sect. 9.1.3.3 for the four lines of the current edge segment.

If one of the conditions is not met, the mode is set to weak filtering. In this case, the edge samples $p_{0,i}$ and $q_{0,i}$, $i = 0, \ldots, 3$ are filtered. For each of the two blocks $\mathbf{B}_q$ and $\mathbf{B}_p$, an additional decision is drawn in this case whether to filter the second edge boundary samples $p_{1,i}$ and $q_{1,i}$, $i = 0, \ldots, 3$, respectively. The condition for filtering are:

$$d_{p0} + d_{p3} < \left\lfloor \frac{3}{16} \cdot \beta \right\rfloor, \tag{9.20}$$

$$d_{q0} + d_{q3} < \left\lfloor \frac{3}{16} \cdot \beta \right\rfloor. \tag{9.21}$$

9.1.3.3 Luma Filtering

In the following, the filtering operation is described for a single line for the vertical edge. The operation for horizontal edges is applied in a transposed fashion accordingly. The operation is applied in the same manner to the four lines of the blocks across the current 4-sample edge segment. The second index to indicate the line is omitted for improved readability here.

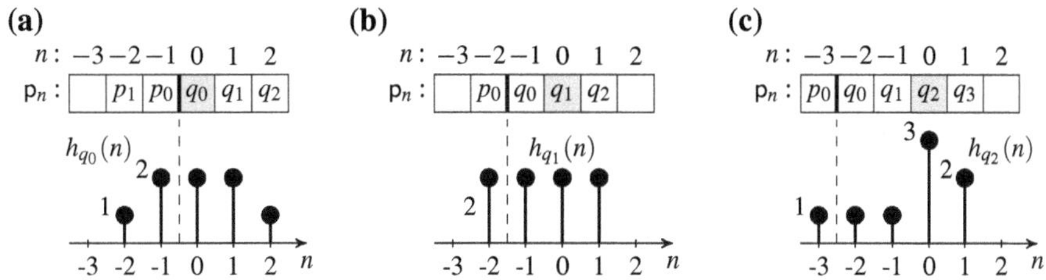

Fig. 9.5 Filtering weights $h_i(n)$, $i \in \{0, 1, 2\}$ for strong deblocking filtering. The sample to be filtered is on $n = 0$. The samples p_i, q_i are numbered by n relative to the one to be filtered

Strong Filtering

If strong filtering is indicated, the samples $p_i, q_i, i \in \{0, 1, 2\}$ are filtered each with a dedicated filter. The filtering weights are illustrated in Fig. 9.5 for the samples in the current block $\mathbf{B}_q$.

Let the samples p_i, q_i be denoted by p_n as shown in Fig. 9.5. The filtered samples q_i', $i \in \{0, 1, 2\}$, are derived by

$$\tilde{q}_i = \text{round}\left\{\frac{1}{8} \cdot \sum_n h_{q_i}(n) \cdot \mathsf{p}_n\right\} \tag{9.22}$$

The result is clipped in order not to exceed the original sample value q_i by $\pm 2t_\text{C}$:

$$q_i' = \min\Big[\max\big[(q_i - 2t_\text{C}), \tilde{q}_i\big], (q_i + 2t_\text{C})\Big] \tag{9.23}$$

The result is then further clipped not to exceed the available value range, $0, \ldots, 2^{B_\text{d}} - 1$ for the video bit depth B_d. For the samples in block $\mathbf{B}_p$, the weights $h_{p_i}(n)$ are derived by mirroring, $h_{p_i}(n) = h_{q_i}(-n)$, and exchanging q_i and p_i accordingly.

Weak Filtering

For weak filtering, only the boundary samples p_0 and q_0 and, if the conditions (9.20) and (9.21) are met, samples p_1 and q_1 are filtered, respectively.

In a first step, the steepness of the edge is determined by deriving Δ_0 as

$$\Delta_0 = \text{round}\left\{\frac{1}{16} \cdot \sum_n h_{\Delta 0}(n) \cdot \mathsf{p}_n\right\}, \quad \text{with } n = -2, \ldots, 1, \tag{9.24}$$

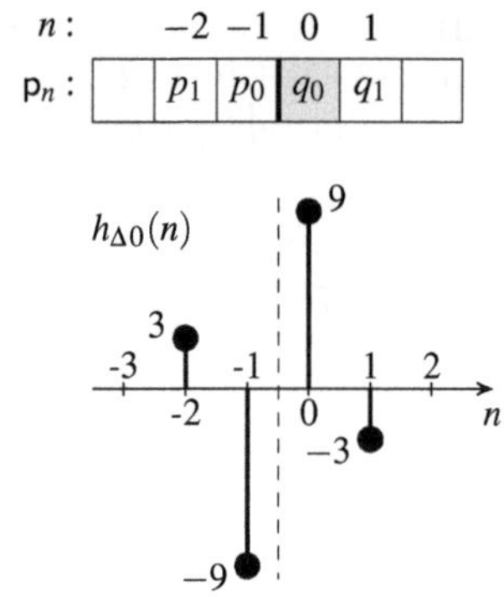

Fig. 9.6 Filtering weights to derive the steepness Δ_0 of the edge between p_0 and q_0

and with $h_{\Delta 0}(n)$ and p_n as shown in Fig. 9.6. If $\Delta_0 < 10 \cdot t_C$, the deblocking filtering operation is applied. Edges with larger values of Δ_0 are considered to be picture content, which hence should not be filtered.[2]

For the actual filtering operation, the value of Δ_0 is clipped to a maximum value of $\pm t_C$. The samples p_0 and q_0 are then modified as

$$p_0^* = p_0 + \Delta_0, \tag{9.25}$$
$$q_0^* = q_0 - \Delta_0. \tag{9.26}$$

As noted above, the secondary samples p_1 and q_1 are only filtered if the conditions (9.20) and (9.21) are met, respectively. For this filtering operation, the value of Δ_0 as derived in (9.24) is updated using a local second order derivative,

$$\Delta_p = \frac{1}{2}\left(\Delta_0 + \frac{1}{2}p_2 - p_1 + \frac{1}{2}p_0\right) \tag{9.27}$$
$$\Delta_q = \frac{1}{2}\left(\Delta_0 - \frac{1}{2}q_0 + q_1 - \frac{1}{2}q_2\right) \tag{9.28}$$

The values of Δ_p and Δ_q are clipped to $\pm\frac{1}{2} \cdot t_C$ and added to the original sample values,

$$p_1^* = p_1 + \Delta_p, \tag{9.29}$$
$$q_1^* = q_1 - \Delta_q. \tag{9.30}$$

Finally, the results of (9.25)–(9.30) are clipped not to exceed the available value range $0, \ldots, 2^{B_d} - 1$ for the video bit depth B_d, accordingly.

[2] Note that this decision is separately drawn for each line. Thereby, the filtering may be switched on or off locally for single lines of the current edge segment.

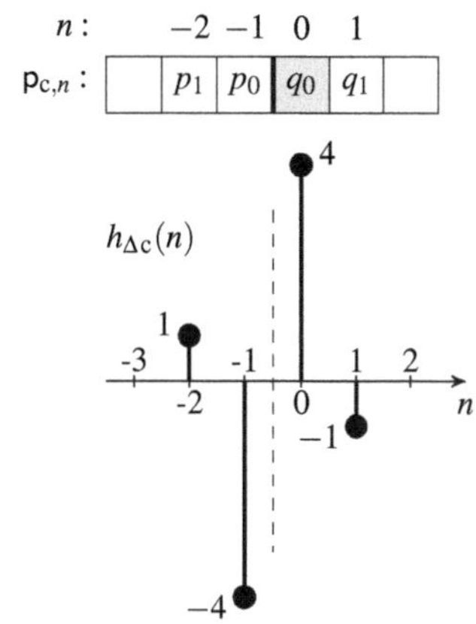

Fig. 9.7 Filtering weights to derive the steepness Δ_c of the chroma edge between p_0 and q_0

9.1.3.4 Chroma Filtering

The chroma deblocking filtering process is only activated for blocks with a boundary strength of $b_S = 2$, i.e. when the block is intra predicted. As with the luma blocks, the chroma deblocking filtering is only performed on edges on the 8×8 grid. For chroma filtering, only a single deblocking filter is specified, regardless of weak or strong filtering for the corresponding luma edge. Only the boundary samples q_0 and p_0 are filtered.

The effective quantization parameter for the derivation of the chroma threshold parameter t_C is derived according to (9.9) with $\mathrm{QP_L}$ derived by

$$\mathrm{QP_L} = \mathrm{round}\left\{\frac{1}{2}\left(\mathrm{QP}_q + \mathrm{QP}_p\right)\right\} + \mathrm{QP_{os,c}}, \tag{9.31}$$

where QP_q and QP_p are the luma quantization parameters and $\mathrm{QP_{os,c}}$ set to the PPS chroma QP offset value $\mathrm{QP_{os,cb}}$ or $\mathrm{QP_{os,cr}}$ for the corresponding chroma component. Note that potential modifications of the chroma quantization parameter on the slice level are not taken into account.

The steepness of the chroma edge is determined by the approximated 1st-order derivative Δ_c as

$$\Delta_c = \mathrm{round}\left\{\frac{1}{8} \cdot \sum_n h_{\Delta c}(n) \cdot \mathsf{p}_{c,n}\right\}, \tag{9.32}$$

with $h_{\Delta c}(n)$ and $\mathsf{p}_{c,n}$ as shown in Fig. 9.7. For the filtering operation, the value of Δ_c is clipped to $\pm t_C$. The samples p_0 and q_0 are then filtered as

$$p_0^* = p_0 + \Delta_c, \tag{9.33}$$
$$q_0^* = q_0 - \Delta_c, \tag{9.34}$$

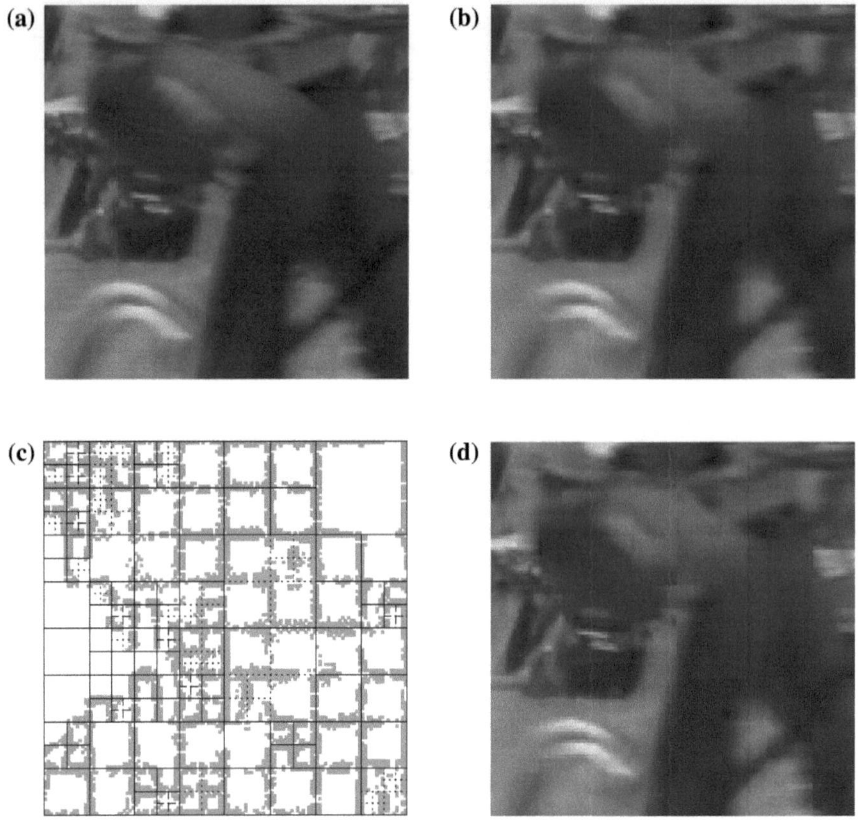

Fig. 9.8 Example for deblocking of 2 × 2 CTBs (CB: *solid lines*, PB: *dashed lines*, TB: *dotted lines*). **a** Original. **b** Reconstruction with deblocking. **c** CB, PB, TB structure, deblocked samples. **d** Reconstruction without deblocking

and the results of (9.33) and (9.34) are clipped not to exceed the available value range.

9.1.4 Deblocking Filter Example

In Figs. 9.8 and 9.9, an example for the application of the deblocking filter is provided. An example area of 2 × 2 CTBs is shown.[3] Besides the original area of the picture from the test sequence and the corresponding decoded reconstruction, the samples along the prediction block and transform block boundaries which are modified by

[3] Frame 208 of the 1080p test sequence ParkScene, HM 11.0, random access configuration of the JCT-VC common testing conditions [4], QP = 32, default HM deblocking filter configuration.

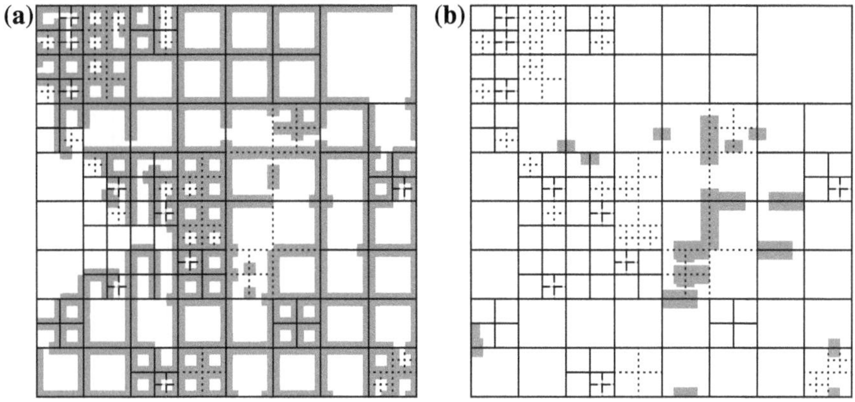

Fig. 9.9 Selection of deblocking filter mode in Fig. 9.8. **a** Weak filtering. **b** Strong filtering

the deblocking filter are visualized. For comparison, the same region without the application of the deblocking filter is shown as well. In Fig. 9.9, the selection of the strong and the weak filtering modes is illustrated for the selected area.

It can be seen from the figure that the deblocking filter does not necessarily cancel out all blocking artifacts. A trade-off between smoothing of the reconstructed picture and the preservation of edges in the content of the picture has to be taken into account. In the example, some visible blocking artifacts at prediction boundaries can be observed as the deblocking filtering decision indicated an edge to be preserved. This is e.g. visible at the chin of the depicted person. Blocking artifacts at other, less extreme prediction block boundaries are reliably detected and removed. Since the deblocking filter operation can be configured in the PPS and in the slice header, the intensity of deblocking artifact removal can be controlled according to the application needs by the encoder.

9.2 Sample Adaptive Offset

Sample Adaptive Offset (SAO) is a sample-based filtering operation which is operated on a CTU basis. It is applied right after the deblocking filter if activated. Two SAO modes are specified: *edge offset* and *band offset*. The former is driven by local directional structures in the picture to be filtered, the latter modifies the intensity values of the samples without a dependency on the neighborhood. Additionally, SAO can be deactivated per CTU. Since SAO directly follows the deblocking filter in the signal path, the two processes can be implemented as successive operations without intermediate storage. Thereby, the required memory access can be reduced [5]. The input to the process is independent from the output, i.e. it is not affected by the filtering operation. Therefore, the SAO process can be parallelized at a large scale. Similar

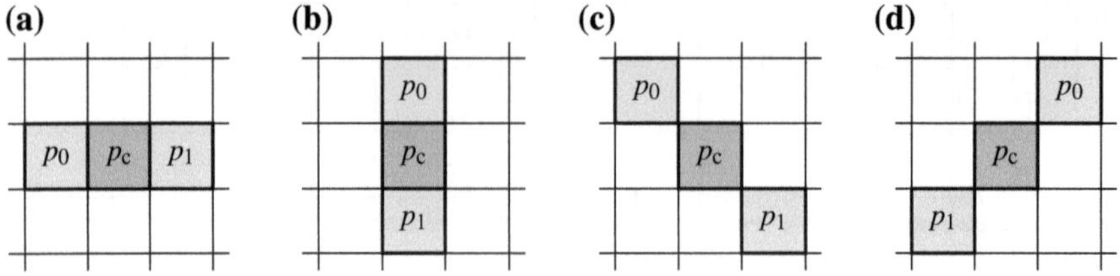

Fig. 9.10 Edge directions for sample adaptive offset. **a** Horizontal. **b** Vertical. **c** Diagonal-down. **d** Diagonal-up

to the deblocking filter, slice or tile boundaries are treated as if they were picture boundaries in the case that filtering across slice or tile boundaries is not enabled.

For each CTU, the applicable SAO mode is indicated.The parameters can be either explicitly signaled or derived from the neighborhood. For reduced signaling cost, the applicable mode and parameters can be inherited from the top or left neighboring CTU in this case. SAO can be applied for luma as well as for the chroma components, where the two chroma components share the same SAO mode. The offset parameters are configured individually for each component nonetheless. For inheritance of the SAO configuration, the parameters of the CTU row have to be stored for potential use in the following row. Since the information only needs to be stored on a CTU basis, the line buffer requirement for SAO is small [5].

While the SAO operation at the decoder side is not very complicated to describe (the specification text requires only about two pages [2]), the operation at the encoder side to derive suitable SAO parameters requires some consideration. In Sect. 9.2.5, a fast SAO parameter derivation algorithm as used in the HM reference software is described [5–7].

9.2.1 Edge Offset

If SAO operates in edge offset mode, the samples of the current coding tree block are modified depending on their relation to two samples from their 8-connected neighborhood. This sample pair represents one of the available edge directions, including horizontal, vertical and diagonal-up and diagonal-down, respectively. The applicable direction is fixed for the CTU and is referred to as the edge offset class of the CTU. As stated above, the two chroma components share the same edge offset class, which may further be different from the luma edge offset class.

In Fig. 9.10, the neighboring samples for the direction of each edge offset class are shown. The current sample under consideration is denoted by p_c, the two neighboring samples are denoted by p_0 and p_1.

Depending on the relation of the neighboring samples in the indicated direction, one of five configurable offset values Δ_{SAO} is selected to be added to the current sample,

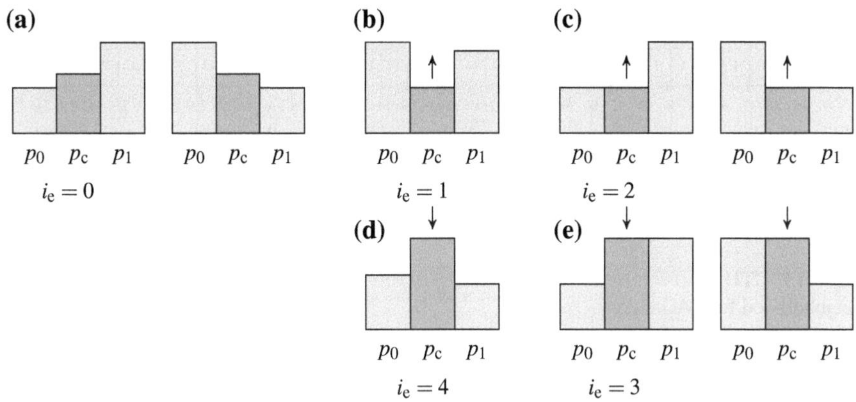

Fig. 9.11 Illustration of the relation of the neighboring sample values for each SAO edge offset index. **a** For $i_e = 0$ no SAO is applied. **b–e** smoothing is only specified for $i_e \neq 0$

$$p_c^* = p_c + \Delta_{SAO}(i_e). \tag{9.35}$$

The applicable offset index i_e is determined by the following steps:

$$i_e' = 2 + \operatorname{sgn}(p_c - p_0) + \operatorname{sgn}(p_c - p_1), \tag{9.36}$$

$$\text{if } i_e' < 2 : i_e = i_e' + 1, \tag{9.37}$$
$$\text{if } i_e' = 2 : i_e = 0, \tag{9.38}$$
$$\text{if } i_e' > 2 : i_e = i_e'. \tag{9.39}$$

The relation of the sample values for each index is visualized in Fig. 9.11. The offset index $i_e = 0$ corresponds to a fixed offset value $\Delta_{SAO}(0) = 0$, i.e. in this case no SAO filtering operation is applied.

As indicated by the arrows in Fig. 9.11, the direction of the SAO offset is predefined for each offset index. For $i_e \leq 2$ only values $\Delta_{SAO}(i_e) \geq 0$ are allowed, for $i_e > 2$, only values $\Delta_{SAO}(i_e) \leq 0$ are allowed. Thereby, only smoothing and no edge sharpening operations are permitted for the SAO filter. Since the offset sign is fixed, only the absolute value of $\Delta_{SAO}(i_e)$ needs to be signaled in the syntax.

9.2.2 Band Offset

In the band offset mode, SAO applies a configurable offset to the sample values which only depends on the sample intensity and not on the neighborhood of the sample. The intensity range is partitioned into three 'bands': (i) low intensity band,

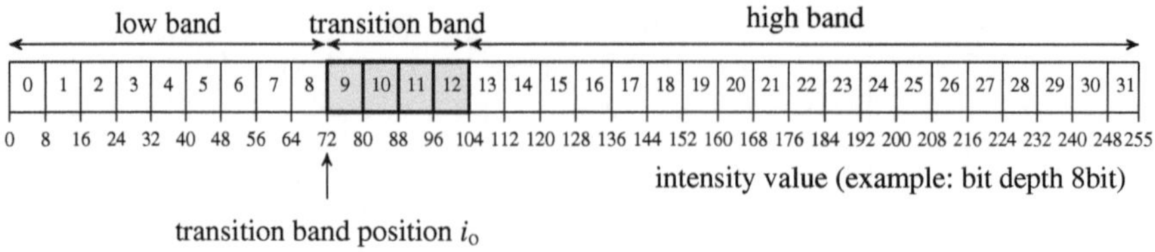

Fig. 9.12 Example for the areas of the band offset mode for SAO band position $i_0 = 9$

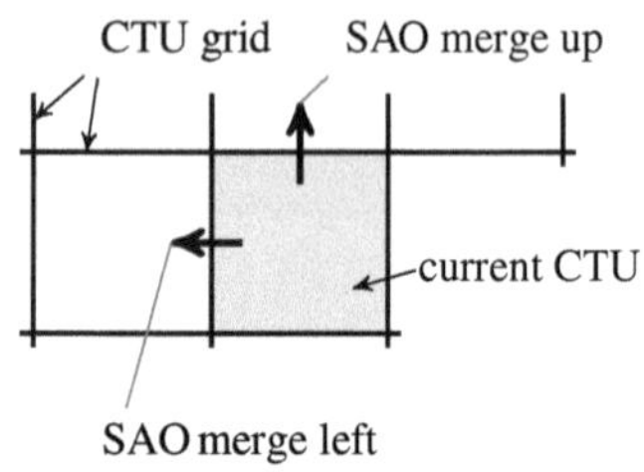

Fig. 9.13 CTU neighborhood for SAO merge either from the left or from the upper neighboring CTU

(ii) transition band, and (iii) high intensity band. The full value range of each component is quantized into 32 bins, represented by the five most significant bits of the sample value. The index i_0 of the lowest bin of the transition band is signaled as the SAO band position in the bitstream. The transition band has a width of four bins. For each of the four bins, an individual offset value for the sample intensity is signaled in the bitstream. These offset values can be positive or negative, allowing for a dedicated adaptation of the intensity values in the transition band for each component. Samples with values in the low or high intensity bands remain unchanged. An example for SAO band positioning is given in Fig. 9.12.

9.2.3 Signaling of SAO Parameters

The signaling of the SAO parameters is performed at the CTU level. A flag in the SPS indicates the activation of SAO for the coded video sequence. If activated, the application of SAO can be separately controlled for luma and the two chroma components at the slice level. For slices with SAO, the corresponding parameters are signaled in a dedicated syntax structure preceding the coding quadtree syntax structure. Since all information at the CTU level is coded with CABAC, this also holds for the SAO parameters.

For an efficient representation of the SAO parameters in the bitstream, the applicable SAO mode with the corresponding parameters can be copied from one of the neighboring CTUs as shown in Fig. 9.13. The SAO merge left flag indicates that the parameters are copied from the left neighbor. If this flag is not set, the SAO merge up flag indicates copy from the upper neighboring CTU. If both flags are not set to true, the SAO parameters for the current CTU are encoded.

The coded SAO parameters comprise the SAO type index which indicates the applicable SAO modes (edge offset, band offset, deactivated) for the luma and the chroma components, respectively. Then, the absolute values of the four offset parameters $|\Delta_{\mathrm{SAO}}(i_{\mathrm{e}})|$ are encoded. In the edge offset mode, the edge offset class is encoded as the next step. The sign of the offset is derived from the edge offset index i_{e} in this case. In SAO band offset mode, the sign of the four offsets and the band position are signaled.

In terms of entropy coding, the SAO merge flags (left and up) as well as the SAO type indices (edge or band offset) are adaptively encoded with context selection. The remaining parameters are coded in CABAC bypass mode. For details on the entropy coding stage, the reader is referred to Chap. 10.

9.2.4 SAO Filter Example

The operation of the two SAO filtering types is illustrated in Fig. 9.14. The samples selected for SAO using band offset and edge offset are shown for two example areas in a coded video sequence.[4] In Fig. 9.14a, b, the application of the SAO band offset filter is illustrated. The samples which are modified by SAO band offset filtering are marked. With band offset, the filter is activated depending on the intensity values of the reconstructed samples. Figure 9.14c, d shows an example for the application of the SAO edge offset filter. In this case, the filter is activated based on the relation of the intensity values of the neighboring samples. It can be seen that in both examples, the SAO filter is applied roughly along edges which are visible in the picture. The CB, PB, and TB partitioning is shown here for reference. As indicated before, the SAO filter operates independently of the underlying coding structure.

In Fig. 9.15, a typical distribution of the SAO modes within a complete picture is shown. In the figure, CTBs with edge mode are marked with a line indicating the applicable edge direction for the CTB. CTBs in band offset mode are marked with a gray square. CTBs with deactivated SAO filter are not further marked. In the example, the edge offset mode is dominantly used. Band offset is mostly applied in background areas.

9.2.5 Encoder-Side Derivation of Sample Adaptive Offset Parameters

The concept of the fast rate-distortion optimized SAO parameter derivation as used in the HEVC reference software is outlined here. The presentation follows Fu et al. [5], who are the main contributors to SAO.

[4] Areas in frames 160 and 250 of the 1080p test sequence BasketballDrive, HM11.0, random access configuration of the JCT-VC common testing conditions [4], QP = 37.

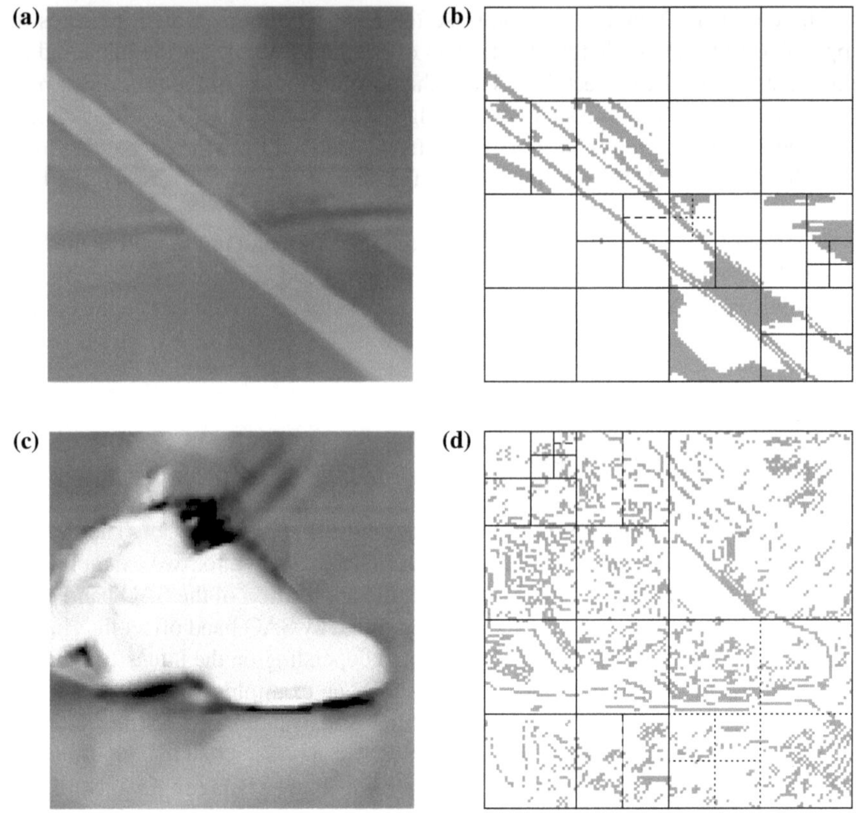

Fig. 9.14 Illustration of the application of SAO band offset and edge offset to the samples of 2×2 CTBs. **a** Reconstructed area. **b** Samples with band offset active. **c** Reconstructed area. **d** Samples with edge offset active

The determination of the SAO parameters is performed for the encoded CTU. I.e., the CUs have been encoded and the reconstructed and deblocked CTBs of the color components are available. The CTBs are input to the SAO filtering determination process.[5]

The rate-distortion optimized application of SAO requires the encoder to determine which of the three available modes—edge offset, band offset, and non-SAO filtering—to use and what offset parameters to apply in case of activated filtering. For CTUs using edge offset, the absolute values of the offsets for each of the four applicable neighborhood conditions need to be found. For CTUs using band offset, the transition band position and the signed offsets for the four transition bands have to

[5] In CTU-based processing, the bottom rows and the right boundary sample columns of the CTU can only be deblocked after the corresponding bottom and right neighboring CTUs have been processed as well, see the discussion in Sect. 9.2.5.2.

Fig. 9.15 Illustration of SAO mode distribution per CTB over a coded picture

be determined. The methods for fast distortion calculation and rate-distortion-based parameter derivation are detailed in the following.

9.2.5.1 Distortion Calculation

For determination of the offset parameter values, the distortion introduced by the corresponding SAO filtering has to be calculated for each tested parameter variation. In [8], the following fast distortion calculation method is proposed for this task.

For the discussion, a CTB using band offset is considered. Let C_b be a set of N_b samples that belong to one band of the 32 bands into which the intensity range is divided for application of the SAO band offset, see the illustration in Fig. 9.16. The squared error introduced by the previous prediction, quantization, and deblocking filtering can be calculated as

$$D_\mathrm{pre} = \sum_{k \in C_\mathrm{b}} \left[b_\mathrm{org}(k) - b_\mathrm{dbl}(k)\right]^2, \tag{9.40}$$

where $b_\mathrm{org}(k)$ and $b_\mathrm{dbl}(k)$ denote the kth original and reconstructed deblocked samples in C_b, respectively. If an offset Δ_sao is applied to the samples in C_b, the squared error becomes

$$D_\mathrm{sao} = \sum_{k \in C_\mathrm{b}} \left[b_\mathrm{org}(k) - (b_\mathrm{dbl}(k) + \Delta_\mathrm{sao})\right]^2. \tag{9.41}$$

The difference between the squared errors can be written as

(a)  (b) 

Fig. 9.16 Illustration of samples belonging to one SAO band. **a** Example luma CTB. **b** Set C_b of samples in one band marked *gray* (band 9, values in range [64,71])

$$\begin{aligned} D_\Delta &= D_{\mathrm{sao}} - D_{\mathrm{pre}} \\ &= \sum_{k \in C_\mathrm{b}} \Delta_{\mathrm{sao}}^2 - 2\Delta_{\mathrm{sao}} \left[b_{\mathrm{org}}(k) - b_{\mathrm{dbl}}(k) \right] \\ &= N_\mathrm{b} \Delta_{\mathrm{sao}}^2 - 2\Delta_{\mathrm{sao}} E_\mathrm{b}. \end{aligned} \tag{9.42}$$

The values D_{pre} and E_b only need to be calculated once for the N_b samples in C_b. Thereby, the variation of the distortion for different values of Δ_{sao} can be determined at limited further cost.

In the HM reference software, the search for the applicable offset value is initialized with

$$\tilde{\Delta}_{\mathrm{sao}} = \left\lfloor \frac{E_\mathrm{b}}{N_\mathrm{b}} \right\rfloor, \tag{9.43}$$

which corresponds to the value where D_Δ is minimized.

It should be noted that with CTB-based encoder processing, the samples at the bottom and right boundary of the CTB cannot be directly deblocked yet as the neighboring CTBs have not yet been processed. Therefore, these sample values may be subject to modification and decision drawn on SAO parameters may not be reliable. In [5], two options are proposed to deal with this issue. For a CTB size of 64×64, the suggested method is to skip the non-deblocked boundary samples in the determination algorithm. For smaller CTB sizes, it is suggested to include the non-deblocked sample values for determination of the SAO parameters.

9.2.5.2 Rate-Distortion Decision

Using D_Δ from (9.42) and the estimated bit cost R_{est} which is needed to encode the corresponding value of the offset parameter Δ_{sao}, the rate-distortion impact of the application of Δ_{sao} can be estimated as

$$J_\Delta = D_\Delta + \lambda \cdot R_{\text{est}}, \tag{9.44}$$

which in turn can be used to decide on the applicable value of the offset parameter. Starting from $\tilde{\Delta}_{\text{sao}}$ in (9.43), offset values $|\Delta_{\text{sao}}| \leq \left|\tilde{\Delta}_{\text{sao}}\right|$ are tested.

Summarizing over the resulting J_Δ values for all bands, the overall rate-distortion impact of SAO on the current CTB can be determined. The approach can be further used to determine the applicable transition band position or the applicable edge offset class. Taking the resulting J_Δ values for edge offset and band offset into account, the decision on the application of SAO for the given CTB can be drawn.

In the HM reference software, an additional slice-level early termination algorithm is applied [9, 10]. For the top hierarchy level (corresponding to temporal sub-layer 0 for configurations with temporal scalability), SAO is always activated on the slice level. Based on the prediction hierarchy of the selected GOP structure, the application of SAO is switched off for a picture for the luma component if less than 75 % of the CTUs in the last picture of the previous hierarchy level have used SAO. For chroma, the threshold is set to 50 %. This method reduces the overall rate-distortion cost for SAO if only few isolated CTUs would be filtered. It also reduces the encoder complexity as the SAO parameter determination for these pictures is skipped.

9.3 Comparison to H.264 | AVC

Since H.264 | AVC only includes the specification of an in-loop deblocking filter, sample-adaptive offset filtering is only applicable as a post filter for codecs following this specification. An overview of the H.264 | AVC deblocking filter is provided in [11]. The HEVC deblocking filter can be seen as an advancement from the H.264 | AVC filter with reduced complexity and enhanced parallelizability as the key development objectives. The filters applied to the samples are the same in HEVC and H.264 | AVC but the control of the filters is different. In H.264 | AVC up to three luma samples and up to two chroma samples on each side of the block boundary can be filtered. In HEVC, chroma samples may only be filtered at the boundaries of intra blocks.

Compared to the H.264 | AVC deblocking filter [12, 13], the filter in HEVC is distinguished by a generally simplified processing structure. The H.264 | AVC deblocking filter operates on a 4×4 edge raster. Due to the maximum applicable filter length, virtually all samples in a picture may be modified by the deblocking filter process. The decisions on the boundary strength b_S are made for each boundary sample pair instead of the 4-sample sections used in HEVC. In H.264 | AVC, five b_S values instead of three are used to control the filter operation.

The H.264 | AVC deblocking filter operation is specified on a 16×16 macroblock basis. As filtering is applied on up to three samples into the respective block and the block grid is 4×4, the deblocking order of vertical and horizontal edges inside the macroblock must be strictly obeyed and parallelization e.g. along neighboring macroblocks is not possible.

The inter-dependency between the edges has a strong impact on the deblocking filter operation if deblocking across slice boundaries is enabled. In this case, H.264 | AVC deblocking must be done after complete decoding of the slices. With the HEVC design, all deblocking besides the 8×8 slice or tile boundary blocks can be included in decoding loop and only the remaining blocks at the slice or tile boundaries need to be revisited upon completion of the decoding process [1].

References

1. Norkin, A., et al.: HEVC deblocking filter. IEEE Trans. Circ. Syst. Video Technol. **22**(12), 1746–1754 (2012). doi:10.1109/TCSVT.2012.2223053
2. High efficiency video coding. ITU-T Rec. H.265 (HEVC). http://www.itu.int/rec/T-REC-H.265/en (2013). Accessed 14 April 2014
3. Information technology—High efficiency coding and media delivery in heterogeneous environments—Part 2: High efficiency video coding. ISO/IEC 23008-2:2013 (HEVC). http://www.iso.org/iso/home/store/catalogue_detail.htm?csnumber=35424 (2013). Accessed 14 Apr 2014
4. Bossen, F.: Common test conditions and software reference configurations. Doc. JCTVC-K1100. 11th Meeting: Joint Collaborative Team on Video Coding (JCT-VC) of ITU-T SG 16 WP 3 and ISO/IEC JTC 1/SC 29/WG 11, Shanghai, CN (2012)
5. Fu, C.-M., et al.: Sample adaptive offset in the HEVC standard. IEEE Trans. Circ. Syst. Video Technol. **22**(12), 1755–1764 (2012). doi:10.1109/TCSVT.2012.2221529
6. JCT-VC. HEVC Reference Software. http://hevc.hhi.fraunhofer.de/svn/svn_HEVCSoftware/ (2014). Accessed 14 Apr 2014
7. Kim, I.-K., et al.: High efficiency video coding (HEVC) test model 10 (HM10) encoder description. Doc. JCTVC-L1002. 12th Meeting: Joint Collaborative Team on Video Coding (JCT-VC) of ITU-T SG 16 WP 3 and ISO/IEC JTC 1/SC 29/WG 11, Geneva, CH (2013)
8. Fu, C.-M., et al.: Sample adaptive offset for HEVC. In: Proceedings of IEEE International Workshop on Multimedia Signal Processing MMSP, vol. 13, pp. 1–5. IEEE, Hangzhou (2011). doi:10.1109/MMSP.6093807
9. Laroche, G., et al. Non-CE1: Encoder modification for SAO interleaving mode. Doc. JCTVC-I0184. 9th Meeting: Joint Collaborative Team on Video Coding (JCT-VC) of ITU-T SG 16 WP 3 and ISO/IEC JTC 1/SC 29/WG 11, Geneva, CH (2012)
10. Alshina, E., et al.: Encoder modification for SAO. Doc. JCTVC-J0044. 10th Meeting: Joint Collaborative Team on Video Coding (JCT-VC) of ITU-T SG 16 WP 3 and ISO/IEC JTC 1/SC 29/WG 11, Stockholm, Sweden (2012)
11. List, P., et al.: Adaptive deblocking filter. IEEE Trans. Circ. Syst. Video Technol. **13**(7), 614–619 (2003). doi:10.1109/TCSVT.2003.815175
12. Advanced video coding for generic audiovisual services. ITU-T Rec. H.264 (AVC). http://www.itu.int/rec/T-REC-H.264/en (2014). Accessed 14 Apr 2014
13. Information technology—Coding of audio-visual objects—Part 10: Advanced video coding. ISO/IEC 14496-10:2012 (AVC). http://www.iso.org/iso/home/store/catalogue_detail.htm?csnumber=61490 (2012). Accessed 14 Apr 2014

Chapter 10
Entropy Coding

The bitstream is constructed of encoded syntax elements. In the case of fixed length codes or variable length codes, syntax elements are represented by unique code words in a one-to-one mapping. With arithmetic coding, no direct connection of the syntax elements to specific bits in the bitstream is given. The different encoding schemes are applied according to the intended properties of the resulting bitstream with respect to the characteristics of the coded syntax elements. The process of extracting syntax elements from the bitstream at the decoder side is called the *parsing process*.

Three coding schemes are used in the HEVC parsing process: fixed-length codes, systematic variable length codes, and context-based adaptive binary arithmetic coding (CABAC). These schemes are employed on the different hierarchy levels of the bitstream according to the intended features of the bitstream. For NAL unit headers and highest-level information on the coded video sequence, fixed-length codes are employed in a byte-aligned syntax structure. Thereby, this information is accessible with very low complex parsing by the decoder and by other systems seeking information on high-level properties of the stream. Information specifying higher-level information like tool configurations and parameter settings is represented using fixed-length or systematic variable length codes in order to balance coding complexity and compression efficiency. The structure of the systematic codes allows for some coarse approximation of the probability distribution of the encoded syntax elements. Syntax elements at the coding block level represent the major part of the coded information. Here, significant effort is invested to achieve the most compact representation of the respective syntax elements, following closely their probability distribution. The effort to approach the symbol entropy as close as possible is expressed in the denotation *entropy coding*. The term is often used to generally address the process of transforming a stream of syntax elements into a bitstream, even if a selected coding scheme is not strictly designed to minimize the bitstream size.

In HEVC, a strict separation between CABAC and non-CABAC coding is employed. The application of the coding schemes is illustrated for a coded slice segment NAL unit in Fig. 10.1. The NAL unit header has a static structure of fixed-length code words which is identical for all NAL unit headers, see also Sect. 5.2.1.

M. Wien, *High Efficiency Video Coding*, Signals and Communication Technology,
DOI 10.1007/978-3-662-44276-0_10

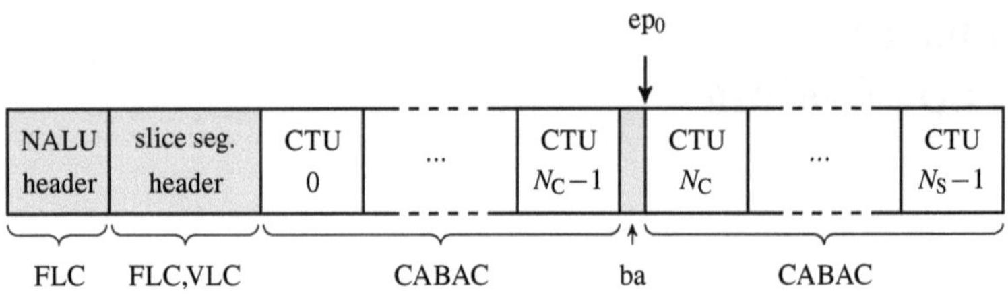

Fig. 10.1 Illustration of the application of fixed-length coding (FLC), variable length coding (VLC), and context-based adaptive binary arithmetic coding (CABAC) in a coded slice segment NAL unit with N_S CTUs and a tile boundary before CTU N_C. Byte alignment (ba) is required to allow to start decoding at the entry point ep_0

Table 10.1 HEVC parsing processes

Descriptor	Parsing process
ae(v)	Context adaptive arithmetic entropy-coded syntax element
b(8)	Byte having any pattern of bit string (8 bits)
f(n)	Fixed-pattern bit string using n bits
se(v)	Signed integer syntax element coded with $C_{eg0}(v)$
u(n)	Unsigned integer using n bits
ue(v)	Unsigned integer syntax element coded with $C_{eg0}(v)$

The slice segment header comprises parameter configurations which are coded using fixed-length and variable length coding. CABAC is applied at the CTU level, with the CTU syntax structure including the CTU SAO parameter setting and the CTU coding quadtree. Within the coded slice segment, CABAC coding is only interrupted for byte alignment if entry points are present, i.e. wavefront parallel processing is enabled or the slice segment comprises multiple tiles.

The small number of six parsing processes are specified for use in the syntax tables of HEVC, one for CABAC and five for non-CABAC parsing. The descriptors for these processes are listed in Table 10.1 below. They are detailed in Sect. 10.1 for non-CABAC and in Sect. 10.2 for CABAC coded syntax elements.

10.1 Fixed- and Variable-Length Coding

10.1.1 Fixed-Length Codes

The parsing process denoted by b(8) is used in the HEVC syntax tables to specify the general syntax structure with the NAL unit header and a NAL unit payload. Without detailing the content of the payload it consists of RBSP bytes (represented as b(8)) for the content and potentially also of emulation prevention bytes in cases where a successive set of RBSP bytes would include a sequence of bits which represent a start code, see Sect. 5.2.1.

Table 10.2 Exp-Golomb code of order 0

v	$C_{\text{eg0}}(v)$
0	1
1,2	0 1 x_0
3,...,6	0 0 1 x_1 x_0
7,...,14	0 0 0 1 x_2 x_1 x_0
	$\vdots$

Syntax elements coded with f(n) have a fixed pre-defined value represented by n bits. Such syntax elements are used in the high-level syntax. An example is the forbidden_zero_bit syntax element which is a f(1) codeword with the fixed value of zero. The parsing process u(n) is used to represent syntax elements with unsigned integer values by n bits in the bitstream. This coding is used in the NAL unit header, the parameter sets and the slice segment header.

10.1.2 Exp-Golomb Codes

For syntax elements that reach large values or syntax elements whose maximum value is not pre-determined by the specification, the Exp-Golomb code of order 0, $C_{\text{eg0}}(v)$, is used. The structure of this code is shown in Table 10.2 for reference.

For syntax elements which have unsigned integer values, the value is decoded from the code word as $v = C^{\text{i}}_{\text{eg0}}(c)$, see (2.4.6). This parsing process is denoted by ue(v) in the syntax tables. For signed values, the bit x_0 is used as the sign flag and the values are reconstructed from the code word c as

$$\begin{aligned} m &= C^{\text{i}}_{\text{eg0}}(c), \\ v &= (-1)^{m+1}\left\lceil \frac{m}{2} \right\rceil. \end{aligned} \tag{10.1}$$

The parsing process for syntax elements with signed value is denoted by se(v).

10.2 CABAC—Context-Based Adaptive Binary Arithmetic Coding

In H.264 | AVC, two entropy coding schemes for block-level information were specified. Context adaptive VLCs (CAVLC) are available in all profiles. CABAC is specified as an alternative (and mostly used) option in Main and High profiles. In HEVC, only CABAC has been selected.

The arithmetic coding engine itself has not changed compared to CABAC as specified in H.264 | AVC. An overview of CABAC in H.264 | AVC was published by Marpe et al. in [1]. For HEVC, the design and use of context dependencies and bypass coding have been revised for high compression efficiency at reduced context sizes and improved parallelizability compared to H.264 | AVC [2].

10.2.1 Process Overview

The binary arithmetic coding engine of CABAC requires transformation of non-binary values into a binary representation for encoding. The transformation process is called *binarization*, the entries of the resulting *bin-string* are called *bins*. The binarization is designed to be prefix-free, i.e. no bin-string is the prefix of a longer bin-string for a given syntax element and the values of the syntax element are thereby uniquely decodable from the reconstructed bin-strings. In case of binary syntax elements (flags), the bin-string is of length 1 and identical to the syntax element. An overview of the binarization schemes applied to HEVC syntax elements is provided in Sect. 10.2.3.

For each coded bin a *context model* is applied. The context model can be adaptive, triggering the adaptive arithmetic coding engine, or static. The static model triggers a non-adaptive coding engine with fixed setting of equal probability for the two binary values '0' and '1'. This engine is called the *bypass* engine. It is used e.g. for coding of sign flags of transform coefficients. Direct writing of such binary values to the bitstream would require the arithmetic coding process to be terminated and restarted afterwards. By using the bypass engine, such binary values can efficiently be included in the arithmetically coded bitstream. In the adaptive coding engine, the context model is updated after each encoding or decoding step. Thereby, the context model adapts to the observed probability distribution in the actual coding process. After processing a complete CTU, the status of all context models can be stored in a buffer to be available for re-initialization of the coding engine later in the decoding process, e.g. for Wavefront Parallel Processing. The encoding and decoding operations with the two arithmetic coding engines are described in Sect. 10.2.2.

A schematic diagram of the CABAC encoding process is shown in Fig. 10.2. The overall CABAC encoding process begins with a binarization stage for non-binary symbols. For each coded bin, either context adaptive coding with an associated dedicated context model or bypass coding with a static probability model is selected. On the context-adaptive branch, the applied context model is updated after coding of each bin. Finally, the coded output bits of the CABAC engine are fed into the bitstream. After the last syntax element of a coded slice segment is reached (or the last syntax element of a row in wavefront parallel processing or a tile) an ending flag is encoded to indicate the termination of the coding process.

On the decoder side, the process of the encoder side is reverted to reveal the coded syntax elements from the bitstream. The context models at the decoder side are initialized in the same manner as the encoder context models. For each syntax element, the bins are successively decoded from the bitstream, using and updating the corresponding context models as appropriate. Since the binarization of the coded syntax elements

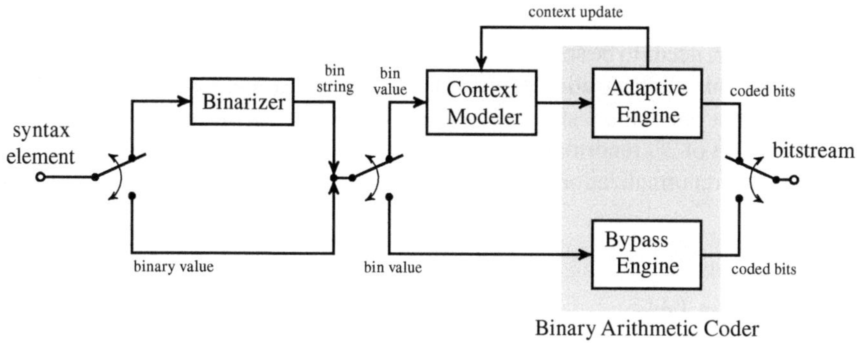

Fig. 10.2 Block diagram of the CABAC encoder as presented in [1]

is designed to be prefix-free, the value of a decoded syntax element is identified as soon as the decoded bin string matches a valid binarization. The decoder then moves on with the first bin of the next syntax element and its associated context model.

10.2.2 Binary Arithmetic Coding

The operation of the binary arithmetic encoding and decoding engines is described in the following. After reviewing the representation of the general engine state and the transition between the applicable probability models, the specific encoding and decoding processes are detailed. It can be seen that the encoding and decoding steps required for processing the value of a bin are closely synchronized.

10.2.2.1 Representation of the Engine State

In binary arithmetic coding the two symbols (0, 1) are encoded. Instead of explicitly addressing the two symbol values, the symbols are characterized to be the *most probable symbol* (MPS) or the *least probable symbol* (LPS), denoting which of the two symbols has the higher probability in the current probability state. In the following, the probability state for a given context model is specified by the probability estimate of the least probable symbol $\tilde{p}_{\mathrm{LPS}}$ and by the value of the most probable symbol $\nu_{\mathrm{MPS}} \in \{0, 1\}$. Since the value and probability state for the other symbol can be derived from $\tilde{p}_{\mathrm{LPS}}$ and ν_{MPS}, the two values are sufficient to characterize the probability state.

Let the current interval range for the engine be denoted by R_{S}. For coding a bin according to its context model, R_{S} is split into two sections according to the probability state representing the most probable and least probable symbol, respectively.

The specification of the arithmetic coding engine allows for implementation in integer arithmetic with a dynamic range of 10 bit for the encoder and 11 bit for the decoder. In the coding process the value of R_s lies within $2^8 \leq R_s < 2^9$. For interval subdivision, R_s needs to be split into two intervals of size $\tilde{p}_{LPS} \cdot R_s$ and $(1-\tilde{p}_{LPS}) \cdot R_s$. In CABAC, this multiplication operation is avoided and replaced by a table-based scheme with a pre-defined state transition table. When the range size drops below the range limit of 2^8, renormalization is applied.

Whenever renormalization is performed, a bit can be written to (or read from) the bitstream.

State Transition Table

For low-complex processing of the arithmetic engine, a context model can only be in exactly one of 64 defined probability states. Each state is characterized by its state index $i_{p_{LPS}}$ and the currently assigned value of the most probable symbol v_{MPS}. The state transition table specifies the transition from one state to another state in case of a context update.[1] The transition rules implemented in this table are schematically shown in Fig. 10.3. The transition after processing an MPS is denoted by $\mathrm{Tr}_{MPS}\{\cdot\}$, the transition after an LPS is denoted by $\mathrm{Tr}_{LPS}\{\cdot\}$. These state transitions apply for the states $i_{p_{LPS}} = 0, \ldots, 62$. The representative probability values $\tilde{p}_{LPSi}$ are indicated in the diagram of Fig. 10.3. The last state $i_{p_{LPS}} = 63$ is not included in the regular state transition table. It is used for terminating the arithmetic coding process, see Sect. 10.2.2.7.

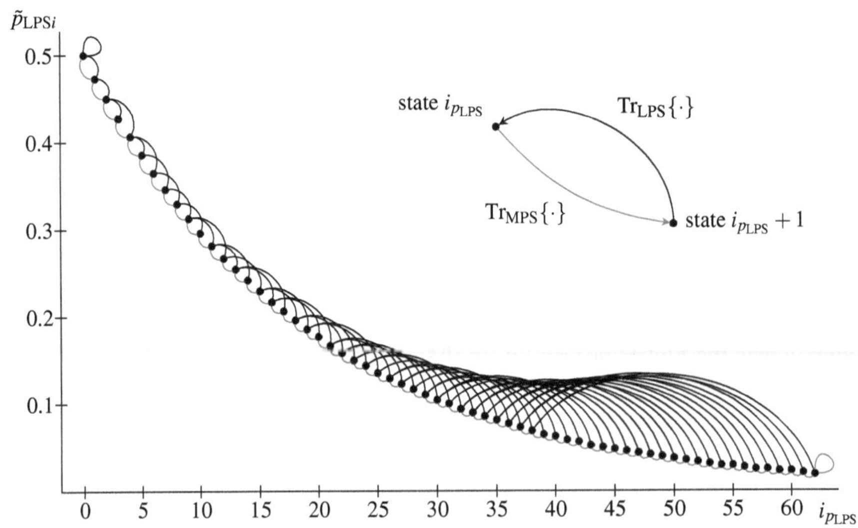

Fig. 10.3 Probability state transition between the available probability states. Transitions after an MPS towards the *right* (*gray*), LPS towards the *left* (*black*) [1]

[1] Table 9-41 in the HEVC specification [3, 4].

The highest value $\tilde{p}_{\mathrm{LPS}i} = 0.5$ is reached for $i_{p_{\mathrm{LPS}}} = 0$. In case of an update after coding an MPS, the index propagates from $i_{p_{\mathrm{LPS}}}$ to the neighboring state $i_{p_{\mathrm{LPS}}} + 1$ to the right, which has a lower LPS probability. For $i_{p_{\mathrm{LPS}}} = 62$, no neighboring state is available. The update after an MPS consequently remains in this state, representing the lowest LPS probability for regular coding. If the LPS has been coded on the other hand, the index is updated towards a lower index value corresponding to an increased LPS probability. Note that the distance between the current state and the updated state is usually large, see Fig. 10.3. Thereby, a faster adaptation is achieved in case of an LPS. If the state is at index $i_{p_{\mathrm{LPS}}} = 0$ and needs to be updated after coding the LPS, the polarity of the LPS changes: The former least probable symbol turns to be the new most probable symbol and the corresponding value v_{MPS} is changed to the new value

$$v_{\mathrm{MPS}} = 1 - v_{\mathrm{MPS}}. \tag{10.2}$$

Table-Based Interval Subdivision

The indexed probability states are used in the process of multiplier-free interval subdivision. For determination of the actual subdivision, the current range R_{s} is quantized to an index with four possible values,

$$i_{R_{\mathrm{s}}} = (R_{\mathrm{s}} \gg 6)\&3, \quad i_{R_{\mathrm{s}}} \in \{0, 1, 2, 3\}. \tag{10.3}$$

In the next step, the state index $i_{p_{\mathrm{LPS}}}$ and the range index $i_{R_{\mathrm{s}}}$ are used for a table look-up, which returns a pre-computed quantized result for the product $\tilde{p}_{\mathrm{LPS}i} \cdot R_{\mathrm{s}}$. The result of this precomputed multiplication depending on $i_{p_{\mathrm{LPS}}}$ and $i_{R_{\mathrm{s}}}$ is visualized in Fig. 10.4. For the special last state $i_{p_{\mathrm{LPS}}} = 63$, the table look-up result is independent of the range index. In this case, the smallest LPS range value of $R_{\mathrm{LPS}} = 2$ is returned for all $i_{R_{\mathrm{s}}}$. As indicated before, this last stage is used for terminating the arithmetic coding process.

10.2.2.2 Encoding with Context Update

The steps of the arithmetic encoding process for a bin value v_{bin} are described. An overview of the process is given in Fig. 10.6.

The state of the arithmetic encoder is characterized by a currently available range R_{s} and the value of the lower bound of this range r_{lb}. When starting the arithmetic encoding process, these two values are initialized to $R_{\mathrm{s}} = 510$ and $r_{\mathrm{lb}} = 0$, indicating that the full number range as specified for CABAC is available. When encoding the value of a bin, the bin value and the context model corresponding to the current bin is passed to the encoder. In the interval subdivision, the MPS corresponds to the lower and the LPS to the higher part of the available range interval. The process of interval subdivision is illustrated in Fig. 10.5.

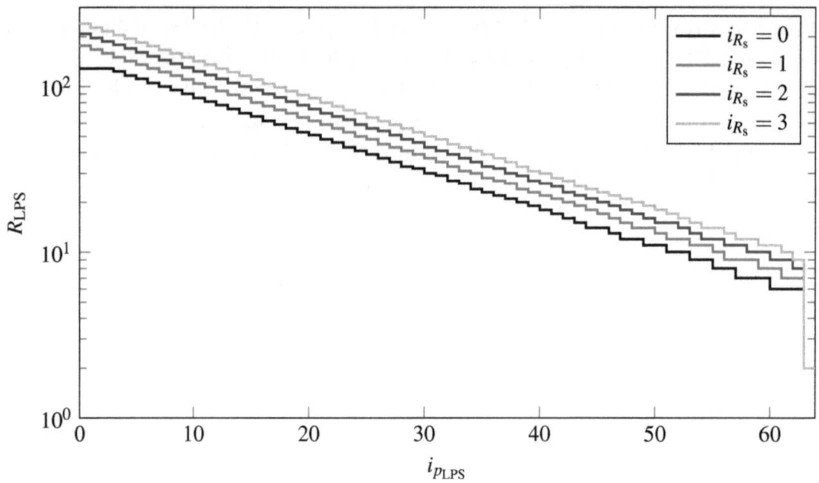

Fig. 10.4 Table-based approximation of $\tilde{p}_{LPSi} \cdot R_s$ depending on the state index $i_{p_{LPS}}$ and the quantized range index i_{R_s}

Fig. 10.5 Interval subdivision of range R_s with shift of the lower bound r_{lb} if the LPS is coded

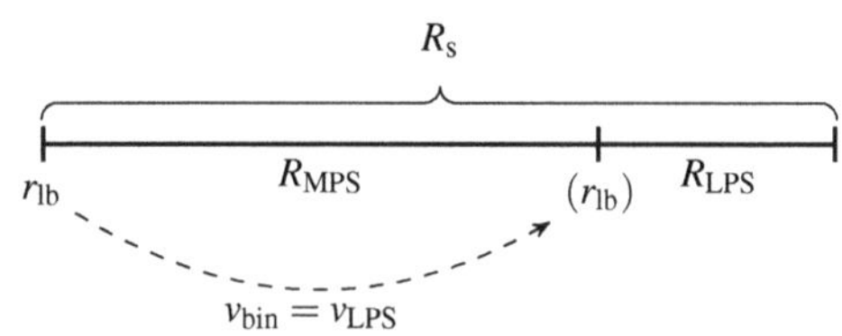

The range R_{LPS} dedicated to the least probable symbol value is derived from the range R_s using the probability state index $i_{p_{LPS}}$ and the range index i_{R_s} according to (10.3). Accordingly, the range dedicated to the most probable symbol value is set to

$$R_{MPS} = R_s - R_{LPS}. \tag{10.4}$$

If the MPS is encoded ($v_{bin} = v_{MPS}$), the range is then set to the new range

$$R_s^* = R_{MPS}. \tag{10.5}$$

The value of the lower bound r_{lb} is not changed. The probability state for the current context model is updated following the MPS path in the state transition table as shown in Fig. 10.3. If the LPS is encoded ($v_{bin} = v_{LPS}$), the range is set to the LPS range and the value of the lower bound is updated,

$$R_s^* = R_{LPS}, \tag{10.6}$$

$$r_{lb}^* = r_{lb} + R_{MPS}. \tag{10.7}$$

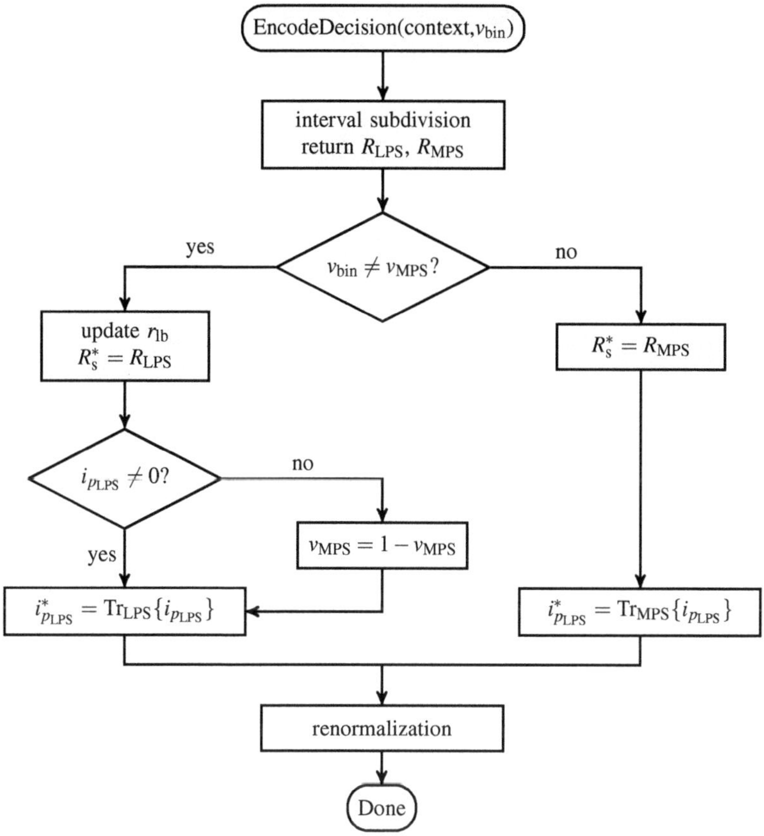

Fig. 10.6 Block diagram of the coding process for encoding a binary decision with context update

The probability state for the current context model is updated following the LPS path in the state transition table as shown in Fig. 10.3. If the probability state index is $i_{p_{LPS}} = 0$, the MPS value is updated according to (10.2). The updated values R_s^* and r_{lb}^* are set to be the current range and lower bound values.

As the last step, a renormalization process is invoked. In this process, bits are written to the bitstream if the available range is below the 8 bit threshold, $R_s < 256$. In this case, the following loop is entered:

- If the lower bound [which may have been updated in (10.7)] is $r_{lb} < 256$, a bit $b = 0$ is written to the stream.
- Otherwise, if the bound is equal to or exceeds the 9-bit threshold, the bound is shifted to $r_{lb}^* = r_{lb} - 512$ and a bit $b = 1$ is written. If the bound is $256 \leq r_{lb} < 512$, no bit can be directly written to the stream. Instead, the bound is shifted to $r_{lb}^* = r_{lb} - 256$ and a counter n_b of outstanding bits to be written is incremented, $n_b += 1$. The next time a bit b is written, it is followed by an according number of inverted

bits with value $(1 - b)$. This method asserts decodability for the given finite precision of the arithmetic coding engine (carry-over control) [5].

- As the last step, both, the range and the lower bound value are scaled according to

$$R_s^* = 2 \cdot R_{LPS}, \tag{10.8}$$

$$r_{lb}^* = 2 \cdot r_{lb}, \tag{10.9}$$

and the loop is re-iterated.

The operation of writing a bit along with potential outstanding bits (if $n_b > 0$) to the bitstream is denoted putBit(b) in the following.

10.2.2.3 Bypass Encoding

In case of encoding a bin value v_{bin} with the bypass engine, a simplified encoding algorithm is employed which omits updating the probability state of a context model. Instead, a fixed probability model is applied with equal probability of 50 % for both bin values, '0' and '1'. Avoiding the probability state update reduces the implementation complexity of this encoding engine compared to the context-adaptive engine. In the design of CABAC, bypass coding has been chosen to arithmetically code bins which have approximately equal probability or do not have great impact on the coding efficiency of the overall coding scheme.

A schematic overview of bypass encoding is given in Fig. 10.7. The order of renormalization and bit encoding is inverted compared to the encoding process with context update as described in the previous section. The bin value of $v_{bin} = 1$ is assigned the upper part of the range, $v_{bin} = 0$ is assigned the lower part. According to the inversion of the processing steps, the value of the lower bound is first scaled,

$$r_{lb}^* = 2 \cdot r_{lb}. \tag{10.10}$$

In case $v_{bin} = 1$, the lower bound is then shifted by the range,

$$r_{lb}^* = r_{lb} + R_s. \tag{10.11}$$

Note that due to the inversion of the processing steps with the scaling for renormalization first and due to equal probability of the two bin values, adding R_s corresponds to adding R_{MPS}. If $r_{lb} \geq 1024$, a bit $b = 1$ is written and the bound is shifted to $r_{lb}^* = r_{lb} - 1024$. Otherwise, if $r_{lb} < 512$, a bit $b = 0$ is written.

Similar to the unscaled intermediate case for the regular encoding process, no bits are written if $512 \leq r_{lb} < 1024$. In this case, the bound is shifted to $r_{lb}^* = r_{lb} - 512$ and the counter n_b of outstanding bits to be written is incremented. These outstanding bits are handled according to the renormalization process described in Sect. 10.2.2.2.

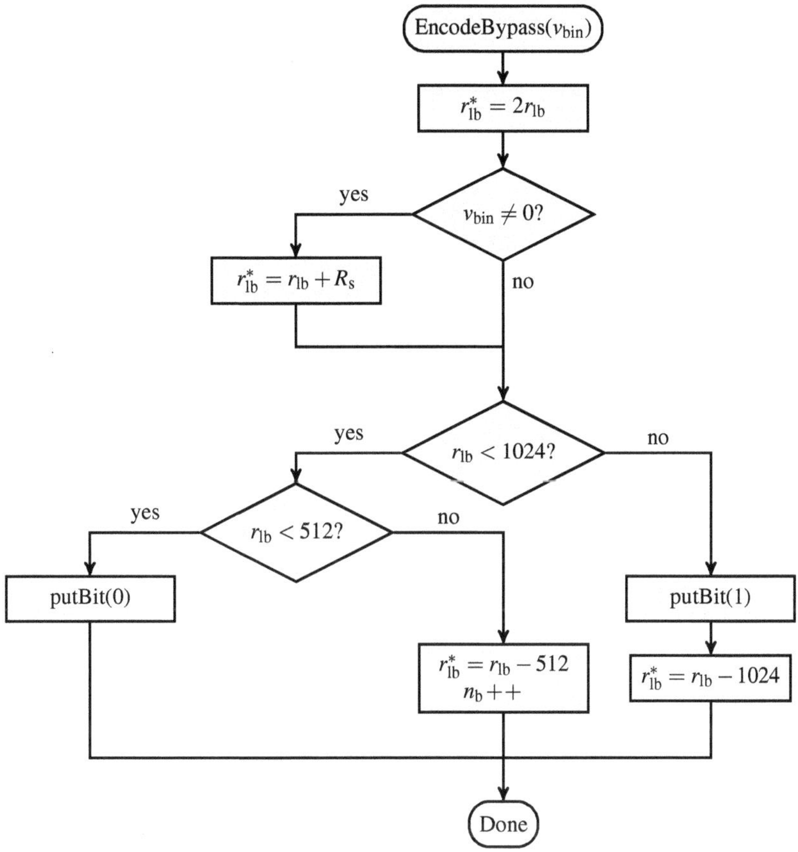

Fig. 10.7 Block diagram of the bypass encoding process

10.2.2.4 Encoding Before Termination

The syntax of HEVC provides several entry points for starting the decoding of parts of the bitstream. On the slice level, special care has to be taken to establish entry points to the bitstream as arithmetic coding is used for the syntax elements here. Accordingly, before an entry point the arithmetic coding engine has to be terminated. The process described in this section applies to flags which are encoded to indicate such a termination point, either at the end of a slice segment, the end of a sub-bitstream, or in the context of PCM coding.[2]

The terminating syntax elements are assigned a dedicated context model (with context index 0) which is associated with the probability state index $i_{p_{LPS}} = 63$, as

[2] HEVC syntax elements `end_of_slice_segment_flag`, `end_of_sub_stream_one_bit`, and `pcm_flag`. In case of PCM coding, byte alignment is performed before the unary coding of the PCM sample values.

indicated in Sect. 10.2.2.1. The bin value $v_{\mathrm{bin}} = 1$ is associated to the LPS, and the interval sub-partitioning results in a fixed LPS range of $R_{\mathrm{LPS}} = 2$.

First, the available range is modified as

$$R_{\mathrm{MPS}} = R_{\mathrm{s}} - 2. \tag{10.12}$$

If the MPS is encoded, the renormalization process as described in Sect. 10.2.2.2 (which writes the bits) is invoked next. If the LPS is encoded, the lower bound of the range is set to

$$r_{\mathrm{lb}}^{*} = r_{\mathrm{lb}} + R_{\mathrm{MPS}}. \tag{10.13}$$

The renormalization process is invoked with $R_{\mathrm{LPS}} = 2$, accordingly. Then, three final bits are written to reach the termination of the coding step. The encoding process before termination is shown in Fig. 10.8.

10.2.2.5 Decoding with Context Update

The decoding process for a context model with probability state update is described. Similar to the encoding process, the state of the arithmetic decoder is characterized by the currently available range R_{s}. Instead of the lower bound of the range, a range offset r_{os} is used in the decoding process. When starting arithmetic decoding, the range is initialized to $R_{\mathrm{s}} = 510$ and the offset r_{os} is set to the integer value represented

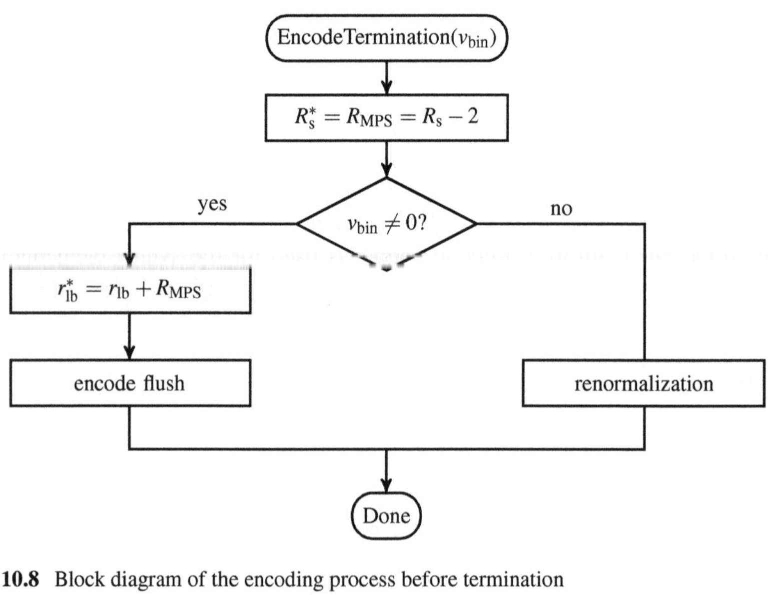

Fig. 10.8 Block diagram of the encoding process before termination

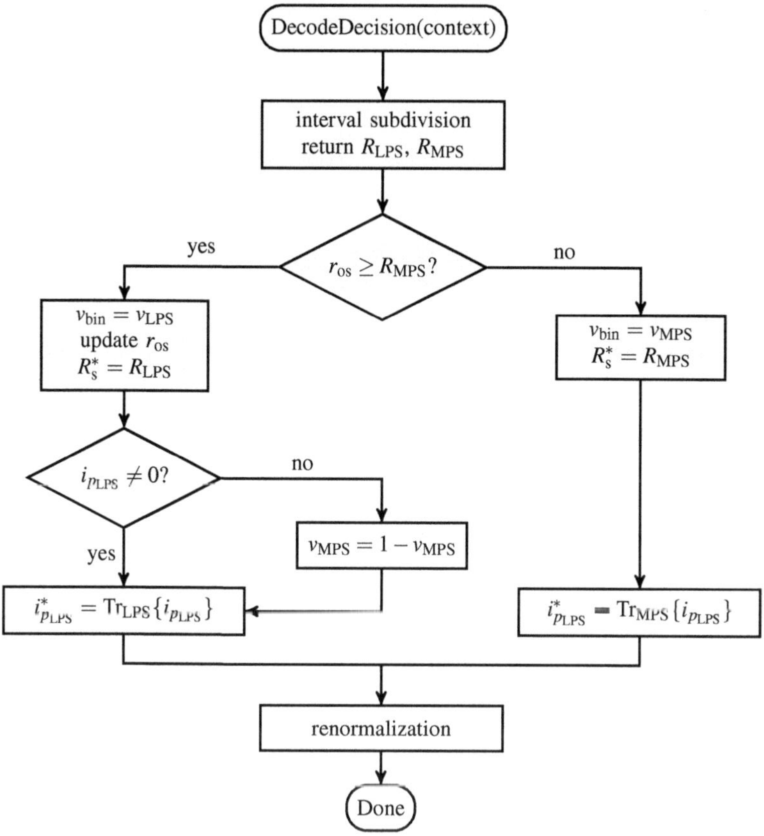

Fig. 10.9 Block diagram of the decoding process for decoding a binary decision

by the first 9 bits to be read. Like with the encoder, the applicable context model is passed to the arithmetic coding engine.

After the bin has been decoded, it is appended to the end of the reconstructed bin string of the given syntax element. If the bin string matches a possible binarization of the syntax element, the value corresponding to the binarization is assigned and the decoder proceeds to the next syntax element.

An overview of the decoding process for a single bin is given in Fig. 10.9.

The LPS range R_{LPS} is derived from the range R_s using the probability state index $i_{p_{LPS}}$ and the range index i_{R_s} according to (10.3). The range dedicated to the most probable symbol value is then set to

$$R_{MPS} = R_s - R_{LPS}. \tag{10.14}$$

If the offset r_{os} is smaller than R_{MPS}, the lower interval and thereby the encoding of the MPS value is detected and the most probable symbol is assigned to the decoded bin, i.e. $v_{bin} = v_{MPS}$. Then, the MPS state transition is performed, as shown in

Fig. 10.3. If the offset r_{os} is larger than or equal to R_{MPS}, the decoded bin value is set to $v_{bin} = v_{LPS}$. The range and the offset values are updated as

$$R_s^* = R_{LPS}, \tag{10.15}$$

$$r_{os}^* = r_{os} - R_{MPS}. \tag{10.16}$$

Thereby, the offset is shifted to the lower bound of the LPS range. If the LPS probability state index is $i_{p_{LPS}} = 0$, the MPS value is updated according to (10.2). For the following steps, the updated values $R_s = R_s^*$ and $r_{os} = r_{os}^*$ are used as the current range and lower bound values.

Renormalization is applied if the new range $R_s < 256$. In this case, the range value is scaled by a factor of 2,

$$R_s^* = 2 \cdot R_{LPS}. \tag{10.17}$$

The offset value is shifted by a factor of 2 as well and the next bit b from the bitstream is read and added to the offset as

$$r_{os}^* = (2 \cdot r_{os}) \oplus b. \tag{10.18}$$

The renormalization process in (10.17) and (10.18) including the bit reading is reiterated until the condition $R_s \geq 256$ is fulfilled.

10.2.2.6 Bypass Decoding

Like on the encoder side, the decoder side processing of the bypass engine inverts the order of renormalization and bin decision. The bypass decoding process is shown in Fig. 10.10.

First, the offset is scaled by a factor of 2 and a bit is added to the offset value according to (10.18). If $r_{os} \geq R_s$, the upper interval is detected which has been assigned to the bin value $v_{bin} = 1$. The offset is shifted accordingly to

$$r_{os}^* = r_{os} - R_s. \tag{10.19}$$

Otherwise ($r_{os} < R_s$), the usage of the lower interval is identified and therefore $v_{bin} = 0$ is assigned. In this case, no further processing of the offset value is required. From the presentation in Fig. 10.10 it becomes obvious that the arithmetic coding process for bypass decoding has a very low number of processing steps and therefore enables a very lightweight implementation.

10.2.2.7 Decoding Before Termination

The dedicated decoding process before termination is summarized in Fig. 10.11. This decoding process is invoked with the context model with state index $i_{p_{LPS}} = 63$.

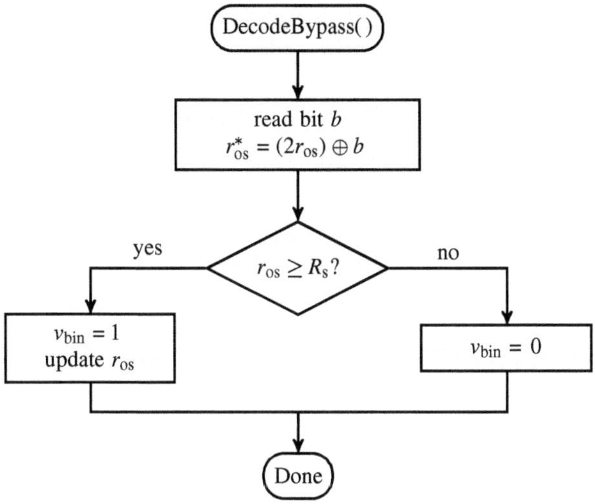

Fig. 10.10 Block diagram of the bypass decoding process

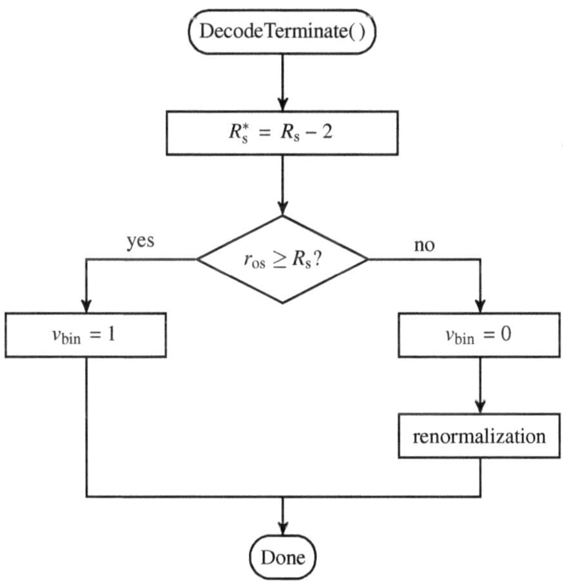

Fig. 10.11 Block diagram of the decoding process before termination

According to the table based interval subdivision, the range associated with the LPS for this special state $i_{p_{LPS}} = 63$ is $R_{LPS} = 2$. For decoding, the MPS range is checked. For this task, the range is modified by fixed LPS range,

$$R^*_s = R_s - 2. \tag{10.20}$$

If the offset r_{os} is larger than R_s, the bin value $v_{bin} = 1$ is decoded and the process is finalized. Otherwise, $v_{bin} = 0$ is decoded and the decoder side renormalization process as described with (10.17) and (10.18) is invoked.

10.2.3 Binarization

The binarization process transforms the values of a non-binary syntax element into strings of binary values. These bin-strings are fed into the arithmetic coding process, which was detailed in Sect. 10.2.2. Each non-binary syntax element has a designated binarization scheme. The binarization must be prefix-free by concept to enable decodability of the respective syntax element.

For the arithmetic encoding and decoding process, the respective bins are ordered by increasing bin index i_{bin} from the left-most bin to the right-most bin. Each bin coding is performed using an assigned context model. The selection of the applicable context models offers a great design choice when developing a CABAC-based entropy coding stage in a video coding system. Criteria for the design of an efficient entropy coding stage at a minimized implementation cost include the following [2]:

- The number of bins to be encoded should be small,
- The number of applied context models should be small as well,
- Specifically, the number of bins coded in bypass mode should be as high as possible (for reduced computational complexity),
- Bins which are coded with an identical context model should be grouped (which is beneficial for simultaneous or parallel processing).

In the development of HEVC, the consideration of these criteria has lead to the established design.

In the following, the binarization schemes applied in HEVC are presented and their features are detailed. While for many syntax elements straight forward binarization schemes are applied (like fixed-length or truncated unary binarization), other syntax elements have been assigned with sophisticated multi-level binarization schemes in response to the aforementioned design criteria.

10.2.3.1 Fixed-Length Binarization (FL)

The fixed-length binarization is applied for syntax elements with unsigned integer value. The binarization corresponds to the binary representation of the value. The length of the binary representation relates to the maximum value to be represented as $l_{FL} = \lceil \log_2(v_{max} + 1) \rceil$. An example for the fixed-length binarization is given in Table 10.3.

Fixed-length binarization is used for context adaptive coding, e.g. for the SAO type index or the prediction mode flag. The binarization is used for bypass coding, e.g. for sign flags for the QP delta, motion vector difference, transform coefficients, or the SAO band position.

Table 10.3 Fixed-length binarization

v	$B_{\mathrm{fl},3}(v)$
0	0 0 0
1	0 0 1
2	0 1 0
3	0 1 1
4	1 0 0
5	1 0 1
6	1 1 0
7	1 1 1
i_{bin}	0 1 2

Table 10.4 Truncated unary binarization

v	$B_{\mathrm{tu},n}(v)$
0	0
1	1 0
2	1 1 0
3	1 1 1 0
⋮	⋮
$n-2$	1 1 1 1 … 1 0
$n-1$	1 1 1 1 … 1 1
i_{bin}	0 1 2 3 … $n-1$

10.2.3.2 Truncated Unary Binarization (TU)

The truncated unary binarization $B_{\mathrm{tu},n}(v)$ corresponds to the truncated unary code $C_{\mathrm{u}}(v)$ described in Sect. 2.4.8.3 with the modification of a given maximum length n. The scheme of the truncated unary code is shown in Table 10.4.

The maximum length n of a code word corresponds to the maximum value that can be represented, $n = v_{\mathrm{max}}$. Knowing the maximum value, the trailing '0' bit can be omitted if n ones have been encoded.

In the specification text, the truncated unary binarization is represented by a truncated Rice binarization with binarization parameter $k = 0$, see Sect. 10.2.3.4. It is used for context adaptive coding, e.g. for the inter merge index, as well as for bypass coding, e.g. for the most probable intra prediction mode.

10.2.3.3 Exp-Golomb Binarization (EG-k)

The k-th order Exp-Golomb binarization $B_{\mathrm{eg}\,k}(v)$ for the value v_{se} of a syntax element is constructed of a prefix and a suffix. The construction of the bin string has been detailed in Sect. 2.4.8.4.

Note that for the purpose of binarization, the polarity of the prefix bits is switched compared to the code words for the EG-0 code, which is used as a VLC code in HEVC. An example for the Exp-Golomb binarization of order $k = 3$ is given in Table 10.5.

Table 10.5 Exp-Golomb binarization of order 3

v	$B_{\mathrm{eg}3}(v)$
$0,\ldots,7$	$0\ x_2\ x_1\ x_0$
$8,\ldots,23$	$1\ 0\ x_3\ x_2\ x_1\ x_0$
$24,\ldots,55$	$1\ 1\ 0\ x_4\ x_3\ x_2\ x_1\ x_0$
$56,\ldots,119$	$1\ 1\ 1\ 0\ x_5\ x_4\ x_3\ x_2\ x_1\ x_0$
$120,\ldots,247$	$1\ 1\ 1\ 1\ 0\ x_6\ x_5\ x_4\ x_3\ x_2\ x_1\ x_0$
$\vdots$	$\vdots$
i_{bin}	$0\ 1\ 2\ 3\ 4\ 5\ 6\ \ldots$

For context adaptive coding, the EG-k binarization is used in combination with other binarization schemes, e.g. for the remaining absolute value of transform coefficients as detailed below. The EG-1 binarization is used for bypass coding of the absolute motion vector difference.

10.2.3.4 Truncated Rice Binarization (TR)

The Truncated Rice binarization (TR) has a binarization parameter k for configuration. Since for $k = 0$ the TR binarization corresponds to a TU binarization, it is used to denote TU binarization in the specification. It is further used as a prefix binarization for the remaining levels of coefficients with level greater than 2, see Sect. 10.2.3.5.

The TR binarization is based on a Golomb-Rice code $C_{\mathrm{gr}k}(v)$ of grade k, see Sect. 2.4.8.4. Here, the code is constructed by a unary coded prefix of n_{p} '1' bits followed by a '0' and k suffix bits.

Let the code be used for unsigned integer values and the suffix be the k-bit binary representation of an integer $0 \leq i < 2^k$. The number of '1' bits in the prefix is denoted by n_{p}. When encoding a value v, the number of prefix '1' bits is determined by

$$n_{\mathrm{p}} = \left\lfloor \frac{v}{2^k} \right\rfloor. \tag{10.21}$$

The suffix is the binary representation of

$$v_{\mathrm{s}} = v - n_{\mathrm{p}} \cdot 2^k. \tag{10.22}$$

The value range that can be binarized with this code is limited by $0 \leq v \leq v_{\max}$. In the context of transform coefficient level coding, the maximum value is set to

$$v_{\max} = 4 \cdot 2^k. \tag{10.23}$$

Thereby, the values $v < v_{\max}$ are represented with a maximum prefix length of $\lfloor \frac{v_{\max}-1}{2^k} \rfloor + 1$ and all available codewords up to this boundary are utilized. For $v = v_{\max}$, the prefix string is coded with $\lfloor \frac{v_{\max}-1}{2^k} \rfloor + 1$ '1' bits and no suffix is appended.

Table 10.6 Truncated Rice binarization: example with order $k=1$ and maximum value 8. Suffix bit $x_0 \in \{0, 1\}$

bit $x_0 \in \{0,1\}$

v	$B_{\text{tr}\,1,8}(v)$
0,1	0 x_0
2,3	1 0 x_0
4,5	1 1 0 x_0
6,7	1 1 1 0 x_0
8	1 1 1 1
i_{bin}	0 1 2 3 4

Note that this specific binarization is only used as a part of the level coefficient binarization if the coded level value v_l is larger than $v_{\max}$. From $v_l = v_{\max}$ on, the TR binarization is followed by an additional binarization string as detailed in the following subsection. An example for the TR binarization with $k = 1$ is given in Table 10.6.

10.2.3.5 Binarization for Coefficient Levels Greater than 2

In the HEVC design, the syntax elements for the remaining absolute level which are to be coded for coefficient levels greater than 2 are bypass encoded in CABAC. Since no context adaptation is available in this case, the design of the binarization must follow the distribution of the remaining absolute levels of the transform coefficients for efficient entropy coding. Starting from the assumption of a geometric distribution for these levels, a parametrized adaptive binarization has been developed [6].

The binarization $B_{\text{rl}\,k}$ for the remaining absolute level v_r of coefficient levels greater than 2 consists of a truncated Rice binarization as described in the previous section and an appended Exp-Golomb binarization for larger coefficient level values.

The binarization parameter k depends on the absolute level values at the previously scanned coefficient positions in the current transform sub-block. It specifies the grade of the Golomb-Rice code and the corresponding maximum value $v_{\max\text{TR}}$ for the TR part as denoted in (10.23). For remaining coefficient level values $v_r \geq v_{\max\text{TR}}$, the binarization consists of a concatenation of the "1111" code word for $v_{\max\text{TR}}$ and

Table 10.7 Binarization $B_{\text{rl}\,k}$ for coefficients with level greater than 2: Example for binarization parameters $k = 0, 1, 2$. The TR prefix is marked bold

v_0	$k=0$	v_1	$k=1$	v_2	$k=2$
0	**0**	0,1	$\mathbf{0x_0}$	0,...,3	$\mathbf{0x_1x_0}$
1	**10**	2,3	$\mathbf{10x_0}$	4,...,7	$\mathbf{10x_1x_0}$
2	**110**	4,5	$\mathbf{110x_0}$	8,...,11	$\mathbf{110x_1x_0}$
3	**1110**	6,7	$\mathbf{1110x_0}$	12,...,15	$\mathbf{1110x_1x_0}$
4,5	**1111**$0x_0$	8,...,11	**1111**$0x_1x_0$	16,...,23	**1111**$0x_2x_1x_0$
6,...,9	**1111**$10x_1x_0$	12,...,19	**1111**$10x_2x_1x_0$	24,...,39	**1111**$10x_3x_2x_1x_0$
10,...,17	**1111**$110x_2x_1x_0$	20,...,35	**1111**$110x_3x_2x_1x_0$	40,...,71	**1111**$110x_4x_3x_2x_1x_0$
...	...	...	...	...	...

Fig. 10.12 Example transform sub-block with coefficient levels

13	2	8	1
10	-5	1	0
4	-3	0	0
-1	0	1	0

Table 10.8 Increment of the binarization parameter k over the transform sub-block scan and binarization pattern length for the TSB in Fig. 10.12

Coefficient	1	0	0	0	1	1	−3	1	8	−5	4	2	10	13
Significance flag	–	0	0	0	1	1	1	1	1	1	1	1	1	1
Coeff. > 1	0	–	–	–	0	0	1	0	1	1	1	1	–	–
Coeff. > 2	–	–	–	–	–	–	1	–	–	–	–	–	–	–
Sign flag	0	–	–	–	0	0	1	1	0	1	0	0	0	0
Remaining level	–	–	–	–	–	–	1	–	7	4	3	1	10	13
TR parameter k	–	–	–	–	–	–	0	–	1	2	2	2	2	3
Binarization length	–	–	–	–	–	–	2	–	5	4	3	3	5	6

a code word from the Exp-Golomb code of grade $k + 1$ for the remaining value $v_r - v_{\max TR}$.

The applicable binarization parameter is initialized to $k = 0$ when starting the inverse scan of each 4×4 transform sub-block. When the absolute level value of the previous coefficient greater than 2 along the scan is $v > 3 \cdot 2^k$, the binarization parameter is increased by $k^* = k+1$ until a maximum of $k = 4$ is reached. Table 10.7 provides the resulting example binarization patterns for $k = 0, 1, 2$.

The combination of the truncated Rice and the Exp-Golomb binarization patterns can be seen as a modified Exp-Golomb code structure of order k where the number of suffix bits is set to k until a prefix length of 4 and the increment of the number of suffix bits is started from the suffix length of 5 on.

By adapting the binarization parameter in relation to the value of the last previously coded coefficient level greater than 2, this bypass-coded binarization scheme can be interpreted as a context adaptive VLC coding scheme embedded into CABAC.

In Fig. 10.12 and Table 10.8, the coded representation of the TSB in Fig. 8.5 is repeated. In Table 10.8, the increment of the binarization parameter and the resulting length of the resulting binarization words is indicated in addition to the flags which are coded in the scans before the remaining level values. Along the inverse scan, the absolute level of the previous scan position is checked for the condition to increase the binarization parameter as described above. In the given example, the maximum parameter is reached for the top-left coefficient in the block.

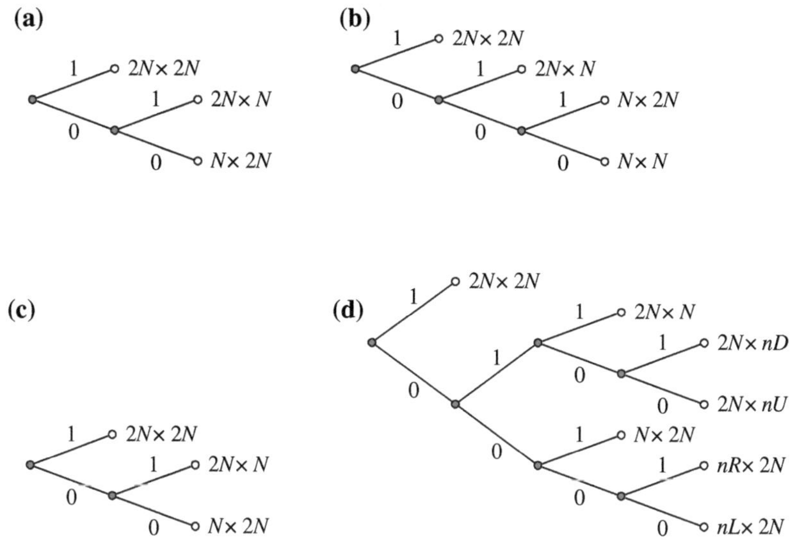

Fig. 10.13 Binarization for the partitioning mode syntax element for inter CUs in dependency of the minimum CU size $N_{C,\min}$ and the usage of asymmetric motion partitioning (AMP). **a** $N_C = 8$. **b** $N_C = N_{C,\min}$ and $N_{C,\min} > 8$. **c** $N_C > N_{C,\min}$ with no AMP. **d** $N_C > N_{C,\min}$ with AMP

10.2.3.6 Binarization for the CU Partitioning Mode

The binarization of the partitioning mode for the current CU depends on the prediction mode of the coding unit and on the size of the corresponding coding block. For intra CUs, the partitioning mode information is only coded if the CU size is equal to the minimum CU size. In this case the partitioning mode syntax element is a flag indicating a $2N \times 2N$ PU size or $N \times N$ partitioning.

For inter CUs, the applicable binarization scheme depends on the minimum CU size $N_{C,\min}$ and on the application of asymmetric motion partitioning as shown in Fig. 10.13.

10.2.3.7 Binarization for the Intra Chroma Prediction Mode

The binarization for the intra chroma prediction mode consists of a prefix and a suffix. The prefix is a flag, indicating that either (in case of a ‘0’) the intra prediction mode for chroma is copied from luma or (in case of a ‘1’) is set to one of DC, Planar, Horizontal, or Vertical prediction.

Table 10.9 Binarization for the CU level delta quantization parameter with the TU prefix is marked bold

v	$B_{\text{dqp}}(v)$
0	0
1	1 0
2	1 1 0
3	1 1 1 0
4	1 1 1 1 0
5	1 1 1 1 1 0
6,7	1 1 1 1 1 1 0 x_0
8,...,11	1 1 1 1 1 1 1 0 x_1 x_0
12,...,19	1 1 1 1 1 1 1 1 0 x_2 x_1 x_0
...	...

10.2.3.8 Binarization for the Inter Prediction Direction Index

In B slices, the number of available intra prediction directions depends on the partition size and shape. For squared prediction blocks, the binarizations '00' for List 0, '01' for List 1, and '1' for bi-prediction from both lists. The binarization can be interpreted as two flags: The first flag deciding on uni- or bi-prediction, and the second flag for List 0 or List 1 prediction in the uni-prediction case.

As indicated in Sect. 7.1.2, 4×4 inter prediction blocks are not allowed in the HEVC specification. For 4×8 and 8×4 prediction blocks, only uni-directional prediction from List 0 or List 1 is allowed. Therefore, the prediction index turns into a flag with one bin.

10.2.3.9 Binarization for CU Delta Quantizer Step Size

The delta QP value as described in Sect. 8.1.7 is represented by an absolute value part $v_{\text{se}} \geq 0$ and a sign flag. The absolute value is binarized using a prefix and, depending on the prefix, a suffix bin string. The binarization method is similar to the binarization for the absolute transform coefficient values greater than 2 in Sect. 10.2.3.5.

The prefix is binarized using a TU binarization with $v_{\max} = 5$. If the value of the syntax element is $v_{\text{se}} > 4$, an EG-0 binarization is appended for the remaining value $v^*_{\text{se}} = v_{\text{se}} - 4$. The binarization is shown in Table 10.9.

For CABAC coding, the TU prefix of the binarization is coded with context adaptivity, using one context for the first bin (indicating delta QP equal to zero) and a second context for the remaining bins of the prefix. The suffix as well as the sign flag are coded in bypass mode.

Table 10.10 Prefix and suffix binarization for the last significant coefficient position for all transform block sizes. For each block size, the respective (0) bin is omitted for the last binarization string

v	$B_{\mathrm{lsc}}(v)$ prefix	suffix
0	0	
1	1 0	
2	1 1 0	
3	1 1 1 (0)	← 4×4
4,5	1 1 1 1 0	x_0
6,7	1 1 1 1 1 (0)	x_0 ← 8×8
8,...,11	1 1 1 1 1 1 0	x_1 x_0
12,...,15	1 1 1 1 1 1 1 (0)	x_1 x_0 ← 16×16
16,...,23	1 1 1 1 1 1 1 1	0 x_2 x_1 x_0
24,...,31	1 1 1 1 1 1 1 1	1 x_2 x_1 x_0 ← 32×32

10.2.3.10 Binarization for the Last Significant Coefficient Position

For coding each of the horizontal and vertical components of the last significant scan position $(x_{\mathrm{lsc}}, y_{\mathrm{lsc}})$, a two-stage binarization scheme is applied. The scheme is presented in Table 10.10. It consists of a TU binarization for the prefix and a FL binarization for the suffix if available. The size of the suffix increases with larger position values. The discussion here details the steps for the horizontal position of the last significant coefficient x_{lsc}. The derivation for the vertical position y_{lsc} follows accordingly.

Let the value of the TU prefix be v_{pre}. If $v_{\mathrm{pre}} \leq 3$, the position of the last significant coefficient is set to

$$x_{\mathrm{lsc}} = v_{\mathrm{pre}}. \tag{10.24}$$

If $v_{\mathrm{pre}} > 3$, a suffix value v_{suf} is coded and the position of the last significant coefficient is calculated as

$$\begin{aligned} k &= \left\lfloor \frac{v_{\mathrm{pre}}}{2} \right\rfloor - 1 \\ x_{\mathrm{lsc}} &= 2^k \cdot (2 + v_{\mathrm{pre}} \& 1) + v_{\mathrm{suf}}. \end{aligned} \tag{10.25}$$

For blocks in vertical scan mode, the applicable scan position is swapped to $(y_{\mathrm{lsc}}, x_{\mathrm{lsc}})$.

The TU prefix of the binarization is coded with context adaptivity. See Sect. 10.2.5.2 for the context selection. The suffix is coded in bypass mode, if present.

10.2.4 Context Initialization

As indicated in Sect. 10.2.2.1, the state of each context is represented by the probability value for the least probable symbol $0 < \tilde{p}_{\mathrm{LPS}} \leq 0.5$ and by the value of the most probable symbol $v_{\mathrm{MPS}} \in \{0, 1\}$. For integer representation, the probability scaled to an integer with 7 bit dynamic range. Since finally the LPS value is to be represented, only half of the range is required.

The contexts are initialized to a probability state p_{LPS} and a most probable symbol value v_{MPS}. Note that the estimated probability $\tilde{p}_{\mathrm{LPS}}$ is represented by the integer value p_{LPS} with $0 < p_{\mathrm{LPS}} < 64$. The initialization of these two variables is performed in dependency on the luma quantization parameter $\mathrm{QP}_{\mathrm{slice}}$ of the current slice. Thereby, variations of the symbol statistics over the range of the QP-driven reconstruction quality and bitrate are taken into account.

The adaptation is specified by an 8 bit initialization value v_{init} which represents the QP-dependent slope value s_{ctx} in the upper four bits and the offset parameter value o_{ctx} in the lower four bits,

$$s_{\mathrm{ctx}} = 5 \cdot \left\lfloor \frac{v_{\mathrm{init}}}{16} \right\rfloor - 45 \tag{10.26}$$

$$o_{\mathrm{ctx}} = 8 \cdot (v_{\mathrm{init}} \& 15) - 16. \tag{10.27}$$

The values for p_{LPS} and v_{MPS} are then derived from a linear equation depending on $\mathrm{QP}_{\mathrm{slice}}$ by

$$p_{\mathrm{ctx}} = \left\lfloor \frac{s_{\mathrm{ctx}}}{16} \right\rfloor \cdot \mathrm{QP}_{\mathrm{slice}} + o_{\mathrm{ctx}}, \tag{10.28}$$

with the integer state value p_{ctx} clipped to the range $1 \leq p_{\mathrm{ctx}} \leq 126$. If the state p_{ctx} is in the lower half of the interval ($p_{\mathrm{ctx}} < 64$), the most probable symbol value is set to

$$v_{\mathrm{MPS}} = 0. \tag{10.29}$$

The probability state p_{LPS} is finally set to the distance of p_{ctx} from a 50 % probability, which corresponds to a value of 63:

$$v_{\mathrm{MPS}} = 0: \quad p_{\mathrm{LPS}} = 63 - p_{\mathrm{ctx}}, \text{ or} \tag{10.30}$$

$$v_{\mathrm{MPS}} = 1: \quad p_{\mathrm{LPS}} = p_{\mathrm{ctx}} - 64. \tag{10.31}$$

The initialization values v_{init} which are used in the HEVC specification are listed in Table 10.11 together with the corresponding slope and offset parameter values s_{ctx} and o_{ctx}. An example for the progression of the p_{LPS} and v_{MPS} over QP is provided in Fig. 10.14.

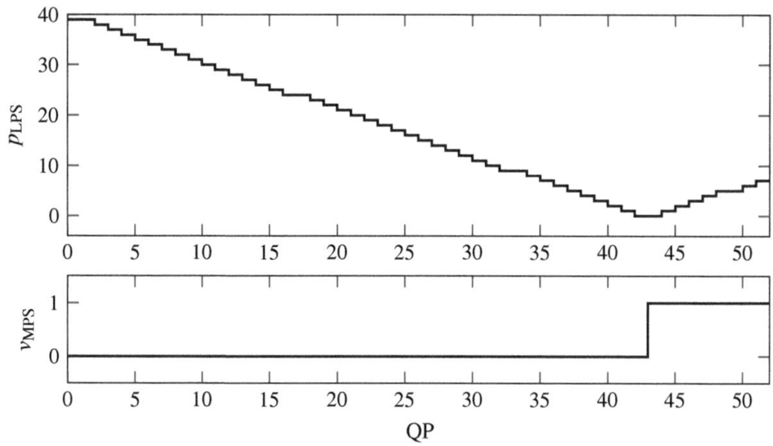

Fig. 10.14 Context initialization for $v_{init} = 197$, $s_{ctx} = 15$, and $o_{ctx} = 24$, which is e.g. used for the CU skip flag syntax element

10.2.5 Context Selection

For each bin of the binarization which is not coded in bypass mode, the applicable context model is selected. The selected model can either be fix or it can be adapted to context conditions. In the specification, the applicable context is indicated by a context index which points to the first initialization value for the syntax element in the context table and a context index increment which indicates the specific context model out of the available set to be applied.

Three sets of context models are specified by the initialization type which depends on the slice type (I, P, or B) of the current slice. One initialization type is used for I slices. For P slices and B slices, the initialization type can additionally be switched by a flag in the slice header.

The applicable context model can further depend on the existence and value of previously coded syntax elements in the local neighborhood. In the following, the context selection processes for these syntax elements with advanced context selection are presented.

10.2.5.1 Context from Left and Above Syntax Elements

The CU split flag of the coding quadtree and the CU skip flag of the current coding unit use information from the previously coded neighboring CUs for context selection. A set of three context models for each applicable initialization type are available.

The location of the left and above neighboring blocks are shown in Fig. 10.15. These blocks are denoted as available if the locations indicated by the samples (x_L, y_L) or (x_A, y_A) are within the same slice and not outside of the picture boundaries.

Table 10.11 Context slope s_{ctx} and offset o_{ctx} for CABAC initialization values v_{init}

v_{init}	s_{ctx}	o_{ctx}
31	−40	104
61	−30	88
63	−30	104
74	−25	64
78	−25	96
79	−25	104
91	−20	72
92	−20	80
93	−20	88
94	−20	96
95	−20	104
107	−15	72
108	−15	80
109	−15	88
110	−15	96
111	−15	104
121	−10	56
122	−10	64
123	−10	72
124	−10	80
125	−10	88
126	−10	96
127	−10	104
134	−5	32
136	−5	48
137	−5	56
138	−5	64
139	−5	72
140	−5	80
141	−5	88
143	−5	104
149	0	24
151	0	40
152	0	48
153	0	56
154	0	64
155	0	72
157	0	88
160	5	−16
166	5	32
167	5	40
168	5	48

(continued)

Table 10.11 (continued)

v_{init}	s_{ctx}	o_{ctx}
169	5	56
170	5	64
171	5	72
179	10	8
182	10	32
183	10	40
184	10	48
185	10	56
194	15	0
196	15	16
197	15	24
198	15	32
200	15	48
201	15	56
208	20	−16
224	25	−16
227	25	8

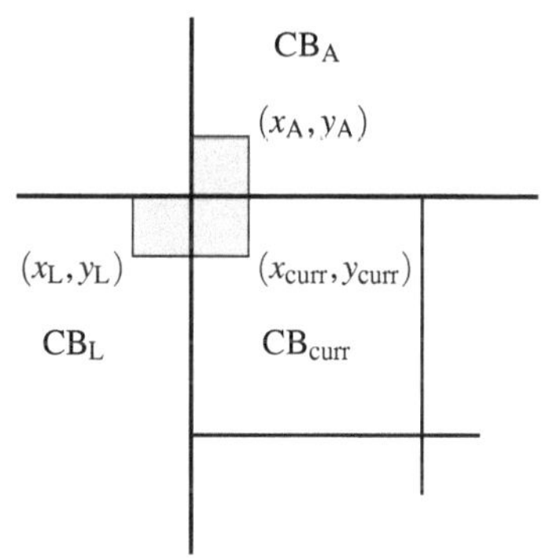

Fig. 10.15 Illustration of a coding block CB_{curr} and the *left* and *above* neighboring blocks CB_L and CB_A

For the CU split flag, the context index is incremented by 1 if the neighboring block is available and its coding tree depth is larger than the coding tree depth of the current block, for both, the left and the above block. For the CU skip flag, the context index is incremented by 1 each if the neighboring block is available and its skip flag is set.

Table 10.12 Context indices i_c for the bins of the last significant luma coefficient prefix binarization

	Bin index								
Transform size	0	1	2	3	4	5	6	7	8
4 × 4	0	1	2						
8 × 8	3	3	4	4	5				
16 × 16	6	6	7	7	8	8	9		
32 × 32	10	10	11	11	12	12	13	13	14

Table 10.13 Context indices i_c for the bins of the last significant chroma coefficient prefix binarization

	Bin index						
Transform size	0	1	2	3	4	5	6
4 × 4	15	16	17				
8 × 8	15	15	16	16	17		
16 × 16	15	15	15	15	16	16	16

10.2.5.2 Last Significant Coefficient Prefix

The column (x) and row (y) positions of the last significant coefficient of a transform block in scanning order is coded with a prefix and a suffix with context adaptive coding for the prefix and bypass coding for the suffix, as detailed in Sect. 10.2.3.10. A separate set of contexts is provided for the x and y positions each, with a total number of $2 \cdot 54 = 108$ context models for this syntax element.

The applicable context index is determined in dependency of the transform block size and the current bin index i_b. With the logarithm of the transform block size $n_S = \log_2(N_S)$, a context offset and shift are determined as

$$o_c = 3 \cdot (n_S - 2) + \left\lfloor \frac{n_S - 1}{4} \right\rfloor, \tag{10.32}$$

$$s_c = \left\lfloor \frac{n_S + 1}{4} \right\rfloor. \tag{10.33}$$

The context index is then determined by

$$i_c = \left\lfloor \frac{i_b}{2^{s_c}} \right\rfloor + o_c. \tag{10.34}$$

The assignment of context indices to the bins of the last significant coefficient binarization is summarized in Tables 10.12 and 10.13. The presentation follows [6].

(a)

0	1	4	5
2	3	4	5
6	6	8	8
7	7	8	

(b)

2	1	0	0
1	1	0	0
0	0	0	0
0	0	0	0

(c)

2	2	2	2
1	1	1	1
0	0	0	0
0	0	0	0

(d)

2	1	0	0
2	1	0	0
2	1	0	0
2	1	0	0

(e)

2	2	2	2
2	2	2	2
2	2	2	2
2	2	2	2

Fig. 10.16 Context selection offset for the significant coefficient flag for 4 × 4 transform blocks and transform sub-blocks of larger transform blocks. **a** 4 × 4 transform block. **b** No neighboring sub-block significant. **c** *Right* neighboring sub-block significant. **d** *Lower* neighboring sub-block significant. **e** Both, *right* and *lower* sub-block significant

10.2.5.3 Coded Sub-Block Flag

The coded sub-block flag is coded for transform blocks larger than 4 × 4. It indicates the presence of non-zero transform coefficients in each 4 × 4 sub block of the transform block. A total number of 72 context models is specified for this syntax element. The coded sub-block flags are coded in the order according to the active scan pattern (diagonal, horizontal, or vertical). The coding order follows the scan in inverse direction. The coded sub-block flag is coded for all but the first and the last sub-blocks of the transform block. For these sub-blocks, the significance is inferred to be given.

For the other sub-blocks along the inverse scan, the applicable context model is selected depending on the coded sub-block flags of the previously coded neighboring blocks below and to the right of the current sub-block, if available. The selection is performed by adding the coded sub-block flags of those available neighbors, leading to three different models to chose from:

- No significant neighboring sub-blocks
- One significant neighboring sub-block (below or to the right)
- Two significant neighboring sub-blocks (below and right)

10.2.5.4 Significant Coefficient Flag

The significant coefficient flag is sent for each coefficient position in a transform sub-block between position (0, 0) and the position of the last significant coefficient along the transform sub-block scan.

Overall, a set of 126 context models are provided for this syntax element. The context model index depends on the initialization type (Intra, P, or B) and on the transform block size. Separate context models are provided for luma and chroma.

Depending on the coefficient position in the current transform sub-block, an offset for the context model index is selected. These offsets are visualized in Fig. 10.16. For the top-left transform sub-block which comprises the lowest frequency coefficients, nine different contexts are used as shown in Fig. 10.16a. For the DC coefficient, a separate context is used. The context selection in the other transform sub-blocks

Table 10.14 Determination of the context set s_c for absolute coefficient level values greater than 1 flags depending on the number n_{lg1} of such previously coded coefficients

s_c	Luma		Chroma	
	$n_{lg1} = 0$	$n_{lg1} > 0$	$n_{lg1} = 0$	$n_{lg1} > 0$
Top-left TSB	0	1	4	5
Remaining TSBs	2	3	4	5

Table 10.15 Determination of the context offset o_c for absolute coefficient level values greater than 1 flags depending on the number n_{lg1} of such previously coded coefficients

o_c	Condition
0	$n_{lg1} \geq 1$
1	No trailing ones (initial)
2	1 trailing one
3	More than 1 trailing one

(if available) depends on the presence of significant coefficients in the bottom or right neighboring transform sub-blocks, see Fig. 10.16b–e. Separate context models are provided in case of 4×4 and 8×8 transform blocks. The higher frequency transform sub-blocks for transform blocks of size 16×16 and 32×32 share a joint set of context models. Note that no significance flag is specified for the highest coefficient of the 4×4 transform block as this position would coincide with the last significant coefficient position, if present.

10.2.5.5 Absolute Coefficient Level Value Greater than 1 Flag

A total set of 72 context models is available for the flag to indicate the absolute level of a coefficient to be greater than 1. Hence, 24 models for each initialization type are available. The context models are organized in sets of four models each. The selection of the set depends on the fact that at least one values greater than 1 has already been encoded in a previous transform sub-block. For luma coefficients, a distinction is further made between the top-left transform sub-block and the remaining sub-blocks (if available).

Within each set of context models, the index of the applicable model is determined on the fact that coefficient levels greater than 1 have been coded before and the number of sequentially coded coefficients with absolute level one (so-called 'trailing ones').

The selection of the applicable context model is summarized in Table 10.14 and 10.15. The context index i_c is determined based on the context set s_c and an offset o_c within the set as

$$i_c = 4 \cdot s_c + o_c. \tag{10.35}$$

The presentation of Tables 10.14 and 10.15 follows [6].

Table 10.16 Context index i_c for absolute coefficient level values greater than 2 flags

	$n_{lg1} = 0$	$n_{lg1} > 0$
Top-left luma TSB	0	1
Remaining luma TSBs	2	3
Chroma TSBs	4	5

10.2.5.6 Absolute Coefficient Value Greater than 2 Flag

For this syntax element, 6 context models are available for each initialization type.

The context index is determined by the current transform sub-block, if it is the top-left luma TSB or another luma TSB or a chroma TSB, and by the number n_{lg1} of previously coded flags for absolute coefficient levels greater than 1. The corresponding context indices are summarized in Table 10.16.

10.3 Comparison to H.264 | AVC

Both specifications apply fixed-length and variable-length coding for the high-level information, including byte alignment for fundamental information on the coded video sequence. On the slice level, H.264 | AVC includes two entropy coding stages, a context adaptive variable length coding scheme (CAVLC) and context adaptive binary arithmetic coding (CABAC) [7, 8]. At the time of specification of H.264 | AVC, the complexity of the arithmetic coding scheme was considered a complexity burden. Therefore, a Baseline profile without arithmetic coding, and a Main profile including CABAC were specified. During the development of HEVC it turned out that the complexity for an efficient CAVLC representation of the slice level syntax elements was not significantly lower than the complexity of the CABAC based scheme. Therefore, the option of two entropy coding schemes was dropped and only CABAC remained as the only entropy coding scheme [9].

While the arithmetic coding engine in HEVC has basically been copied from H.264 | AVC, the design for the context adaptation and for bypass coding has been optimized. The design goal for HEVC was a small required set of context tables and reduced parsing and context dependencies to enable high throughput processing. For a detailed comparison of the CABAC coding schemes in H.264 | AVC and HEVC, the reader is referred to Sze and Budagavi [2].

References

1. Marpe, D., et al.: Context-based adaptive binary arithmetic coding in the H.264/AVC video compression standard. IEEE Trans. Circ. Syst. Video Technol. **13**(7), 620–637 (2003). doi:10.1109/TCSVT.2003.815173
2. Sze, V., Budagavi, M.: High throughput CABAC entropy coding in HEVC. IEEE Trans. Circ. Syst. Video Technol. **22**(12), 1778–1791 (2012). doi:10.1109/TCSVT.2221526
3. High efficiency video coding. ITU-T Rec. H.265 (HEVC). http://www.itu.int/rec/T-REC-H.265/en (2013). Accessed 14 Apr 2014
4. Information technology—High efficiency coding and media delivery in heterogeneous environments—Part 2: High efficiency video coding. ISO/IEC 23008-2:2013 (HEVC). http://www.iso.org/iso/home/store/catalogue_detail.htm?csnumber=35424 (2013). Accessed 14 Apr 2014
5. Rissanen, J., Langdon, G.G.: Universal modeling and coding. IEEE Trans. Inform. Theory **27**(1), 12–23 (1981). doi:10.1109/TIT.1981.1056282
6. Sole, J., et al.: Transform coefficient coding in HEVC. IEEE Trans. Circ. Syst. Video Technol. **22**(12), 1765–1777 (2012). doi:10.1109/TCSVT.2012.2223055
7. Advanced video coding for generic audiovisual services. ITU-T Rec. H.264 (AVC). http://www.itu.int/rec/T-REC-H.264/en (2014). Accessed 14 Apr 2014
8. Information technology—Coding of audio-visual objects—Part 10: Advanced Video Coding. ISO/IEC 14496-10:2012 (AVC). http://www.iso.org/iso/home/store/catalogue_detail.htm?csnumber=61490 (2012). Accessed 14 Apr 2014
9. Sullivan, G.J., Ohm, J.-R.: Meeting report of the seventh meeting of the Joint Collaborative Team on Video Coding (JCT-VC), Geneva, CH. Doc. JCTVC-G1100. 7th Meeting: Joint Collaborative Team on Video Coding (JCT-VC) of ITU-T SG 16 WP 3 and ISO/IEC JTC 1/SC 29/WG 11, Geneva, CH (2011)

Chapter 11
Profiles, Tiers, and Levels

In this Chapter, the profiles, tiers and levels specified in the base version of HEVC are presented. These include the three Main profiles targeting at 8-bit and 10-bit YCbCr 4:2:0 video and at 8-bit YCbCr 4:2:0 still image coding. The ongoing development of further profiles is related to extensions such as scalable video coding, multiview video coding, and range extensions beyond 10-bit YCbCr 4:2:0 video. These extensions are briefly presented in Chap. 12.

It is important for a decoder to be able to identify if an incoming bitstream is within or beyond its processing capabilities. This includes the usage of tools, buffering capabilities, as well as the picture size which scales the computational effort. The specification of profiles, tiers, and levels provides the means to convey this information. A profile defines the set of coding tools which may be used to encode a video sequence into a bitstream. The encoder is not required to use all coding tools available in the profile but may select the coding tools which it considers to be suitable. The levels indicate restrictions on parameters which determine decoding and buffering capabilities. These include the maximum picture size, the coded and decoded picture buffer sizes, the number of slice segments and tiles in a picture, as well as the maximum sample rate and maximum bitrate. It further sets a requirement for the minimum compression ratio which must be met by a bitstream. The concept of tiers has been introduced to enable the differentiation between different application spaces which require different available bitrate ranges. Correspondingly, the maximum bitrate and the maximum CPB size differ between tiers. Two tiers of levels are specified in HEVC. The Main tier targets consumer applications, the High tier is designed for professional applications.

A decoder can claim to be conforming to the specification at some profile, tier, and level. In order to fulfill this claim, it must be capable of decoding any bitstream conforming to the specification which indicates the corresponding profile, tier, and level. Furthermore, it must be able to decode any conforming bitstream of the same profile which has the same or a lower level in the same or a lower tier. The indication of the applicable profile, tier, and level is given by a corresponding syntax structure which is specified in the VPS as well as the SPS. While the VPS syntax structure specifies the parameters for the overall stream including all layers and temporal

M. Wien, *High Efficiency Video Coding*, Signals and Communication Technology,
DOI 10.1007/978-3-662-44276-0_11

sub-layers, the SPS syntax structure is dedicated to a single coded video sequence. In the case of single-layer coding as specified in the first version of HEVC, the VPS and SPS syntax structures are identical. In the upcoming extensions, which include multiple layers (scalable layers or views), an SPS is activated for each layer. The SPS for the layers may be identical or diverging, depending on the scenario. For example for spatial scalability, a dedicated SPS for each picture resolution of the scalable stream is required.

Additionally, flags for general profile compatibility are provided. These serve as a means to indicate the conformance not only to the one indicated profile but also to potentially other profiles (which include a subset of the coding tools of the indicated bitstream profile). Thereby, interoperability points between different conforming devices can be established.

11.1 Profiles

Conceptually the definition of profiles for a video coding specification is driven by application requirements. Since profiles constitute general compatibility points of the specification, they provide a strong means to structure the available tool set but also for easy identification outside of the video coding specification itself, e.g. for branding and advertisement. The less profiles exist, the more devices will allow for interoperability. New profiles with or without new tools may be defined to support a new application space and to provide an explicit name to address the specification by its profile. On the other hand, if the market indicates a strong use of a certain profile with additional constraints and restrictions on the applicable tools, dedicated profiles may be defined to address this application space.[1]

The three profiles specified in the first version of HEVC, called the Main, the Main 10, and the Main Still Picture profile. The profiles can be characterized by a hierarchical structure as visualized in Fig. 11.1. The profiles target at different application spaces each as detailed in the following subsections. The common constraints of these profiles are listed here:

- The profiles only support the representation of YCbCr 4:2:0 video.
- The size of coding tree blocks is constrained to be one of 16×16, 32×32, and 64×64.
- The application has to decide to either use tiles for modification of the CTU scanning order, or to enable wavefront parallel processing, or not to use any of them. It is not allowed to used both tools at the same time.
- If tiles are used, the minimum tile size is 256×64 samples.
- The maximum number of CABAC coded bits in a CTU of size $N \times N$ shall be less than or equal to $5 \cdot b_{\text{raw}}/3$, where b_{raw} is the number of bits needed to represent

[1] For example in H.264 | AVC, a subset of the Baseline profile was specified as a separate Constrained Baseline profile in 2009. The main difference is that the constrained profile does not include slice groups and arbitrary slice ordering, see H.264 | AVC Annex A [2, 3].

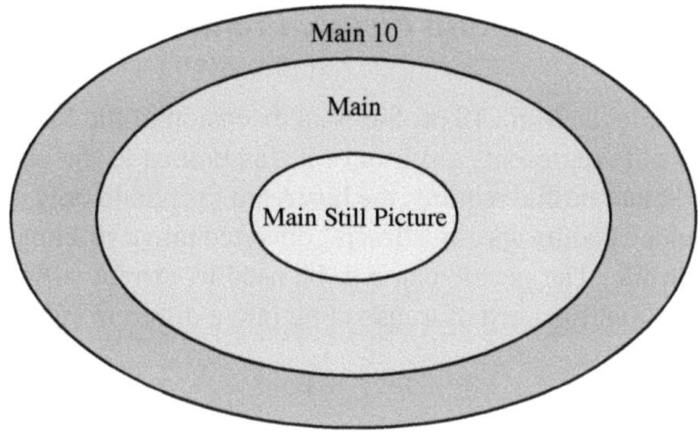

Fig. 11.1 Hierarchy of the main profiles in HEVC

the uncompressed samples, that is $b_{\mathrm{raw}} = (N^2 + N/2) \cdot B_{\mathrm{d}}$ for YCbCr 4:2:0 video with bit depth B_{d}.

Further constraints which have been considered not to be negotiable or dependent on a profile definition are included in the specification text. An example is the restriction of a minimum block size of 4×8 or 8×4 for motion compensation using uni-prediction and a minimum block size of 8×8 for bi-prediction. In contrast to H.264 | AVC, motion compensation with bi-prediction is allowed in both existing profiles which include inter prediction, i.e. Main and Main 10.

11.1.1 Main Profile

The Main profile only allows for $B_{\mathrm{d}} = 8$ bit video bit depth. Since a bit depth of 8 bit is employed by many applications of today ranging from video communication over streaming to broadcast, this profile is expected to find broad use. The fixed definition of $B_{\mathrm{d}} = 8$ bit limits the burden of memory requirements and may therefore be well-suited for mobile and lightweight Internet applications.

11.1.2 Main 10 Profile

The Main 10 profile allows for a video bit depth in the range of $8 \leq B_{\mathrm{d}} \leq 10$. Thereby, the application range for the specification is extended towards high-quality applications and professional use. Since the bit depth is increased, the memory consumption increases accordingly.

11.1.3 Main Still Picture Profile

While the Main 10 profile is an extension of the Main profile, the Main Still Picture profile represents a subset of it. In contrast to the other two profiles, the Main Still Picture profile requires the bitstream to contain only one single picture. Thereby, the video coding specification is converted into a still image coding specification for this profile. The profile can e.g. be used to extract stills from a coded video sequence without the need of transcoding into a different still picture format.

11.2 Tiers and Levels

As stated above, the tier concept is introduced to enable a differentiation between the level set for consumer applications and the level set for professional applications. Both tiers share identical constraints on parameters like the picture size, or the number of slice segments, tiles, etc. The goal of the definition of tiers is to differentiate in the maximum allowed bitrates and correspondingly, in the required maximum size of the coded picture buffer S_{CPB}. The tiers and levels as specified in HEVC are listed in Table 11.1. As can be seen from the table, the High tier is only specified for level 4 and higher. Level 4 is the first level to support full HD resolution, see also Table 11.2. The maximum size of the coded picture buffer S_{CPB} is specified to have different values for the Main and High tiers. The same holds for the maximum bitrate r_{b}. The maximum allowed bitrate for the High tier is 2.5–4 times higher than the maximum bitrate for the Main tier.

The maximum picture size is specified by the maximum number of luma samples N_{pix}. The picture width N_{P} and height N_{L} are further bound by

$$N_{\mathrm{P,max}}, N_{\mathrm{L,max}} \leq \sqrt{8 \cdot N_{\mathrm{pix}}}.$$

Thereby, the maximum picture aspect ratio is 8:1 or 1:8, respectively. This maximum aspect ratio allows for side-by-side coding of all commonly used picture formats. This capability may be useful e.g. for side-by-side coding of multiple videos. The maximum picture aspect ratio is illustrated in Fig. 11.2.

Comparing the values for S_{CPB} and for r_{b} in Table 11.1, it can be seen that for levels ≥ 2, the coded picture buffer size S_{CPB} corresponds to the maximum allowed bitrate. Thereby, the buffer is required to be able to store at least 1 s of incoming coded video [4]. For level 1, the minimum buffer size is larger.

Fig. 11.2 Illustration of the maximum allowed picture aspect ratio of 8:1 in HEVC

Table 11.1 Tier and level limits

Lev.	N_{pix}	S_{CPB}		N_{Sl}	T_{row}	T_{col}	r_{pix}	r_b		cr
		MT	HT					MT	HT	
1	36864	350	–	16	1	1	552960	128	–	2
2	122880	1500	–	16	1	1	3686400	1500	–	2
2.1	245760	3000	–	20	1	1	7372800	3000	–	2
3	552960	6000	–	30	2	2	16588800	6000	–	2
3.1	983040	10000	–	40	3	3	33177600	10000	–	2
4	2228224	12000	30000	75	5	5	66846720	12000	30000	4
4.1	2228224	20000	50000	75	5	5	133693440	20000	50000	4
5	8912896	25000	100000	200	11	10	267386880	25000	100000	6
5.1	8912896	40000	160000	200	11	10	534773760	40000	160000	8
5.2	8912896	60000	240000	200	11	10	1069547520	60000	240000	8
6	35651584	60000	240000	600	22	20	1069547520	60000	240000	8
6.1	35651584	120000	480000	600	22	20	2139095040	120000	480000	8
6.2	35651584	240000	800000	600	22	20	4278190080	240000	800000	6

N_{pix} maximum number of luma samples per picture
S_{CPB} maximum CPB size (1,000 bits), MT = Main tier, HT = High tier
N_{Sl} maximum number of slice segments per picture
T_{row} maximum number of tile rows per picture
T_{col} maximum number of tile columns per picture
r_{pix} maximum sample rate (sample/s)
r_b maximum bitrate (1,000 bits/s), MT = Main tier, HT = High tier
cr minimum compression ratio

The maximum number of slices and tile rows and columns is specified by N_{Sl}, T_{row}, and T_{col}, respectively. Note that tiles are enabled only from level 3 on, which supports VGA and standard definition resolutions, see Table 11.2.

For the *decoded* picture buffer, the maximum size S_{DPB} depends on the number of luma samples $N_P \cdot N_L \leq N_{pix}$ in the picture. The value of S_{DPB} can be in the range of 6 to 16 pictures as illustrated in Fig. 11.3. This value corresponds to the maximum possible size of the reference picture set. The maximum number of reference pictures to be used for prediction in a reference picture list is constrained to be equal to or less than 8.

11.3 Syntax Structure

The syntax structure for the signaling of profile, tier, and level employs unsigned fixed length codes only. The syntax elements for the indication of the general profile space and profile, the general tier, and the general level are available at fixed bit positions in the first bytes of the syntax structure. Thereby, this information can be easily parsed by applications and network elements besides the decoder itself.

Table 11.2 Maximum picture rates (Hz) for common video picture sizes of $N_P \times N_L$ samples and a CTU size of 64×64 samples

Level	N_P	N_L	1	2	2.1	3	3.1	4	4.1	5	5.1	5.2	6	6.1	6.2
QCIF	176	144	15.0	100.0	200.0	300.0	300.0	300.0	300.0	300.0	300.0	300.0	300.0	300.0	300.0
CIF	352	288	–	30.0	60.0	135.0	270.0	300.0	300.0	300.0	300.0	300.0	300.0	300.0	300.0
Q720p	640	360	–	–	30.0	67.5	135.0	272.0	300.0	300.0	300.0	300.0	300.0	300.0	300.0
VGA	640	480	–	–	–	50.6	101.2	204.0	300.0	300.0	300.0	300.0	300.0	300.0	300.0
4CIF	704	576	–	–	–	40.9	81.8	164.8	300.0	300.0	300.0	300.0	300.0	300.0	300.0
QHD	960	540	–	–	–	30.0	60.0	120.8	241.7	300.0	300.0	300.0	300.0	300.0	300.0
XGA	1024	768	–	–	–	–	42.1	85.0	170.0	300.0	300.0	300.0	300.0	300.0	300.0
720p	1280	720	–	–	–	–	33.7	68.0	136.0	272.0	300.0	300.0	300.0	300.0	300.0
1080	1920	1080	–	–	–	–	–	32.0	64.0	128.0	300.0	300.0	300.0	300.0	300.0
4K	3840	2160	–	–	–	–	–	–	–	32.0	64.0	128.0	256.0	300.0	300.0
8K	7680	4320	–	–	–	–	–	–	–	–	–	–	32.0	64.0	128.0

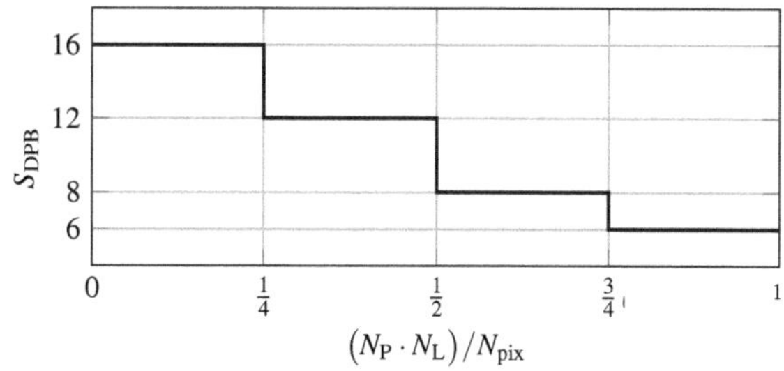

Fig. 11.3 The maximum DPB size over the normalized number of samples in the picture

The specification allows for the indication of 32 different profiles. While this number range appears to be sufficient for a single-layer specification, the number of planned and not yet foreseen extensions may easily cover this number range. In order to prevent the potential issue of a complete coverage of the profile number range, an dedicated indication for the profile space has been specified. This syntax element allows for the indication of four dedicated profile spaces and thereby prevents potential overfilling of the profile number range.

Besides the general profile, tier, and level parameters, the syntax structure further includes the option to indicate specific parameters for each temporal sub-layer. This information can be very useful e.g. if a decoder is not capable of decoding the profile and level of the bitstream at full temporal resolution but might be capable of decoding at reduced temporal resolution (requiring the extraction of the temporal sub-layer from the full bitstream). The sub-layer parameter signaling allows for an assessment of such possibilities.

The syntax structure includes multiple reserved zero syntax elements which allow for specification of additional flags and parameters in future extensions of the specification.

References

1. High efficiency video coding. ITU-T Rec. H.265 (HEVC). http://www.itu.int/rec/T-REC-H.265/en. Accessed 14 Apr 2014
2. Advanced video coding for generic audiovisual services. ITU-T Rec. H.264 (AVC). http://www.itu.int/rec/T-REC-H.264/en (2014). Accessed 14 Apr 2014
3. Information technology—Coding of audio-visual objects—Part 10: Advanced video coding. ISO/IEC 14496–10:2012 (AVC). http://www.iso.org/iso/home/store/catalogue_tc/catalogue_detail.htm?csnumber=61490 (2012). Accessed 14 Apr 2014
4. Sullivan, G.J., et al.: Overview of the high efficiency video coding (HEVC) standard. IEEE Trans. Circ. Syst. Video Technol. **22**(12), 1649–1668 (2012). doi:10.1109/TCSVT.2221191

Chapter 12
Extensions to HEVC

In this chapter, a survey of the currently ongoing activities for extension of the HEVC specification is provided. As these activities are beyond the scope of this book and have not been finalized by the time of writing, the presentation is kept at a higher level. Pointers to relevant standardization documents and overview papers are provided for further reading.

The extension work currently includes support for more color formats and higher bit depths, support of spatial and fidelity scalability, and support of multiview and 3D-video coding. As discussed in the previous chapters, the development of these extensions has been foreseen from the beginning of the standardization phase of version 1 of HEVC [1]. Hence, provisions have been implemented throughout the relevant parts of the specification to enable a smooth integration of the envisaged modifications. In the context of HEVC, the range extensions are often referred to by the short form *RExt*. The scalable extension of HEVC is referred to as Scalable High Efficiency Video Coding (SHVC). The extensions for multiview and 3D-video are referred to as Multiview High Efficiency Video Coding (MV-HEVC) and 3D High Efficiency Video Coding (3D-HEVC), respectively. While the range extensions and the scalable extensions are worked on in the JCT-VC, the multiview and 3D extensions are developed in a separate dedicated joint collaborative team, called the Joint Collaborative Team on 3D Video Coding Extension Development (JCT-3V) [2]. This group also works on further multiview and 3D extensions for H.264 | AVC.[1]

In terms of specification, the scalable, multiview, and 3D extensions share the concept of a layered coding approach. The different layers which may be included in a bitstream are addressed by the NAL unit header layer identifier l_{id}, which is required to be equal to 0 for bitstreams conforming to the existing version 1 profiles of the specification. While in SHVC, the layers represent the video sequence in different spatial resolutions or different reconstruction fidelity, the MV-HEVC and 3D-HEVC

[1] Besides the topics mentioned above, the JCT-3V work includes H.264 | AVC multiview plus depth coding and H.264 | AVC multiview frame compatible video coding. This part of the JCT-3V is out of scope of this book. An overview of the ongoing standardization activities in the field of 3D video coding is provided in [3].

M. Wien, *High Efficiency Video Coding*, Signals and Communication Technology,
DOI 10.1007/978-3-662-44276-0_12

extensions use the layer identifier to differentiate between different views which are provided for the same scene. The layer identifier is further used for auxiliary pictures, e.g. carrying depth information or alpha planes. In order to achieve a clean and unified design, general layer-related modifications of the base specification which are common to the mentioned layered extensions are harmonized and collected in a joint annex of the specification. The modifications which are specific to only one of the layered extensions are consequently held in a separate dedicated annex, each. Thereby, an unnecessary duplication of specification text is avoided. The draft of the annex for the general modifications for layered coding is discussed in Sect. 12.2. The dedicated modifications and tools for each extension are described in the following Sects. 12.4 and 12.5.

Scalable and multiview extensions as well as the range extensions have been specified for previous standards, particularly including H.264 | AVC. The most relevant differences in the realization of the extension functionality are highlighted in each of the corresponding sections.

12.1 Range Extensions

As with H.264 | AVC and previous video coding specifications, dedicated profiles for professional applications are being sought by industry. This application space is targeted at very high reconstruction quality and uses different color formats. Consequently, it is particularly driven by increased bit depth and video formats with higher sample resolution for the chroma components for highest fidelity of the reconstructed video. HEVC profiles supporting such requirements are being developed. By the time of writing, the draft for the corresponding amendment of HEVC is planned to be finalized in 2014 [4]. The status of the RExt work can be reviewed in the corresponding JCT-VC output document. By the time of writing this is draft 5 of the extension specification [5].

Here, the profiles under consideration are presented and the most significant additional coding tools are highlighted.

12.1.1 Proposed RExt Profiles

The planned profiles for the range extensions partially follow the nested structure of the existing Main, Main 10, and Main Still Image profiles. The draft profiles are listed and briefly described below:

Main 12 This profile extends the Main profile to a maximum of 12 bit video bit depth.

Main 4:2:2 12 This profile extends the Main 12 by support of the 4:2:2 chroma format in addition to the 4:2:0 format specified for the already standardized Main profiles. See Fig. 2.6 for the location of the chroma samples relative to the luma samples for the two formats.

Main 4:4:4 10 This profile extends the Main 10 profile to support the 4:4:4 and 4:2:2 chroma formats in addition to 4:2:0. In the 4:4:4 chroma format, the chroma samples are collocated with the luma samples. Note that this format is also suitable e.g.for coding of RGB video. A flag can be set to indicate that the three color components are coded as separate color planes.

Main 4:4:4 12 This profile extends the Main 4:4:4 10 profile to a maximum of 12 bit video bit depth.

12.1.2 Proposed RExt Tools

Besides adaptation in the high-level syntax (e.g.to include constraints flags for the supported chroma formats) and an adaptation of the binarization for the transform coefficients for higher bit depths, the tools described in the following subsections are the most relevant new coding tools which are currently discussed for inclusion in the range extension profiles. The tools are described according to Draft 5 of the range extensions specification [5].[2] The document includes a list of changes for RExt and references to the corresponding proposal documents. Thereby, the draft serves as a suitable starting point for further study. As the draft is still under development at the time of writing, the final design and specification of the tools included in the RExt extensions may still be subject to change.

12.1.2.1 Residual Quadtree

The chroma components in 4:2:2 format share the vertical resolution of the luma component at have half the horizontal luma resolution. In order to support the 4:2:2 chroma format, the residual quadtree is kept unchanged for the three color components, but for chroma it carries two stacked $\frac{N}{2} \times \frac{N}{2}$ transform blocks with each corresponding $N \times N$ luma transform block. The scheme is illustrated in Fig. 12.1. Otherwise, the coding of the transform blocks remains unchanged compared to the non-RExt profiles. For the 4:4:4 chroma format, no dedicated additional modifications are required as the chroma prediction and transform block sizes match the luma block sizes in this case.

---

[2] This draft includes the intra block copy scheme which has been removed from the draft at the Valencia meeting of JCT-VC in March/April 2014. It is therefore not further detailed here.

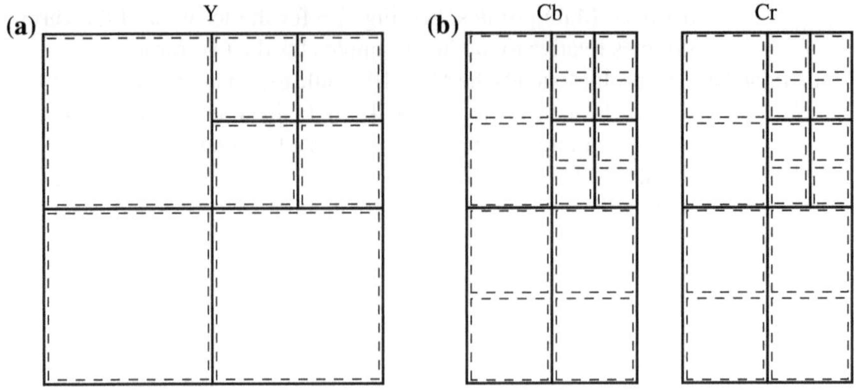

Fig. 12.1 Example residual quadtree in YCbCr 4:2:2 video. **a** Luma transform blocks. **b** Corresponding chroma transform blocks (RQT: *solid*, TBs: *dashed*)

12.1.2.2 Residual DPCM Coding

For very high quality video coding, the application of transform skip and transform bypass are considered to be more frequently applied, especially when coding (computer generated) graphical content. In both coding modes, the residual signal is not processed by an inverse transform stage for reconstruction. Instead, the sample values are directly processed. See Chap. 8 for a discussion of the two modes.

In order to increase the compression performance of coding the non-transformed residual signal, the residual DPCM coding mode has been proposed. In this mode, a DPCM is applied either along the horizontal or along the vertical direction. For intra blocks using this tool, the horizontal or vertical DPCM prediction direction is derived from the direction of the intra prediction mode. For inter blocks, the DPCM prediction direction is explicitly signaled in the bitstream.

12.1.2.3 Inter-Component Residual Prediction

In the case of 4:4:4 chroma format, the option for residual prediction from the luma residual to the chroma residual is proposed. The amplitude of the luma residual signal is scaled by a factor $\alpha \in \{0, \pm 1, \pm 2, \pm 4, \pm 8\}$ for construction of the residual signal as

$$\mathbf{R}_\mathrm{C} = \mathbf{R}'_\mathrm{C} + \alpha \cdot \mathbf{R}_\mathrm{Y}, \tag{12.1}$$

where $\mathbf{R}_\mathrm{C}$ denotes the chroma residual block, $\mathbf{R}'_\mathrm{C}$ is the residual block reconstructed from the decoded and scaled transform coefficients, and $\mathbf{R}_\mathrm{Y}$ is the luma residual block.

Inter-component residual prediction aims at additional decorrelation of the transmitted residual signal. The tool reportedly provides gains especially for RGB sequences [6].

12.1.3 Comparison to H.264 | AVC

In H.264 | AVC, the High 4:2:2 and the High 4:4:4 profiles serve a comparable application space as the discussed RExt profiles. The extensions were referred to as Fidelity Range Extensions (FRExt) at the time. The High profiles including High 4:2:0 of H.264 | AVC specify three coding tools in addition to the coding tools of the H.264 | AVC Main profile [7, 8]. These tools were already previously proposed for inclusion into the specification but were not considered in the profiles of the first version of H.264 | AVC. Adaptive block-size transforms were introduced, enabling the application of an 8×8 transform in addition to the 4×4 transform for residual coding.[3] Furthermore, quantizer weighting matrices and a transform skip operation were included. These tools are not dedicated specifically to range extensions applications but generally provide improved coding performance.

In HEVC, adapted (or extended) versions of the tools have been incorporated into the Main profiles from the beginning. In H.264 | AVC, only lossless residual coding for 4:4:4 content is available as a dedicated RExt coding tool in the High 4:4:4 profile. An overview of the fidelity range extensions for H.264 | AVC is provided in [10].

12.2 Common Specification Structure for Multi-Layer Video Coding Extensions

In terms of specification, the common aspects of the layered extension schemes are collected in a new joint annex for the common layer-related processing in the extensions of the HEVC specification. At the time of writing, this annex is referred to as Annex F "Common syntax, semantics and decoding processes for multi-layer video coding extensions". The aspects related only to multiview coding are specified in Annex G "Multiview coding" [11]. These are discussed in Sect. 12.3. The aspects related only to the scalable extensions are specified in Annex H "Syntax, semantics and decoding processes for scalable extension" [12]. SHVC is discussed in Sect. 12.3. In the current drafting stage, 3D-HEVC is worked on as Annex H "3D High Efficiency Video Coding" [13]. Since SHVC will be finalized before the 3D extensions, the number "H" for the 3D Annex will be subject to change in the further standardization process. The 3D extension is reviewed in Sect. 12.3.

---

[3] As a modification of the original proposal by the author [9], the transform block size is signaled in the bitstream [7, 8].

The JCT-VC and the JCT-3V plan to align the finalization of the different parts of the extension specification such that the new annexes for the SHVC and MV-HEVC extensions can be jointly published as a new integrated edition of the HEVC specification. Since the development of 3D video is not at the same stage as the work on SHVC and MV-HEVC, it is likely that this extension will be added as an amendment in a later version of the specification.

12.2.1 Definition of Layers

Each layer included in the coded video sequence is addressed by a unique layer identifier l_{id} in the NAL unit header. [4] The 6 bit l_{id} syntax element in the NAL unit header conceptually allows for 64 different layers which may be addressed. The NAL units with $l_{id} = 0$ are required to form a bitstream conforming to one of the version 1 profiles of HEVC. The maximum value $l_{id} = 63$ is reserved for potential future extensions, e.g. if the number of views to be encoded exceeds the available number range. The VPS is extended to specify the interpretation of l_{id} in the context of the given bitstream. For multiview coding, l_{id} is used to indicate the view which is encoded. For scalable coding, l_{id} indicates the dependency identifier which is used for definition of the inter-layer prediction dependencies. The VPS syntax structure enables future extensions of the interpretation of l_{id} including a combination of view index and dependency identifier. This design allows for future flexibility, e.g. for a combination of multiview coding with scalability. Such options are not further explored in the current draft extensions.

In multiview video coding, the different *views* of the video sequence are represented by dedicated values of l_{id}. Typically, a view with a dedicated view identifier in a multiview video sequence is assigned to the video sequence from one camera in a multi-camera recording. The view with $l_{id} = 0$ is referred to as the *base view* or the *independent view*. The other views are referred to as *dependent views*. For scalable video coding, the bitstream includes multiple scalable layers which provide either an enhanced reconstruction fidelity or an increased picture resolution. As detailed in Chaptercha:struc, temporal scalability is already inherently provided in HEVC. In the context of scalable video coding, the layers are referred to as the *base layer* and the *enhancement layers*. The base layer is conforming to a profile of HEVC version 1. With the required value of $l_{id} = 0$, it represents the lowest available layer in the bitstream. The enhancement layers are organized in a hierarchical fashion and may or may not depend on lower layers.

Generally, a layer which is used for prediction in a non-base layer is referred to as a *reference layer*. The reconstructed picture at the reference layer is referred to as the reference layer picture. Accordingly, the layer which is to be output is referred to as the *target layer*. The specification differentiates between direct and indirect reference

---

[4] Recall that in order to differentiate the temporal from other scalability dimensions, the NAL unit header temporal identifier t_{id} refers to a *temporal sub-layer* instead of a layer.

layers. A direct reference layer is a layer which another layer predicts from. If the direct reference layer itself predicts from another layer, this layer is referred to as an indirect reference layer relative to the current target layer. The use of layers as reference layer is indicated in the VPS. While for scalable coding, only one layer (which usually would be the highest available layer) is the target layer, multiple target layers are available with multiview coding.[5]

12.2.2 Proposed Joint Tools

The common syntax, semantics, and decoding processes for multi-layer video coding extensions includes various modifications and extensions which are required to specify multi-layer bitstreams and the meaning of the layers. Here, some relevant elements of the Annex are presented.

12.2.2.1 High-Level Syntax

Several definitions which were originally specified for a single layer bitstream are modified to embrace the multi-layer extensions, see [12], Clause F.3. The most relevant modifications concern the definitions of the coded picture and the access unit. A coded picture in the multi-layer context is a coded picture with a given value of l_{id}. An access unit comprises all coded pictures with the same output time, including all associated VCL and non-VCL NAL units [11]. Thereby, multiple coded pictures are included in a multi-layer access unit. The coded pictures in an access unit are associated with the same value of POC for identification.

The video parameter set is extended to include a description of the presence and the relation between all layers in the coded video sequence as well as the high-level features of each of the layers. The VPS is required to have a layer identifier l_{id} equal to 0 which corresponds to the base layer. Given its nature as the highest available parameter set, the VPS applies to all layers comprised in the bitstream. For each layer, dedicated sequence and picture parameter sets are activated. Both, the SPS and the PPS contain changes which apply for enhancement layer NAL units in order to enable parameter inference from a reference layer. The SPS further includes a description of the relation between the spatial resolutions of the reference layer and the target layer, as well as the location of the reference layer picture relative to the target layer picture. Thereby, the geometrical relation of the pictures in the different layers is completely specified. This is needed for spatial scalability in order to appropriately configure the reference layer resampling process. In multiview coding as well as in scalable

[5] It should be noted that reference from another layer is not required. A bitstream may have multiple independent layers. For multiview coding, this enables multiple independent views. For scalable video coding, this corresponds to a simulcast transmission of independent versions of the same video sequence.

coding with fidelity scalability only, the picture resolution remains unchanged for all layers.

The hypothetical reference decoder is extended such that a buffer model for the complete stream as well as dedicated buffer models for the layers are supported. Besides the coded picture buffer for the full bitstream, a bitstream partition buffer (BPB) holding only the layer-related part of the bitstream is specified for this task.

12.2.2.2 Inter-Layer Prediction

Prediction between different layers in a bitstream is handled by the reference picture sets and reference picture lists.[6] For this task, an inter-layer reference picture is simply placed in the reference picture list of a target layer picture. Thereby, the reconstructed sample arrays and the associated motion and partitioning information of the reference layer can be accessed for inter-layer prediction. The process of reference picture list construction is specified in the dedicated Annexes for multiview and scalable coding. Note that no changes to the CTB decoding process compared to the base layer HEVC specification are included. Besides the modification of the reference picture list and the necessary integration with other layers, each layer can be decoded by a virtually unchanged HEVC decoder. This design is targeted at a straight forward extension of existing HEVC decoder implementations for support of the layer-based profiles.[7]

The inclusion of inter-layer reference pictures is constrained to comprise pictures from other layers only from within the same access unit. This corresponds to the conventional approach in scalable video coding where lower layers of the same picture are used for prediction of the target layer. In multiview coding, the constraint limits the prediction to either temporal prediction in the same view or inter-view prediction at the same output time. The same concept has been applied in H.264 | AVC multiview video coding and can be considered a complexity-performance trade-off [14].

12.2.2.3 Auxiliary Pictures

In the specification 'regular' coded pictures are denoted as *primary pictures* in distinction from so-called *auxiliary pictures*. By definition, an auxiliary picture does not have a normative effect on the decoding process of primary pictures [11]. Auxiliary pictures have been present in previous video coding standards such as H.262 | MPEG-2, MPEG-4 Visual or H.264 | AVC. They can be used e.g. for coding of alpha planes which enable indication of transparency of picture regions for mixing applications. They can also be used for coding of depth pictures which carry depth information

[6] Note that l_{id} is also used to identify auxiliary pictures such as depth information in 3D video coding, see below. The applicability of prediction between layers depends on the picture type at each layer.

[7] It should be noted that for scalable coding this scheme may imply a complexity burden as further detailed in Sect. 12.4.

for a primary picture. An auxiliary picture is identified by a dedicated value of l_{id}, where the respective value and the interpretation of this value are specified in the VPS. Thereby, the processing of auxiliary pictures is seamlessly integrated with the processing of primary pictures.

Currently, only alpha planes and depth pictures are specified as auxiliary picture types. The specification allows for a total number of 256 different auxiliary picture types. While most of the type indices are reserved for future use of ITU-T and ISO/IEC, a set of 16 indices remains unspecified. The interpretation of these types is out of the scope of the specification and can thereby be used for other applications. Auxiliary pictures of different type are not necessarily required to be associated with a primary picture. This allows e.g. for coding of pictures from different sources in auxiliary pictures.[8] Alpha planes carry transparency information for primary pictures. Alpha plane auxiliary pictures are therefore required to be associated with the corresponding primary pictures.

In the context of 3D video coding, depth maps are used in the prediction process of primary pictures. Therefore, the concept of auxiliary pictures must be adapted for 'true' 3D video coding, see Sect. 12.5.

12.3 Multiview Coding

The required extensions of the specification to support multiview coding are covered by Annex F to the largest extent. Annex G "Multiview coding" particularly includes the missing specification of the construction of the inter-layer reference picture set as well as the construction of the applicable reference picture lists. By the time of writing, the proposed amendment for multiview coding is at draft 6 [11].

As indicated in the previous section, the layers to be used for reference are specified in the VPS. The pictures with the corresponding values of l_{id} are included in a inter-layer reference picture set. Within the inter-layer reference picture set, the pictures are organized according to their distance from the target layer. The reference picture list to be used for inter prediction in the current picture is then constructed from the pictures which are available from the regular 'temporal' reference picture set and the inter-layer reference picture set. By default, the inter-layer reference pictures are inserted into the reference picture list after the pictures marked as 'Current ShortTerm Before' for List 0 and 'Current ShortTerm After' for List 1.[9] As for single-layer coding, the order of the pictures in the reference picture lists can be modified as intended by the encoder.

---

[8] For example, Note 4 in the Annex mentions the application of auxiliary pictures for range sensing cameras.

[9] See Sect. 4.3.2 for the HEVC reference picture list construction.

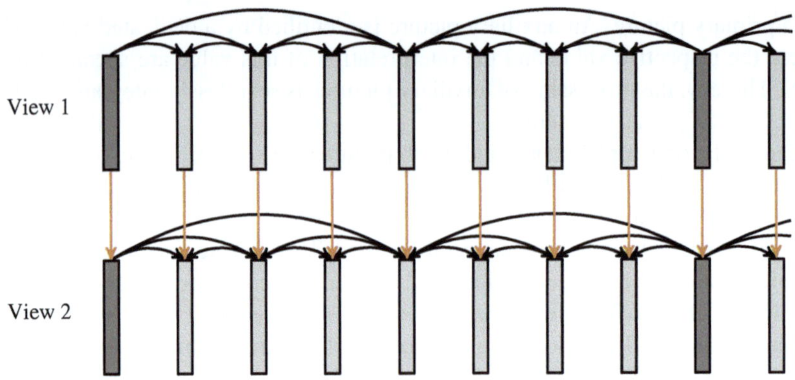

Fig. 12.2 Example coding structure for multiview coding with two views (stereo) with View 1 being the independent view

An illustration of the coding structure of a two-view (stereo) video sequence is provided in Fig. 12.2. In the example, both views have the same temporal coding structure. In View 2, which is the dependent view, additional inter-layer prediction from the independent View 1 is enabled for improved compression performance.

12.3.1 Proposed Multiview Profile

The current draft for the multiview extension includes the definition of a stereo profile only. The profile is called the Stereo Main profile and is based on the Main profile. Thereby, 8 bit YCbCr 4:2:0 video is supported. The profile allows for coding of two views (the independent view and one dependent view) of the same spatial resolution.

Further profiles may be specified in future amendments of the specification. These might allow for encoding of more than two views. Furthermore, the inclusion of auxiliary pictures may be enabled to support transmission of a depth map for each view. Note that in this case and according to the definition of auxiliary pictures, prediction of the primary pictures from an auxiliary picture is not allowed.

12.3.2 Comparison to H.264 | AVC

Conceptually, the multiview coding scheme specified for HEVC is very similar to the scheme specified in H.264 | AVC. The high-level syntax approach which has been developed for H.264 | AVC [14] is transferred to the HEVC specification. In the current extensions for HEVC, this approach is further extended to scalable coding.

12.4 Scalable Extension

Scalable High Efficiency Video Coding (SHVC) supports coding of a video sequence with multiple layers of different spatial resolution or reconstruction fidelity. By the time of writing, the proposed amendment for the scalable extensions is at draft 4 [12].

Following the basic structure provided by the common multi-layer video coding extension in Annex F, the decoding process for scalable layers does not include changes below the slice level. The relation between the scalable layers in terms of resolution and relative location is specified in the VPS. In order to enable spatial scalability, the different layers have different (increasing) spatial resolution. For fidelity scalability, the resolution of the video sequence does not change from layer to layer. This case can be interpreted as spatial scalable coding with a 1:1 ratio between the resolutions of the spatial layers.

12.4.1 Proposed SHVC Profile

The current draft includes the specification of a single scalable profile [15]. It is called the Scalable Main profile and is consequently based on the Main profile. Thereby all layers are required to be in YCbCr 4:2:0 format with a component bit depth of 8 bit.

12.4.2 Proposed SHVC Tools

Like in the multiview coding extension, the scalable extension in Annex H "Syntax, semantics and decoding processes for scalable extension" includes the specification of the construction of the inter-layer reference picture set as well as the construction of the applicable reference picture lists. Inter-layer reference pictures can be inserted into the applicable reference picture list in a very similar fashion as described in Sect. 12.3. Since it can be assumed that the motion in the different scalable layers of the coded video sequence is aligned, prediction from inter-layer reference pictures is constrained to solely apply zero-motion vectors. An illustration for inter-layer and temporal prediction in the case of spatial scalability with two layers is provided in Fig. 12.3. The prediction block from the inter-layer reference picture is scaled according to the relation of the resolutions between the two layers.

12.4.2.1 Resampling for Inter-Layer Reference Pictures

Depending on the size and the potential location offset of the reference layer picture in relation to the target layer picture, the scaling ratio for interpolation of the reference layer picture to the target layer picture size is derived. The resampling process derives

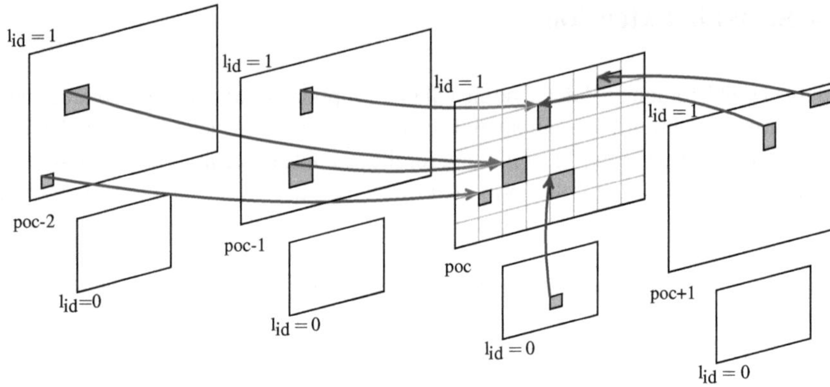

Fig. 12.3 Illustration of a coding structure with spatial scalability using inter-layer prediction

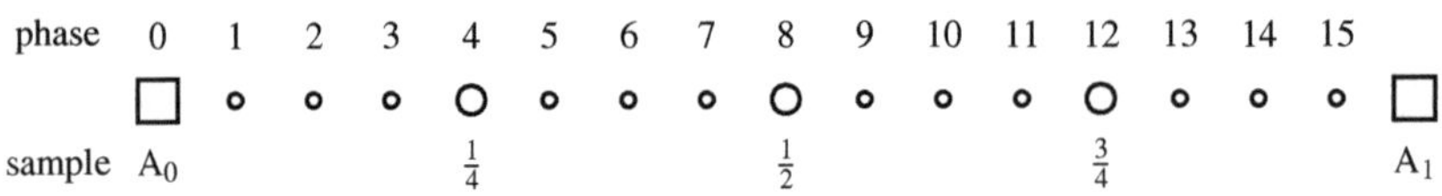

Fig. 12.4 Illustration of the 16 horizontal phase locations for inter-layer resampling

a (sub-sample) reference layer sample location and uses an interpolation operation to derive the sample value at target layer resolution. Motion vectors in the inter-layer reference picture are scaled according to the horizontal and vertical layer scaling. Like all other operations in the HEVC decoding process, the required operations are specified in integer arithmetic.

The location of the samples in the inter-layer reference picture is derived at 1/16th sample precision in horizontal and vertical direction. The sub-sample locations are denoted as horizontal and vertical phases where phase 0 corresponds to the left (top) neighboring sample. The assignment of the phase locations is shown in Fig. 12.4 for the horizontal case with the full-sample locations A_0 and A_1.

For each of the 16 phase locations, a dedicated interpolation filter is specified. The interpolation filters have been derived using the same method as for the sub-sample interpolation filters for motion compensated prediction, see Sect. 7.3.3. In fact, the filter coefficients are identical for the corresponding half-sample and quarter-sample phase locations in horizontal and vertical direction. For illustration, the SHVC luma inter-layer resampling filters are shown in Fig. 12.5. The left integer position (phase 0) is obviously derived without further interpolation. The phase positions up to the half-sample position at phase 8 are depicted. Phase 4 corresponds to the luma quarter-sample interpolation filter. The remaining seven phase positions are derived by mirroring the respective filters shown in the figure.

For the chroma samples, the resampling process is operated accordingly. The existing four-tap inter sub-sample interpolation filters are extended to 16 phase locations in the same fashion.

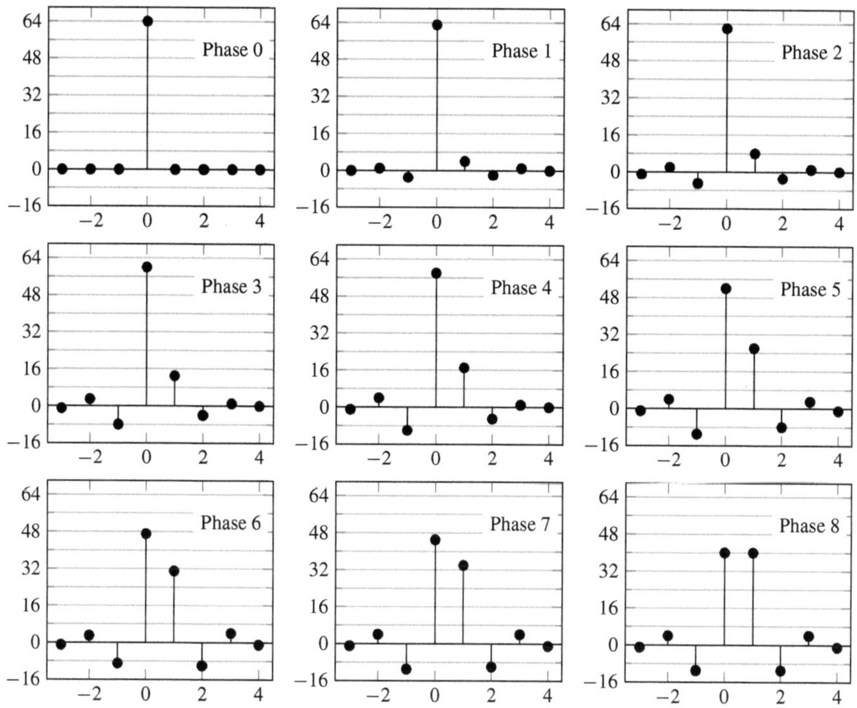

Fig. 12.5 SHVC resampling interpolation filters for the first eight phases (without scaling)

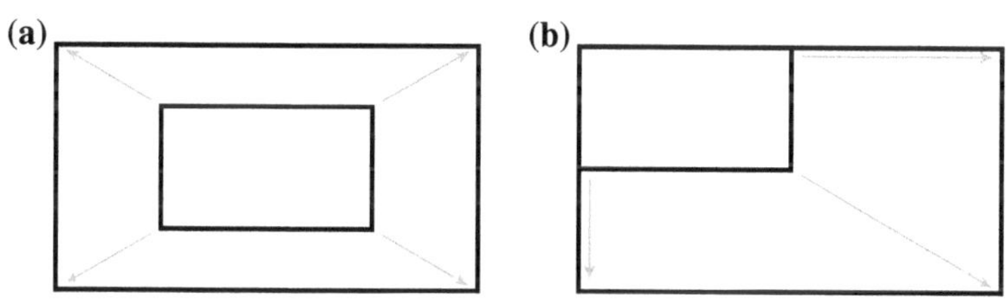

Fig. 12.6 Resampling phase alignment. **a** *Center* alignment. **b** *Top left* alignment

A cross layer phase alignment flag is further specified in the VPS. It is used to indicate if the location of the luma sample grids in the layers of the bitstream are aligned at the center of the picture or at the top left corner of the picture. This information is used to derive the correct phase positions for the resampling operation. The phase alignment is illustrated in Fig. 12.6.

12.4.2.2 Coding Standard Scalability

In the draft scalable extensions of the HEVC specification, an option is provided to support coding standard scalability with an H.264 | AVC base layer. The presence of such a base layer is indicated by a flag in the VPS. If present, the decoding process for the base layer according to the H.264 | AVC specification is performed. The reconstructed video as well as the motion information are then made available to the enhancement layer in the same way as with an HEVC base layer.

In the current draft, the transport of the H.264 | AVC base layer is not specified and is thereby left to the systems layer. There are options discussed to encapsulate H.264 | AVC NAL units in dedicated HEVC NAL units or to encapsulate the HEVC enhancement layer NAL units in H.264 | AVC NAL units with a previously reserved NAL unit type each [16, 17]. These approaches would allow for joint transport of all layers in a single bitstream.

12.4.3 Comparison to H.264 | AVC

The scalable extension of H.264 | AVC is called Scalable Video Coding (SVC) [7]. Like the scalable extension for HEVC, it supports multiple layers of different spatial resolution for spatial scalability or different reconstruction fidelities for fidelity scalability (also referred to as quality scalability). Both types of scalability can be combined according to the application needs. Furthermore, it includes an integrated concept for temporal scalability, which has only been supported in a less straight-forward way in the H.264 | AVC base specification. In order to address the dimensions of scalability, a three-byte NAL unit header extension was added to the original one-byte NAL unit header. Thereby, the SVC NAL unit header enabled indication of the reconstruction point in terms of dependency layer, quality, and temporal level.[10] An overview of SVC is provided in [18].

In SVC, scalability is enabled with a single-loop decoding concept using key pictures. For key pictures, multiple motion compensation loops may be used. For non-key pictures, only one motion compensation loop is required to reconstruct the target layer of the scalable bitstream, potentially using motion information from the reference layers. For fidelity scalability, SVC supports two different approaches: Like with spatial scalability, *coarse-grain scalability* allows for switching between different layers at each start of a new coded video sequence. As a second option, *medium-grain scalability* is specified. Here, single NAL units with quality refinement information may be dropped while reconstruction is still possible. Thereby, adaptation to bitrate requirements of an application can be achieved with low-complex operations on NAL units in the bitstream.

[10] In H.264 | AVC, temporal scalability is expressed in terms of *temporal levels* instead of the term temporal sub-layers, which is used in HEVC.

In contrast to SVC, the main focus of the SHVC design is on a "simple solution" for the specification of the scalable extension of HEVC [19]. While the computational complexity of this approach may be higher compared to a single-loop approach, specifically in the case of quality scalability [20, 21], the specification complexity is low. The specification extensions required for enabling the scalability features do not modify the signalling and decoding process below the slice level. Thereby both, the scalable extension and the multiview extension follow the approach chosen for multiview coding in H.264 | AVC. This design should thereby enable an easy extension of existing hardware or software implementations as the core decoder specification for each layer remains unchanged [14].

12.5 3D Video Coding

As an extension to conventional multiview coding, 3D video coding integrates the representation of depth information in the bitstream. In this context, the sequence of coded pictures of a view is referred to as the *texture view*. The corresponding depth information is referred to as the *depth view*. Based on the provided depth information, additional views besides the originally coded ones can be synthesized. This enables the receiver e.g. to adapt the depth impression according to the application needs or to render various intermediate views which are required for support of auto-stereoscopic displays. Also, the adaptation of the view point is possible [22, 23]. By the time of writing, the specification of 3D video coding is at draft 2 [13]. The reference software and the test model are at version 7 [24], which reflects the ongoing activity since the MPEG Call for Proposals in March 2011 [25] and the formation of the JCT-3V in April 2012 [2]. In distinction from the original HM reference software for HEVC, the reference software is referred to as HTM. The finalization of the specification is planned for mid 2015. By the time of writing, prospective profiles have not been drafted yet.

In order to achieve maximum compression efficiency, 3D video coding employs joint coding of depth and texture information, allowing for inter-view prediction under consideration of depth as well as inter-component prediction between texture and depth pictures. The available concepts for 3D coding with depth are visualized in Fig. 12.7. In the figure, texture and depth coding for a single picture is shown. For simplified representation, inter prediction along the temporal direction is left aside in the figure. Temporal inter prediction within each view (texture and depth) is available in addition to the shown inter-view and inter-component prediction in this scheme.

Figure 12.7a shows the simplest case where no prediction between texture and depth pictures is used. The texture and depth component are coded independently. This case corresponds to the realization of 3D video coding using MV-HEVC or H.264 | AVC MVC using auxiliary pictures for the depth. In Fig. 12.7b, prediction from the texture to the depth picture within the same view is depicted. For example partitioning and motion vector information for the depth component can be predicted from the texture component for improved compression efficiency.

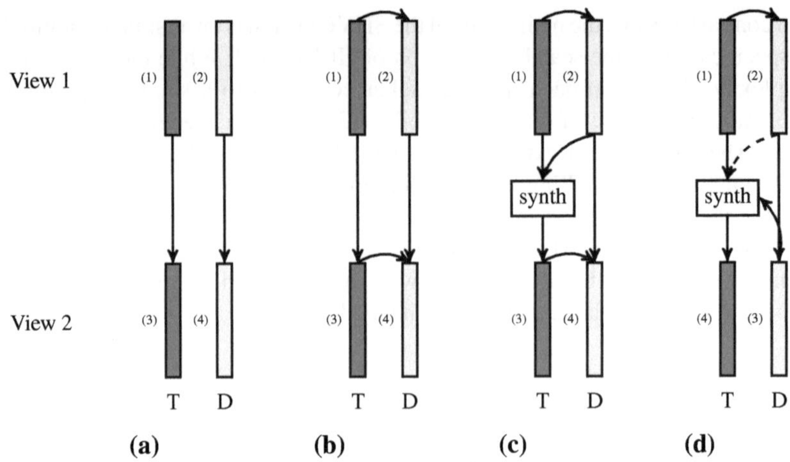

Fig. 12.7 Access unit prediction structures for 3D video coding with texture pictures (T) and depth pictures (D), coding order indicated in parentheses. **a** Multiview plus depth. **b** Inter-component prediction. **c** View synthesis prediction (used in HTM). **d** View synthesis prediction with dependent view depth first

Figure 12.7c extends the prediction to employ both, texture and depth information from the base view for prediction of the texture in the dependent view. The information from the base view is projected to the dependent view according to the available depth information of the base view. This process is referred to as view synthesis prediction in the HTM. The consideration of the depth information in the prediction process leads to an improved compression performance in the enhancement view [23]. Finally, Fig. 12.7d shows a modified approach for view synthesis prediction, where the order of the texture and depth coding in the dependent view is modified. This enables the utilization of the dependent depth view information for prediction of the dependent texture view. Since coding of the texture information covers the largest portion of the bitstream, using the corresponding depth information for prediction of the texture can have a further beneficial impact on the compression [26]. By January 2014, 3D-HEVC is reported to provide 24 % rate savings over MV-HEVC plus depth coding under JCT-3V testing conditions [27].

References

1. Sullivan, G.J., et al.: Standardized extensions of high effciency video coding (HEVC). IEEE J. Sel. Top. Signal Process. **7**(6), 1001–1016 (2013). doi:10.1109/JSTSP.2013.2283657
2. Terms of reference of the Joint Collaborative Team on 3D Video Coding Extension Development (JCT-3DVE). ITU-T and ISO/IEC JTC 1. Geneva. http://www.itu.int/en/ITU-T/studygroups/com16/video/Documents/ToR-JCT-3V-TDPLEN-053216.pdf (2012). Accessed 14 Apr 2014
3. Rusanovksyy, D., et al.: White Paper on State of the Art in 3D Video (Draft 2). Doc. JCT2-A1006. 1st meeting: Joint Collaborative Team on 3D Video Coding Extension Development (JCT-3V) of ITU-T SG 16 WP 3 and ISO/IEC JTC 1/SC 29/WG 11, Stockholm, SE (2012)

4. Chiariglione, L.: MPEG Time Line. Doc. N14204. 107th meeting: MPEG, San Jose, USA. http://mpeg.chiariglione.org/sites/default/files/timeline/w14204.zip (2014). Accessed 14 Apr 2014
5. Flynn, D., et al.: High effciency video coding (HEVC) range extensions text specifcation: Draft 5. Doc. JCTVC-O1005. 15th meeting: Joint Collaborative Team on Video Coding (JCT-VC) of ITU-T SG 16 WP 3 and ISO/IEC JTC 1/SC 29/WG 11, Geneva, CH (2013)
6. Pu, W., et al.: RCE1: Descriptions and Results for Experiments 1, 2, 3, and 4. Doc. JCTVC-O0202. 15th meeting: Joint Collaborative Team on Video Coding (JCT-VC) of ITU-T SG 16 WP 3 and ISO/IEC JTC 1/SC 29/WG 11, Geneva, CH (2013)
7. Advanced video coding for generic audiovisual services. ITU-T Rec. H.264 (AVC). http://www.itu.int/rec/T-REC-H.264/en (2014). Accessed 14 Apr 2014
8. Information technology—Coding of audio-visual objects—Part 10: Advanced video coding. ISO/IEC 14496-10:2012 (AVC). http://www.iso.org/iso/home/store/catalogue_tc/catalogue_detail.htm?csnumber=61490 (2012). Accessed 14 Apr 2014
9. Wien, M.: Variable block-size transforms for H.264/AVC. IEEE Trans. Circ. Syst. Video Technol. **13**(7), 604–613 (2003). doi:10.1109/TCSVT.2003.815380
10. Sullivan, G.J., et al.: The H.264/AVC advanced video coding standard: overview and introduction to the fidelity range extensions. In: Tescher, A.G. (ed) Proceedings SPIE5558 Conference on Applications of Digital Image Processing XXVII, vol. 5558. SPIE. Denver, CO (2004). DOI:10.1117/12.564457
11. Tech, G., et al.: MV-HEVC Draft Text 6. Doc. JCTVC-F1004. 6th meeting: Joint Collaborative Team on 3D Video Coding Extension Development (JCT-3V) of ITU-T SG 16 WP 3 and ISO/IEC JTC 1/SC 29/WG 11, Geneva, CH (2013)
12. Chen, J., et al.: High efficiency video coding (HEVC) scalable extension Draft 4. Doc. JCTVC-O1008. 15th meeting: Joint Collaborative Team on Video Coding (JCT-VC) of ITU-T SG 16 WP 3 and ISO/IEC JTC 1/SC 29/WG 11, Geneva, CH (2013)
13. Tech, G., et al.: 3D-HEVC Draft Text 2. Doc. JCTVC-F1001. 6th meeting: Joint Collaborative Team on 3D Video Coding Extension Development (JCT-3V) of ITU-T SG 16 WP 3 and ISO/IEC JTC 1/SC 29/WG 11, Geneva, CH (2013)
14. Vetro, A., et al.: Overview of the stereo and multiview video coding extensions of the H.264/MPEG-4 AVC standard. Proc. IEEE **99**(4), 626–642 (2011). doi:10.1109/JPROC.2010.2098830
15. Sullivan, G.J., Ohm, J.-R.: Meeting report of the 15th meeting of the Joint Collaborative Team on Video Coding (JCT-VC), Geneva, CH. Doc. JCTVC-O1000. 15th meeting: Joint Collaborative Team on Video Coding (JCT-VC) of ITU-T SG 16 WP 3 and ISO/IEC JTC 1/SC 29/WG 11, Geneva, CH (2013)
16. Boyce, J.: Specification text and profile for SHVC with AVC base layer. Doc. JCTVC-O0143. 15th meeting: Joint Collaborative Team on Video Coding (JCT-VC) of ITU-T SG 16 WP 3 and ISO/IEC JTC 1/SC 29/WG 11, Geneva, CH (2013)
17. Samuelsson, J., Sjöberg, R.: AHG15: AVC and HEVC encapsulation for hybrid codec scalability. Doc. JCTVC-O0190. 15th meeting: Joint Collaborative Team on Video Coding (JCT-VC) of ITU-T SG 16 WP 3 and ISO/IEC JTC 1/SC 29/WG 11, Geneva, CH (2013)
18. Schwarz, H., et al.: Overview of the scalable video coding extension of the H.264/AVC standard. IEEE Trans. Circ. Syst. Video Technol. **17**(9), 1103–1120 (2007). doi:10.1109/TCSVT.2007.905532
19. Sullivan, G.J., Ohm, J.-R.: Meeting report of the 13th meeting of the Joint Collaborative Team on Video Coding (JCT-VC), Incheon, KR. Doc. JCTVC-M1000. 13th meeting: Joint Collaborative Team on Video Coding (JCT-VC) of ITU-T SG 16 WP 3 and ISO/IEC JTC 1/SC 29/WG 11, Incheon, KR (2013)
20. Feldmann, C., et al.: Single-loop SNR scalability using binary residual refinement coding. In: Proceedings of IEEE International Conference on Image Processing ICIP'13 (2013). DOI:10.1109/ICIP.2013.6738326
21. Wien, M., et al.: JCT-VC AHG report: Single-Loop Scalability (AHG16). Doc. JCTVC-O0016. 15th meeting: Joint Collaborative Team on Video Coding (JCT-VC) of ITU-T SG 16 WP 3 and ISO/IEC JTC 1/SC 29/WG 11, Geneva, CH (2013)

22. Müller, K., et al.: 3-D Video Representation using Depth Maps. Proc. IEEE. **99**(4), 643–656 (2011). doi:10.1109/JPROC.2091090
23. Müller, K., et al.: 3D high-effciency video coding for multi-view video and depth data. IEEE Trans. Image Process. **22**(9), 3366–3378 (2013). doi:10.1109/TIP.2013.2264820
24. Zhang, L., et al.: Test Model 7 of 3D-HEVC and MV-HEVC. Doc. JCT3VG1005. 7th meeting: Joint Collaborative Team on 3D Video Coding Extension Development (JCT-3V) of ITU-T SG 16 WP 3 and ISO/IEC JTC 1/SC 29/WG 11, San Jose, CA, USA (2014)
25. Video and Requirements. Call for Proposals on 3D Video Coding Technology. Doc. N12036. 96th meeting: MPEG, Geneva, CH (2011)
26. Jäger, F.: Depth-based Block Partitioning for 3D Video Coding. In: Proceedings of International Picture Coding Symposium PCS '13. IEEE, Piscataway, San Jose, USA (2013). doi:10.1109/PCS.2013.6737770
27. Baroncini, V., et al.: Viewing Report for Comparison of 3D-HEVC and MVHEVC with depth coding. Doc. JCT3V-G0243. 7th meeting: Joint Collaborative Team on 3D Video Coding Extension Development (JCT-3V) of ITU-T SG 16 WP 3 and ISO/IEC JTC 1/SC 29/WG 11, San Jose, CA, USA (2014)

Index

M. Wien, *High Efficiency Video Coding*, Signals and Communication Technology,
DOI 10.1007/978-3-662-44276-0

MIX
Papier aus verantwortungsvollen Quellen
Paper from responsible sources
FSC® C105338

If you have any concerns about our products,
you can contact us on
ProductSafety@springernature.com

In case Publisher is established outside the EU,
the EU authorized representative is:
Springer Nature Customer Service Center GmbH
Europaplatz 3, 69115 Heidelberg, Germany

Printed by Libri Plureos GmbH
in Hamburg, Germany